Deconstructing
Digital Capitalism
and the Smart Society

ALSO BY MEL VAN ELTEREN
AND FROM MCFARLAND

*Managerial Control of American Workers:
Methods and Technology from the 1880s to Today* (2017)

Labor and the American Left: An Analytical History (2011)

*Americanism and Americanization:
A Critical History of Domestic and Global Influence* (2006)

Deconstructing Digital Capitalism and the Smart Society

Invasive Platforms, Unchecked Monopolies, Humane Alternatives

MEL VAN ELTEREN

McFarland & Company, Inc., Publishers
Jefferson, North Carolina

ISBN (print) 978-1-4766-9609-6
ISBN (ebook) 978-1-4766-5548-2

LIBRARY OF CONGRESS CATALOGING DATA ARE AVAILABLE

Library of Congress Control Number 2024058449

© 2025 Mel van Elteren. All rights reserved

No part of this book may be reproduced or transmitted in any form or by any means, electronic or mechanical, including photocopying or recording, or by any information storage and retrieval system, without permission in writing from the publisher.

Front cover images © Shutterstock

Printed in the United States of America

McFarland & Company, Inc., Publishers
Box 611, Jefferson, North Carolina 28640
www.mcfarlandpub.com

To my brothers,
Walter and Henk

Acknowledgments

Words are inadequate to express my gratitude to my dear wife and soulmate, Nancy A. Schaefer, for her encouragement and unflagging support. I am deeply indebted for her careful editing of the manuscript and critical remarks on the content itself, which led me to rethink—and subsequently revise—specific parts of my argument. Of course, any weaknesses in the book are my own.

Table of Contents

Acknowledgments   vi
Introduction   1

1. Contemporary Capitalism and the Rise of Data Colonialism   7
2. The Internet: From a Government-Owned Network to a Privatized One   25
3. Moving Towards Surveillance Advertising Led by Google   37
4. Identifying the Distinctive Features of Various Types of Online Platforms   59
5. The Wider Array of Smart Tech Devices and Systems   79
6. The Smart Self and Algorithmic Management Expanded and Intensified   96
7. Smart Homes and Smart Cities: Hype About Benefits Versus Drawbacks for Users   112
8. Smart Technologies, City Surveillance and Predictive Policing   127
9. Moves to Rein in Capitalism Interlocked with Data Colonialism   150
10. Taking Up the Challenges of Regulating Artificial Intelligence   170
11. Proposals to Deprivatize the Internet and Foster Platform Socialism   188
12. Competing Digital Empires and the Future of Platform Capitalism Globally   207

Chapter Notes   223
Bibliography   255
Index   275

Introduction

There has been much public reflection on the use of digital technology following the 2013 Edward Snowden revelations of pervasive surveillance and data-gathering by U.S. security services in close cooperation with major tech firms, coupled with the 2016 interferences in the American presidential election associated with the Facebook-Cambridge Analytica scandal erupting in 2018. This has led to growing public awareness and discontent with the privacy violations, monopolies and strategies of Big Tech companies. Called "techlash," this development has become an enduring focal point of U.S. media and politics. Key factors that have contributed to the backlash against Big Tech are concerns about data processing and security issues, as well as the harmful practices of major players dominating the market for search engines, social media, online shopping, short-term rental and gig economy platforms. Recent publications have shed more light on the logic of Big Tech companies in colonizing the everyday lives of citizens and using data for commercial and political aims. Their practices include the harvesting of user data for purposes such as personalized advertising and various types of consumer manipulation; surveillance and control of workers through digital management systems; and governmental surveillance and policing at various levels. These practices further encompass biases in big data and automated discrimination in using artificial intelligence systems; the creation of internet bubbles; and the spread of misinformation and "fake news." There are also the negative effects on the mental health of users affected by cyber bullying, hate speech and extremist content, as well as the addictive qualities of smartphones and social media. Last but not least, one should mention the significant rise of a wide range of criminal activities that involves the use of digital technology to commit fraud, identity theft, data breaches, the spread of computer viruses, and scams, which is extended to other malicious acts. Cyber criminals exploit vulnerabilities in computer systems and networks to gain unauthorized access, steal sensitive information, disrupt services, and cause financial or reputational harm to individuals, organizations, and governments.

Yet no matter how much bad press and regulatory activity tech companies may have received, they continue to make products today that customers apparently want to buy. And despite the techlash, many people still continue to use digital products and services passionately, and often very uncritically. Observers even witness an over-reliance on smart technology, particularly among the younger generation. As smartphones, tablets, desktop computers, laptops and other digital gadgets related to the internet have become omnipresent and widely used, there is little inkling among the larger population to abandon any of the digital devices and services they have become accustomed to. A broad consensus exists that this ensemble of technologies has made life easier, more convenient, more interactive and more enjoyable. Moreover, despite their discontent over the bad practices of big online platforms, even the fiercest tech critics continue to rely on the same technology and cling to their devices to access online platforms as these are operating now.

At this time, critics consider privacy breaches, monopolistic practices and deployment of surveillance technologies as the main problems of big online platforms. Internet reformers suggest the answers to these questions reside in more—and better—regulations. It must be acknowledged, though, that such approaches exclude non-capitalist alternatives regarding the ways consumers might lead their digital lives. The issues of privacy, data and size are indeed important, but they are secondary to a deeper set of concerns about platform ownership and control, and, importantly, who benefits from the current status quo. While the issues are various and complex, they cannot be seen separately from the fact that the internet is owned by private firms and is run for profit. Yet in addition to a focus on the political-economic context, it is also necessary to consider the extent to which negative ramifications exist for users that are inherent to the technologies in question rather than the result of their capitalist setup.

This book addresses these issues and offers an in-depth analysis of digital capitalism and its multiple harmful practices as it has become increasingly intertwined with various forms of online surveillance, behavior modification and algorithmic management; that is, the delegation of managerial functions to algorithmic and automated systems in platform economies. It also looks at possible ways to change the current state of affairs, which is dominated by the legacies of privatization. In the 1990s, the U.S. government handed the private sector a network that was created at large public expense. Corporations took over the internet's physical structure and profited from selling access to it. But privatization went further than that and ascended to the upper layer of the internet's infrastructure where the system is used and experienced. This is where the real money came to be made by monetizing activity; that is, what people did once they got online. At that point, in the 2000s, the digital platforms emerged, including Google, Amazon, Facebook, Uber, and so on.

While the early internet had nodes overseas, today's internet is truly international—it is better thought of as a set of linked internets, centered on the major economic powers: the United States, the European Union, and China. Nevertheless, this book focuses primarily on the United States (and secondarily on the EU), which is justifiable, given the historical and continuing centrality of American institutions and businesses in the internet's existence. Yet one must keep in mind that the growth of platform capitalism, logistics, and the "Internet of Things" in the context of rapidly expanding uses of artificial intelligence as occurs in the U.S. is a core development and policy goal in China too. This constellation of factors has emerged as a significant arena of commercial and geopolitical competition between the West (particularly the U.S.) and China, with India an important player in the longer term.

The first chapter begins with an examination of Shoshana Zuboff's analysis in her seminal work *The Age of Surveillance Capitalism* (2019), which influenced much of the debate initially, and identifies major weaknesses in understanding the current state of digital capitalism. This arguably sets the stage for a more appropriate, political-economic approach to understanding "surveillance capitalism" in terms of the military-industrial-communications complex as it developed after World War II, first outlined by John Bellamy Foster and Robert McChesney. Next, the chapter both broadens and sharpens the analytical perspective further to capture the essence of what is going on by incorporating Nick Couldry and Ulises Mejias' insightful approach in terms of "data colonialism."

Chapter 2 takes up this historical thread, giving a concise overview of the origins of the internet and its noncommercial first period beginning in the 1970s, followed by an analysis of its transition in the 1990s from a government-owned computer network to a network owned and run by private corporations. This privatization did not involve a simple transfer of ownership from the public sector to the private but rather a more complex movement whereby corporations programmed the profit motive into every level of the network. This demanded proper hardware, software, entrepreneurship and new legislation, a process that took decades, touching on all the network's many pieces.

Chapter 3 describes how the groundwork was laid for the further commercialization of the internet and in particular, interactive digital advertising. Marketers rely on detailed forms of surveillance in order to customize advertising appeals, products, and pricing schemes that extract the maximum value from consumers. The drive toward customization, differentiation, and specialization remained limited for many years by the costs of information gathering as well as the lack of the technological infrastructure necessary for customization. This all changed with the arrival of networked, interactive communication devices and programmable manufacturing equipment. The chapter traces the emergence of surveillance advertising, which first became manifest in the monetizing of surveillance data by the leading ad network DoubleClick that utilized the technical "affordances" (that is, the perceived action possibilities inherent to the design) of web-based communication systems. Moreover, this chapter outlines the trajectory of "proto-platformization" in the late 1990s, which was a key moment in the "digital enclosure" of the internet. The chapter concludes with a closer look at how Google began to deploy its search engine for surveillance advertising and later upgraded its operations by acquiring DoubleClick and merging the two business models to form the prevailing prototype of such platforms.

Chapter 4 examines the origins, distinctive features and further development of major types of digital platforms other than that of Google. These include Facebook as a pioneer of social media enclosures; Amazon as an online marketplace for goods cum universal logistics system and provider of cloud services; Uber and its many "gig economy" imitators that all deploy algorithmic management; and Airbnb, the online platform for short-term rental accommodations. This chapter seeks to explain what these platforms have in common so that their owners and investors profit from rentier relations. Revenue is generated via the ownership and control of the platform under conditions of limited competition, which allows them to charge transaction and usage fees. Data collected on the platform also enables profit generation through advertising revenue. Further, at the same time as they centralize profits for shareholders, platforms outsource labor to "independent" contractors and service providers to externalize the costs and risks of doing business. In addition, many platforms benefit from network effects that enable them to scale rapidly, dominate markets and gain a monopoly over their sector.

The next two chapters broaden the scope to the wider array of data-driven, network-connected and automated systems that involve digital devices and technologies centered on three "smart" spaces: the smart self, the smart home and the smart city, respectively. Chapter 5 opens with a theoretical framework regarding politics, power and profit and the way these components of an emerging "smart society" are channeled through and changed by technology. It then looks at the devices and technologies associated with the smart self, including those of the so-called Quantified Self movement and credit scoring systems utilized both in the U.S. and China.

Chapter 6 continues with an examination of algorithmic management in relation to the smart self. It scrutinizes various systems of digital managerial control of workers in connection with fundamental changes in the features of work—from job ladder to job to discrete, dispersed tasks—which finds its nadir in digital Taylorism as epitomized by online "microwork." This chapter also pays attention to potential worker resistance and counterstrategies to the management practices to which microworkers are subjected.

Subsequently, Chapter 7 zooms in on home automation systems, internet-connected home security surveillance systems and smart home medical devices, as well as the involvement of insurance companies and law enforcement departments with smart home systems. This is followed by an overview of the smart city trend globally and a description of various smart city forms, the features they have in common and their political-economic driving factors. The chapter then discusses the current state of affairs in the U.S., focusing in particular on the serious drawbacks of the prevalent technocratic smart city concept. It concludes with an outline of democratic and "public good" alternatives that have been developed in some European cities, which could be a source of inspiration for similar initiatives in the U.S. and elsewhere.

Chapter 8 delves into highly pervasive city surveillance and control, specifically in relation to smart policing. It looks at the rise of militarized policing, both physical (the occupying army model of policing) and virtual (the intelligence agency model of policing) that coexist and are regularly deployed together. It analyzes in more detail the usual shifts in police surveillance to predictive policing, recognizing that similar changes and continuities can be found within a variety of institutions adopting new data sources and technologies. The chapter also takes a look at community resistance and countermovements against both city surveillance and smart policing.

Chapter 9 shines the spotlight on contemporary public policy measures aimed at tempering capitalist digital enclosure and the relentless drive to commodify information. It examines the antitrust measures and other regulations that the European Union has sought to implement, along with the United States, albeit to a much lesser extent there until quite recently. These regulations deal with the negative effects of online platforms controlled by Big Tech companies. They involve EU legislation such as the General Data Protection Regulation Act (GDPR), which has inspired similar privacy and data protection legislation by the State of California, followed by several other U.S. states (but not the federal government). They also include the EU Digital Markets and Digital Services Acts, which are unlikely to be adopted as models by the U.S. government at any level. This chapter also addresses the latest changes that have taken place on the antitrust front in America, with significant moves towards prosecuting Big Tech companies for monopolistic practices, particularly in the high-profile cases brought against Google and Amazon.

Chapter 10 concentrates on recent moves toward the regulation of the use of artificial intelligence, as sophisticated AI systems are beginning to penetrate much deeper into important sectors of society. This involves first and foremost the EU Artificial Intelligence Act, finalized in December 2023, which is the world's first comprehensive legal framework on artificial intelligence with the potential to influence international norms and standards for AI. In addition, the chapter addresses related EU regulatory measures regarding the labor rights of digital gig workers and a ban on harmful addictive techniques deployed in online games, social media, streaming services and online marketplaces. Secondly, relevant developments in the U.S. are taken into consideration. These revolve around President Biden's executive order of October 2023, which contains directives issued by the White House regarding safe use of AI in the face of potentially harmful consequences of the deployment of the latest generation of (high-risk) AI models. The Biden administration had also been pressuring lawmakers for AI legislation, but a heavily polarized U.S. Congress has made little progress in passing effective legislation so far. Finally, the chapter refers to a potential global movement towards a geopolitically broadened and strengthened regulatory framework, building on the new EU AI regulations.

Chapter 11 takes a critical look at proposals for far-reaching institutional reforms of the digital ecosystem. These encompass deprivatization of the internet and the creation of forms of "platform socialism" that would build partly on existing practices of platform cooperativism. The proposed "platform socialism" is an ideal and regulative idea (in philosophical terms) that may never fully materialize, although its proponents argue that attempts to approximate it can and should be made. These involve, first and foremost, economic "socialization," understood in this context as the organization of the digital economy through the social ownership of digital assets, and democratic control over the infrastructure and systems that govern people's digital lives.

This chapter also discusses the fact that some effects of digital tools deemed detrimental to users are inherent to these technologies—particularly their design and behavioral requirements—rather than to their capitalist arrangements. Tackling these problems demands an approach that goes beyond any socialist reform attempt.

The final chapter analyzes the three dominant digital powers, the U.S., China and the EU, in terms of their twenty-first-century forms of colonialism and contrasting strategies regarding the regulation of tech companies' practices. It further discusses the practical and policy implications of a proposed decolonial approach to data and technology that transcends socialist reform, but thus far has not led to any powerful movement activism. It is duly recognized that socialist reform attempts as mentioned here would face an immense uphill battle, given the current dynamics of global politics. The chapter concludes with a closer look at the various battles about the regulation of tech companies' deployment of digital technologies involving the U.S. market-driven regulatory model, the Chinese state-driven model, and the EU rights-driven model. It then considers the possibilities there might be for a global shift to the EU model, which aims to contain the excesses of platform capitalism and is associated with greater economic fairness and a more human-centric digital society.

1

Contemporary Capitalism and the Rise of Data Colonialism

The starting point of our critical analysis, Shoshana Zuboff's 2019 book *The Age of Surveillance Capitalism*, is probably the most prominent example of several recent books that aim to analyze the massive changes unleashed by the rise of new, data-driven intensive business models within the broader context of contemporary capitalism.[1] It has drawn considerable interest in academic circles and the quality press, tying the concept of "surveillance capitalism" to her name as a scholar.[2] While this social psychologist/philosopher undoubtedly made a valuable contribution to the discussion, Zuboff was not the first or only scholar to come up with the concept. The British sociologist Frank Webster already posited the arrival of surveillance-powered "social Taylorism" in the late 1980s, according to the prominent researcher of the political and social implications of digital technology, Evgeny Morozov.[3] Indeed, proof is found in Webster's co-authored book, *Information Technology: A Luddite Analysis*.[4] The concept of surveillance capitalism was also used elsewhere, in a 2014 essay by John Bellamy Foster and Robert W. McChesney in *The Monthly Review*.[5] They approached the notion of surveillance capitalism from a very different angle than Zuboff has done. Rather than focusing on the economic changes of the last couple of decades as in Zuboff's case, they instead present and define surveillance capitalism in terms of much broader changes dating all the way back to the inception of a post–World War II "Pentagon capitalism" that came to be known as the military-industrial complex, to which we shall return later.

Surveillance capitalism, Zuboff claims, was born with the internet and with the proliferation of digital and networked technologies more generally. Simply stated, her thesis is: Today capitalism is increasingly surveillance-based, driven by companies who make money not only by knowing our behaviors, but also by attempting to influence or configure those behaviors in ways that maximize money-making opportunities. As Zuboff puts it in her terms,

> Surveillance capitalism unilaterally claims human experience as free raw material for translation into behavioral data. Although some of these data are applied to service improvement, the rest are declared as a proprietary be*havioral surplus*, fed into advanced manufacturing processes known as "machine intelligence," and fabricated into *prediction products* that

anticipate what you will do now, soon, and later. Finally, these prediction products are traded in a new kind of marketplace that I call *behavioral futures markets*. Surveillance capitalists have grown immensely wealthy from these trading operations, for many companies are willing to lay bets on our future behavior.[6]

So at the heart of Zuboff's account is the idea of prediction. Prediction is surveillance capitalism's modus operandi, one could say. If, as Zuboff contends, companies are better able to predict what we will do in a particular context, they can better adapt commercial offerings to suit and profit from those predicted behaviors. And what surer method of accurate prediction is there than to intervene at its source and shape the behaviors in question? In order to achieve this goal, machine processes are configured to intervene in the state of play in the real world among real people and things in real time. These interventions are designed to enhance certainty by doing things through nudging, tuning, herding, manipulating, and modifying behavior in specific directions. At its extreme, prediction—probability—becomes certainty. Hence Amazon's leading idea that based on our viewing and purchasing habits, it can ship us books it knows we will like before we ourselves realize it.[7] One should add that, at the current juncture, behavioral, psychological, and neurological insights from the interdisciplinary field of "*neuro*liberalism" are being used to deliberately shape and govern human conduct online.[8] This is done in particular through digital nudging—the deployment of design, information and interaction elements to guide (or perhaps even steer) user behavior in digital choice environments. The latter are user interfaces—such as web-based forms and ERP (Enterprise Resource Planning) dashboards—that require people to make judgments or decisions. The design of the choice environment in which information is presented can exert a subconscious influence on the outcome. In other words, what is chosen often depends upon how the choice is presented, such that the "choice architecture alters people's behavior in a predictable way."[9] Automated recommender systems, which are nowadays a pervasive part of users' online experience, can be seen as digital nudges, because they determine different aspects of the choice architecture for users.[10]

Zuboff focuses mainly on Google, Facebook and Microsoft, which she sees as the prototypical examples by means of which the DNA of surveillance capitalism is best examined. She emphasizes that surveillance capitalism is a human creation, not the result of technological inevitability. It was pioneered and elaborated through trial and error at Google, "invented" around 2001 as the solution to financial emergency in the wake of the dot-com bust, when the fledgling company faced the loss of investor confidence. As investor pressure mounted, Google's founders Larry Page and Sergey Brin abandoned their declared antipathy toward advertising. Instead they embarked on a course that boosts ad revenue by using their exclusive access to user data logs in combination with their considerable analytical capabilities and computational power to generate predictions of user click-through rates, taken as a signal of an ad's relevance. This meant in practice that Google would both repurpose its growing collection of behavioral data, now capitalized on as a behavioral surplus, and develop methods of secret surplus capture that could uncover data which users intentionally chose to keep private—as well as to infer extensive personal information that users did not or would not provide. This surplus would then be analyzed for hidden meanings that could predict click-through behavior. The surplus data thus formed the basis for new predictions markets involving targeted advertising.[11]

Surveillance capitalism quickly spread to Facebook and later to Microsoft, while in Zuboff's telling, a company like Amazon, with its new emphasis on "personalized" services and third-party revenue, was clearly veering towards it as well. At her time of writing, Zuboff still thought differently about Apple, to whom surveillance capitalism was allegedly a constant challenge, both as an external threat and as a source of internal debate and conflict. Apple's iPods/iTunes innovations leveraged the new capabilities of digital technologies to overturn the consumption experience that had prevailed in the age of mass production. This change of direction depended on a few key elements. Digitalization made it possible to rescue valued assets—in this case songs—from the institutional spaces in which they were trapped. The costly institutional procedures that General Motors' CEO Alfred Sloan had described in relation to mass production and consumption in the 1920s and after were eliminated in favor of a direct route to listeners. In the case of the CD, for instance, Apple bypassed the physical production of the song along with its packaging, inventory, storage, marketing, transportation, distribution, and physical retailing. The combination of the iTunes platform and the iPod device made it possible for listeners to continuously reconfigure their songs at will. Thus, Apple was among the first to attain formidable commercial success by catering to a new society of individuals and their demands for individualized consumption. In Zuboff's view, Apple's business strategy implied trustworthy relationships of advocacy and reciprocity embedded in an alignment of commercial operations with consumers' genuine interests.[12]

Surveillance capitalism started with advertising, but then became more general. The Big Tech companies have used the knowledge about their users not only for ever more precisely targeted and subtle ads, but also to sell to the companies, politicians, governments, secret services and any other agency willing to pay for it. Today, surveillance capitalism is no longer restricted to companies or even to the internet sector. It has spread across a wide range of products, services, and economic sectors, including insurance, retail, health care, finance, entertainment, education, transportation, and more, generating whole new ecosystems of suppliers, producers, customers, market makers and market players.

Driven by competition over prediction products, surveillance capitalists first learned that the more surplus, the better the prediction, which led to economies of scale in supply efforts. Then they learned that the more varied the surplus, the higher its predictive value. This new drive toward economies of scope led them from the desktop to mobile, out into the world, to monitor people's drive, hike or run, shopping, search for a parking space, bodily conditions and face, and always their location. And ultimately, they understood that the most predictive behavioral data can be derived from what Zuboff calls "economies of action" as systems are increasingly designed to intervene in the state of play and actually modify behavior, steering it toward desired commercial outcomes. The experimental development of this new means of behavior modification can be seen in Facebook's infamous emotional contagion experiment[13] and the augmented reality game *Pokémon Go* developed at Google. In this phase of surveillance capitalism's evolution, the means of production became subordinate to an increasingly complex and comprehensive means of behavior modification. Thus, surveillance capitalism has created a new form of power that Zuboff terms "instrumentarianism." It knows and shapes human behavior toward others' ends. It reaches its goal through the automated medium of an increasingly ubiquitous architecture of "smart" networked devices, things, and spaces.[14]

A Limited Perspective

There is much to commend Zuboff's approach. She offers a penetrating and persuasive account of how platforms like Facebook and Google mine online behavior so as to better match advertisements with users' interests, and how much data such companies possess and exploit. If users' behavior can be accurately predicted and they are shown the "right" ads, then they check through more than they otherwise might, and Facebook or Google—or whichever company it happens to be—earns more money ("click-through rates" having become the standard pricing metric). Given this business model, the incentive to orient the user toward clicking through—via a range of techniques that Zuboff refers to collectively as "behavioral modification," or that behavioral economists have termed "nudging"—is clear enough.

What seems much less clear is Zuboff's contention that in view of these developments in the spheres of collection and use of user data and attempts to nudge consumers toward behaving in ways that suit capitalists, surveillance capitalism amounts to a decisively new economic system. Lacking a robust, theoretically and historically grounded conception of capitalism, as Morozov points out, Zuboff's text follows a narrative trajectory, in which she begins by choosing a prior stage that she calls "advocacy-oriented capitalism," and then proceeds to roll out the *deus ex machina* of information technology, big data and machine learning. She concludes that the current stage of capitalism, "surveillance capitalism," is a stark departure from the previous one, and that drastic changes in information technology explain the transition. She turns to recent history only very selectively, mostly to shore up her presentist two-stage scheme.[15]

Zuboff contrasts the drive to extract behavioral surplus with something she calls the "behavioral value reinvestment cycle." This is actually just a feedback loop; for Zuboff, it is a consumer-friendly way of using data to improve the customer's experience of a product or service. "Behavioral value reinvestment" is something done by a different type of capitalism—"advocacy-oriented capitalism." As the name suggests, it is exemplified by companies that leverage the possibilities of the digital revolution to "empower" consumers.

Google is the archetype of Zuboff's theory. In its early years, Google had the potential to become Zuboff's favorite "advocacy-oriented" firm; service improvement was its only incentive in gathering data. Once it embraced personalized advertising, things changed. Now Google wanted more user data to sell ads, not just to improve services. The data it gathers in excess of the objectively determined need to serve users is Zuboff's behavioral surplus. In other words, the behavioral reinvestment turns into behavioral surplus only when some objective limit of user monitoring needed to improve the service is exceeded. This important threshold is introduced in *The Age of Surveillance Capitalism* but never explicitly theorized, as Morozov stipulates.[16] As a capitalist firm, Google wants to maximize that surplus, expanding in depth—penetrating ever deeper into people's minds and life worlds—but also in breadth, by offering new services in new domains and diversifying its "surveillance assets."

Advocacy-oriented capitalism, meanwhile, is far more democratic in Zuboff's understanding—it uses data to better respond to consumer needs and demands, not to shape and determine those needs. So for her, surveillance capitalism is a "rogue mutation" of a more normal, socially just version of capitalism. However, it is a fallacy to think that surveillance capitalism is anything new. As Matthias Crain adroitly puts it, "…this position disregards historical continuities to focus only on what is new. Although the magnitude of contemporary commercial surveillance is certainly mind-bending, the system reflects enduring structural imperatives within a capitalist political economy depending on perpetual growth."[17]

Zuboff's analysis obscures the extent to which the logic of surveillance advertising is in fact a continuation of a logic of extraction and commodification inherent to capitalism itself. The point is that surveillance capitalism is not a radical departure. It represents an actualization or intensification of tendencies that have always been at the heart of capitalism, which, until recently, could not materialize for a variety of reasons associated primarily with the technical infrastructures of product and service delivery.[18]

In some ways behavior modification aims have always been integral to capitalism in general and advertising in particular. This was certainly not the first time that companies had profited from personal data. Long before the internet became a mass medium, marketing firms, insurance companies, and others have been using computers to collect and analyze information about consumers, as a pioneering scholar of surveillance studies, Oscar H. Gandy, Jr., has documented.[19] Still, the mainstreaming of the internet significantly enlarged the surface area for such surveillance.

A clear foreshadowing of today's state of affairs can be found as early as the 1960s, when the Simulmatics Corporation sold advertising agencies datafied behavioral science approaches and had visions of a "mass culture model," which would be used to "collect consumer data from companies across air media—publishing houses, record labels, magazine publishers, television networks, and moviemakers—in order to devise a model that could be used to direct advertising and sales by way of a meta-media-and-data corporation that sounds rather a lot like Amazon."[20] By 1991, Equifax had compiled granual personal information including purchase habits on 120 million Americans into a digital marketing base sold as a CD-ROM. So indications were there for decades that surveillance capitalism was the natural continuation of industrial capitalism.[21]

Advocacy-oriented capitalism, Zuboff's preferred alternative, is as capitalist as the surveillance variant is.[22] It features private appropriation of user feedback, even if firms purportedly pursue such appropriation in the name of service improvement. Why, then, is advocacy capitalism considered to be superior to surveillance capitalism? Partly, because in the absence of advertising, it is deemed to be free from the power imbalances of unequal exchange, making the relationship between companies and consumers one of "reciprocity."

However, the idea that "normal" forms of capitalism are founded on some degree of reciprocity (particularly between management and workers) is at best naïve if not disingenuous. While Zuboff wants to present capitalism as a kind of impartial technology that revolves around the changing needs of consumers, the truth is that capitalism has been shaped by power and struggle. Industrial capitalism was not a polite agreement, but rather an (at times very precarious) equilibrium between different parties that was achieved through the exercise of political action.

This also calls into question Zuboff's bold conclusion: "The struggle for power and control in society is no longer associated with the hidden facts of class and its relationship to production but rather by the hidden facts of automated engineered behavior modification."[23] To the contrary, capitalist elites (investors, especially venture capitalists) still play crucial roles in financing and (often indirectly) controlling the state of play. Zuboff herself mentions the relevance of the ownership of the means of behavior modification for surveillance capitalism, and refers to the rise of shareholder capitalism and its impact on the management of corporations. Moreover, the scientists, engineers, social scientists and managers involved choose (consciously or not) to advance and operate the digital systems in question for the interests of surveillance capitalists. They could make a different choice, as some indeed have done or may very well do in the future. In this context, "production" should also be understood in a much broader sense than traditional manufacturing to include all kinds of services and highly skilled white-collar work.

Under surveillance capitalism, consumers become subjected to imperatives that are not theirs, their autonomy is undermined, and so forth. However, to contend that the absence of behavioral surplus means that the relationship between Zuboff's prototypical (and idealized) advocacy firm Apple and its consumers is free from the dynamics of unequal exchange is to ignore all the ways in which Apple regularly pushes its customers around. This even included deliberately not supporting iPhone repairs from third-party shops, at least until August 2019, when Apple launched its Independent Repair Provider Program in the U.S., which was expanded globally two years later. Since "advocacy" firms are defined only by their refusal to appropriate behavioral surplus, such habitual exercises of corporate power are not registered by Zuboff's theory.[24]

The myth of a virtuous, well-functioning advocacy-oriented capitalism quickly breaks down when applied to the real world. Zuboff herself lists a number of Apple's negative externalities—practices that are common to many capitalist firms: "extractive pricing policies, offshoring jobs, exploiting its retail staff, abrogating responsibility for factory conditions, colluding to depress wages via illicit noncompete agreements in employee recruitment, institutionalized tax evasion, and a lack of environmental stewardship."[25] In fact, Apple is becoming a full-blown surveillance-capitalist firm targeting massive behavioral surpluses. Apple tracks its users for many purposes and so reflects the trend of data extraction, except that its business model does not generally depend on the sale of this data. Apple receives substantial amounts of money from Google for allowing it privileged access to iPhone users. When it came to light that Apple's iOS and MacOS systems do collect information on user location and search activity, Apple's subsequent Privacy Policy of December 2017 stated that collected data "will not be associated with [a user's] IP address." Nevertheless, iPhone features still support the surveillance needs of marketers.[26] Further, at the introduction of its new iPhones 14 and 14 Pro, in September 2022, Apple left a striking policy change undiscussed, which the *Financial Times* and *Bloomberg* had recently reported. Apple wanted to expand its advertisement network barely one and a half years after it had set high thresholds for other advertisement sellers. For that purpose, it had opened over 200 job vacancies for the development of its own advertising platform, from AI specialists to professionals in commercial functions. Needless to say, this expansion contradicts Apple's image as a privacy advocate. The company campaigned for years against the "free" services of especially Facebook, which often mishandled user data. This culminated in stricter privacy rules, which since June 2021 apply to the App Store. "App Tracking Transparency" (ATT) is the rule that obliges apps to ask users for permission to follow their behavior on the iPhone, which most users refuse. In the eyes of advertising businesses, Apple is thus cutting off competitors and grabbing advertising dollars itself as the advertisements appear mainly in the App Store.[27]

To make the behavioral surplus of *users* so crucial to the theory as Zuboff does is to conclude that the extraction of surplus from all the other parts of value production does not matter, or perhaps even does not exist.[28] *The Age of Surveillance Capitalism* offers a thorough examination of how advertising-supported firms have incentives to extract ever more data, harming users, democracy, and much else in the process. What is missing is an account of how value—all of it, not just those parts accruing to behavioral surplus—is produced in the digital economy.[29]

As the political scientist James Muldoon argues, Zuboff's book is limited in its scope to the business model of advertising platforms and says little about other forms of exploitation in the digital economy. The overarching framework of "behavioral surplus = bad/service improvement = good" fails to notice the ways in which an extractive business model that commodifies human life is at the heart of both industrial and digital capitalism. When this framework is focused on other online platforms, the dichotomy begins to break down. For example, to what extent does ride-hailing company Uber's use of consumer data to improve its service constitute a form of advocacy-oriented capitalism? If the company uses the data to aggressively push out competitors, underpay workers and control drivers' behaviors (as has been proven), is this merely advocacy-oriented capitalism gone wrong or would it constitute full-blown surveillance capitalism? The problem lies in the fact that Zuboff's book does not recognize how broader aspects of the critique of surveillance capitalism could be applied to capitalism itself. The problem with surveillance capitalism is not simply the new techniques of surveillance but the competitive capitalist economy that drives them.[30]

Lastly, Zuboff's dating of the "discovery" of behavioral surplus as value generator to Google's venture into advertising does not reveal the geopolitical conditions that made this possible. Why did companies like Google and Facebook emerge in the U.S. to conquer the rest of the world? In-depth historical investigations point to carefully planned efforts—begun during the Cold War and undertaken in Washington, Wall Street, Hollywood, and only later, Silicon Valley—to facilitate the "global free flow of information," which is basically an euphemism for the global expansion of data-intensive U.S. businesses.[31]

For a deeper, more comprehensive understanding of the emergence of today's surveillance capitalism it is necessary to take its political economy history into account. John Bellamy Foster and Robert McChesney have made a Marxist attempt of this kind, as laid out next.

Surveillance Capitalism's Military-Industrial Roots and Its Entwinement with U.S. Intelligence Agencies

In their 2014 essay in *Monthly Review*, an independent socialist magazine, Foster and McChesney approach the notion of surveillance capitalism from a completely different angle than Zuboff. Rather than focusing on the economic changes of the last couple of decades, they present and define the concept in terms of much broader changes dating all the way back to the Second World War. In their view the evolution of the U.S. economy since the 1940s had required different forms of surveillance. This was facilitated by the close relationships between various institutions, with the military-industrial complex as the hub persisting throughout numerous structural shifts and changes in emphasis.

The authors explain how surveillance capitalism's inception arose from the postwar architecture that combined the massive promotion of sales in the form of a corporate marketing revolution based in Madison Avenue with the creation of a permanent warfare state. The latter was dedicated to an imperial control of world markets and to fighting the Cold War with its central base in the Pentagon. This constellation was buttressed by arms and fictional nuclear preparedness on the one hand, and a highly organized system of customer surveillance designed to encourage consumption on the other.

Crucially, under the model of the military-industrial complex as promoted by then–Army Chief of Staff (and later president) Dwight D. Eisenhower immediately after the war, the U.S. technological, scientific research and industrial capacity were to become "organic parts" of the U.S. military structure in conditions of national emergency, effectively giving the civilian economy a dual-use purpose. In a 1946 memorandum, Eisenhower noted: "The future security of the nation ... demands that all those civilian resources which by conversion or redirection constitute our main support in time of emergency be associated with the activities of the Army in time of peace."[32] The model became a permanent feature of the U.S. economy, giving birth to a sprawling military-technological-corporate complex, about which Eisenhower notably expressed his own second thoughts, ambivalence and even fear in his 1961 farewell address to the nation. Civilian industry, science, and academia were deployed alongside an exorbitant and ever-expanding war budget to finance the Defense Department's perpetual state of conflict with Cold War enemies and unruly populations, making the world safe for the unchallenged reign of U.S. monopoly capitalism (imperialism) and pump priming the economy whenever an additional surge of government spending ("military Keynesianism") was required.³³

According to Foster and McChesney, the military-industrial complex and the corporate marketing system constituted the two principle surplus-absorption mechanisms until the financial crisis of the 1970s, when a third vector of surplus-absorption was added—that of financialization, which supplemented the system as the previous mechanisms waned. They explain further:

> Like advertising and national security, [financialization] had an insatiable need for data. Its profitable expansion relied heavily on the securitization of household mortgages; a vast extension of credit-card usage; and the growth of health insurance and pension funds, student loans, and other elements of personal finance. Every aspect of household income, spending, and credit was incorporated into massive data banks and evaluated in terms of markets and risk. Between 1982 and 1990 the average debt load of individuals in the United States increased by 30 percent and with it the commercial penetration into personal lives.³⁴

The authors give a concise history of the military-industrial complex's surveillance of citizens, mentioning the crucial moment of the emergence of the "Army Files" (also known as CONUS) scandal in 1970–1971, whereby the Army had been spying on, and keeping files on, over seven million U.S. citizens through the use of over 1,500 plainclothes agents. The dossiers included a vast "subversives" file on civil rights demonstrators, anti-war protestors and others involved in "civil disturbances" or dissent within the Army. It was because of that scandal that Americans came to hear about ARPANET, the precursor of the internet, where these secret files were kept and where the limitless storage of data posed a threat to a healthy democracy and U.S. citizens' right of privacy.³⁵

The extent of Silicon Valley's integration with the U.S. government was revealed to the public in 2013, when the former NSA contractor and whistleblower Edward Snowden provided evidence proving that the U.S. National Security Agency (NSA) and Federal Bureau of Investigation (FBI) had direct access to the internal servers of nine major tech firms—AOL, Apple, Facebook, Google, Microsoft, PalTalk, Skype, YouTube and Yahoo—each of which provided direct access through major internet service providers to the NSA as part of the latter's secret projects such as Boundless Information and PRISM. Snowden provided documentary evidence that the NSA, in its own words, had managed to gain "direct access"—that is, independent of all intermediaries—to practically all data circulating on the internet within the U.S. sphere. It also gained access to data from mobile phones emanating from hundreds of millions of Americans as well as populations abroad. Foster and McChesney explicated:

> These monopolistic corporate entities readily cooperate with the repressive arm of the state in the form of its military intelligence and police functions. The result is to enhance enormously the secret national security state, relative to the government as a whole. Edward Snowden's revelations of the NSA's PRISM program, together with other leaks, have shown a pattern of a tight interweaving of the military with giant computer-internet corporations, creating what has been called a "military-digital complex."³⁶

Although much of the debate over the national security state arising from the Snowden revelations focused simply on the dangers of corporations colluding with the U.S. government (in addition to direct state spying), Foster and McChesney went on to debunk the prevailing assumption that massive corporate-controlled data banks were themselves secure, or that firewalls prevented personal information from circulating out of control. Rather, at the core of the system of capitalist surveillance was the selling of people's information to the highest bidder as an accumulation strategy. In the new surveillance complex, corporations, the military and domestic security forces had become increasingly interdependent.[37]

In the June 2018 issue of *Monthly Review*, the editors (John Bellamy Foster and Brett Clark) pointed out that Shoshana Zuboff defined surveillance capitalism more narrowly as a system in which corporate surveillance of the population is used to acquire information that can be monetized and then sold. Their persuasive critique was that this view effectively divorced surveillance capitalism from class analysis, and from the overall political-economic structure of capitalism. Moreover, it largely dodged the issue of the symbiotic relations between the military and private corporations—involving marketing, finance, high tech, and defense contracting—which was the primary focus of Foster and McChesney's 2014 analysis.[38] Nevertheless, even their approach does not go far enough regarding the reality of contemporary digital capitalism, as will become clear next.

The Heart of the Matter: Data Colonialism in Its Historical Context

To understand the full scope of current developments it is necessary to take a much broader perspective that focuses on what Nick Couldry and Ulises Mejias call "data colonialism" in their book, *The Costs of Connection* (2019). They argue that data colonialism can only be understood against the longer history of colonialism and capitalism. What is at issue here is "the systematic attempt to turn all human lives and relations into inputs for the generation of profit. Human experience, potentially every layer and aspect of it, is becoming the target of profitable extraction. We call this condition *colonization by data*, and it is a key condition of how capitalism is evolving today."[39] According to Couldry and Mejias, there were four key components to historical colonialism:

> the appropriation of *resources*; the evolution of highly unequal social and economic *relations* that secured resource appropriation (including slavery and other forms of forced labor as well as unequal trading relations); a massively unequal *global distribution of the benefits* of resource appropriation, and the spread of *ideologies* to make sense of this all (for example, the reframing of colonial appropriation as the release of "natural" resources, the government of "inferior" peoples, and the bringing of "civilization" to the world).[40]

In describing the transformations taking place today as data colonialism, they use the term "colonialism" because it captures poignantly major structural phases within human history and specifically within capitalism. Colonialism has not been the standard interpretation of what is changing in contemporary capitalism. Yet it is becoming increasingly clear that capitalism's current growth cannot be captured simply in terms of ever-more ambitious business integration or the ever-expanding exploitation of workers. David Harvey and others have characterized today's developments as increasing waves of "accumulation by dispossession," a feature characteristic of capitalism throughout its history.[41] But even this fails to grasp how the axis of capitalism's expansion has transformed through a shift in the supposed "raw material" that capitalism aims to get under its control.

What makes the current moment distinctively colonial, Couldry and Mejias argue, is the discovery of new forms of "raw material." Data colonialism appropriates for profitable exploitation a resource that did not emerge to be appropriated until some twenty-five years ago. Regarding access, contemporary businesses tend to approach data as if it was "just there," freely available for extraction and the release of its potential for humankind. In the history of colonialism, a similar claim was expressed in the legal doctrine of *terra nullius*, land such as the territory now known as Australia that supposedly belonged to "no one" (*nullius*).[42] In other colonies, the legal justification for appropriation could be more complicated. For example, in British North American colonies, John Locke's argument prevailed that common land was not meant by God to go undeveloped forever,[43] although even Locke wrote at times as if America was a *terra nullius*. In Spanish American colonies there was the fiction of the *Requerimiento*.[44] This Monarchical Edict of 1513 was read by conquistadores to newly found subjects—indigenous peoples—who, before attacking them, were abruptly informed that their lands belonged to the spiritual leader of the explorers, someone called the pope, successor to Saint Peter and leader of the Roman Catholic Church. The edict declared that the authority of God, the pope, and the king, was embodied in the conquistadores and then declared the native peoples as vassals subordinate to that authority.[45] After that, the way was clear for plunder, enslavement, and extremely brutal retaliation in the case of any act of indigenous resistance.

Zuboff, too, compares the basic features of "surveillance capitalism" to colonial conquest, especially that of the first Spaniards under Columbus who landed on the Caribbean islands. In her telling, this conquest unfolded in three stages: legalistic measures to provide the invasion with a cloak of justifications, a declaration of territorial claims, and the founding of a town to legitimate the declaration. Likewise, the first surveillance capitalists simply declared our private experience to be theirs for the taking, for translation into data for private ownership and then proprietary knowledge.[46] However, although Zuboff's surveillance capitalism thesis acknowledges colonial parallels to the actions of Big Tech corporations like Google, it never offers a developed notion of what might be colonial about contemporary data practices. At most, it recognizes in colonial history a striking echo. Moreover, the global and historical span of colonialism does not fit well with Zuboff's core argument that surveillance capitalism is a recent and deviant version of contemporary U.S. capitalism that, while disturbing, can be reined in to restore capitalism to an unproblematic trajectory.[47]

It should also be recognized that there were clear connections between colonialism abroad and internal colonialism, notably in Britain in the form of the enclosure movements. Starting from the early sixteenth century and gathering steam from the late eighteenth century onward, debates about the nature of enclosure and "improvement" of English commons and "wastes" drew upon the colonial experience overseas, both in terms of the language used and in making direct comparisons. (In this context, the word "waste" was used to describe all commons and waste lands.) Conceptual similarities were drawn between "planting" colonies overseas and planting farmers on wastes and commons at home. Enclosure was not something forged in the fields and wastes of England (and Wales and Scotland), and imposed on Britain's colonies. Rather, enclosure became a two-way process in which the language of British settler colonialism was deployed not only to describe the enclosures of commons and wastes in the British Metropole, but also to encourage and justify the practices in question.[48] From the later decades of the eighteenth century, commons and wastes were explained as empty spaces, *terra nullius*, the claim made as part of the discourse of British overseas colonialism. The idea of wastes being empty spaces was not altogether new though. It belonged to a discourse used to explain and justify the enclosure of large swaths of fens and forests from at least the turn of the seventeenth century, but by this time the influences and modes of making claims had shifted. As part of this process the peoples that inhabited commons and wastes were defined as "uncivilized" and unable to manage the land; natives of the land but with no ownership title. One can see this reflected in John Locke's belief that those who lived in forests and woods were "irrational, untaught," which itself was an established notion, deeply rooted in popular culture.[49]

The challenge of funding and running Britain's North American colonies and their subsequent loss after the American War of Independence gave rise to new models of how Britain's lands might be more profitably managed. Conversely, this provided the inspiration, example and justification for Britain's subsequent colonial endeavors. As E.P. Thompson notes, the concept of exclusive property in land, which was central to enclosure, was carried across the Atlantic, to the Indian subcontinent, and into the South Pacific, by British colonists, administrators and lawyers.[50] Thompson's point is that, in exporting the legal tools of enclosure from the Metropole, local customs in territories under British colonial rule were made to yield—or were even entirely reinvented—to embrace British conceptions of private property of land, and the primacy of exclusive title was considered a necessary condition for "improvement."

Arguments for the enclosure of wastes were made as an alternative to Britain's overseas imperialism. According to influential thinkers like Arthur Young, chair of the agricultural committee of the Royal Society in the 1780s, and others, this countermovement did away with the need to establish colonies beyond the bounds of the Metropole. Such costly speculations (and violences) were deemed both unnecessary (given the millions of acres of waste at home) and a betrayal of the needs of the British people and the economy. Thus, Young's schemes to colonize English waste were put forward as an alternative to the (failed) example of the North American colonies.[51]

Today's equivalent of territorial colonialism (both abroad and internal in the Metropole) is data as the "exhaust" of life processes, as it was called in a 2011 World Economic Forum report.[52] Data is assumed just to be out there for the taking. In the digital era, conquest by declaration can be clearly traced in the six critical declarations that Google conjured out of thin air when, as a young company, it first asserted them.[53] These declarations laid the foundation for the wider project of data colonialism and its particular form of dispossession—defended at any cost—as each declaration builds on the one before it. If one falls, they all fall, just like dominoes:

- We claim human experience as raw material free for the taking. On the basis of this claim, we can ignore considerations of individuals' rights, interests, awareness, or comprehension;
- On the basis of our claim, we assert the right to take an individual's experience for translation into behavioral data;
- Our right to take, based on our claim of free raw material, confers the right to own the behavioral data derived from human experience;
- Our rights to take and to own confer the right to know what the data disclose;
- Our rights to take, to own, and to know confer the right to decide how we use our knowledge;
- Our rights to take, to own, to know, and to decide confer our rights to the conditions that preserve our rights to take, to own, to know, and to decide.[54]

Thus, the first platform created by capitalists/data colonialists unilaterally declared people's private experience to be theirs for the taking, for translation into data for private ownership and their proprietary knowledge. They relied on misdirection and rhetorical concealment, with secret declarations users could neither understand nor contest. Or, put differently, first of all there was the arrogant appropriation of user's behavioral data—viewed as a free resource, there for the taking. Then the use of patented methods to extract or infer data even when users had explicitly denied permission, followed by the use of technologies that were opaque by design and fostered user ignorance.

The *Requerimientos* of our digital era are known as end-user license agreements (EULAs) or statements of rights and responsibilities (SRRs). As several authors have demonstrated through close readings of exemplary versions released by Google and Facebook, these documents specify how newly discovered resources and their presumed owners are to be treated by the present-day colonizers.[55] Whereas historical colonialism expanded by appropriating for exploitation geographical territory and the resources that conquest would bring, data colonialism expands by appropriating for profit ever more layers of human life itself. Thus, human life is quite literally annexed to capital. This goes far beyond exploitation of digital labor, on which much of the debate on contemporary capitalism is focused.[56]

Moreover, internet users cannot be seen as workers in that capacity. They do not receive wages, and their work is deployed to lower production costs, increase production and so forth. From a Marxist perspective, it is therefore "hard to make the case that what they do is labor, properly speaking."[57] Thus internet users cannot be considered part of the working class, nor can one say they are exploited in the traditional sense. But it can be argued that media consumers are exploited not because they literally work for media companies but because, as an audience, they become a commodity sold to advertisers without themselves benefiting at all from this arrangement.[58]

Data colonialism appropriates many specific aspects of human life: from work to school to health treatment to self-monitoring, as well as from basic forms of sociality to routine economic transactions, plus the framework of decision-making and direction of what is called "governance." Further, as ever more of people's activities and even inner thoughts take place in contexts in which they automatically are prepared for appropriation as data, there is, in principle, no limit to how much of human life can be appropriated and exploited—a clear reminder of the expansionary potential of capitalism that Marx envisioned, albeit actualized in circumstances he could not have foreseen.[59]

The new data colonialism occurs against the background of centuries of capitalism, and is taking familiar aspects of the capitalist social and economic order to a new and more integrated stage; so new in fact to be as yet unnamed, according to Couldry and Mejias. In their argument they mention three further aspects of data colonialism and its relation to capitalism's development. The first is that none of this would have been possible without the radical changes in communication infrastructures over the past thirty-five years, specifically the embedding of computer systems at so many levels in human life. This involves first and foremost the emergence of the revolutionary new technological infrastructure for connecting humans, things and systems, known generally as the internet. The second point is that technologies work, and have consequences for human life, only by being woven into what people do, where they find meaning, and how their lives are interdependent. In other words, data colonialism requires the creation of a new social and economic order that is potentially as enduring as the order that enabled capitalist market societies to thrive from the nineteenth century onward. The third point concerns how the power relations, generated by this emerging order, function. Data colonialism captures not only physical resources but also people's very resources for knowing the world. This means that economic power and cognitive power (the power over knowledge) are converging as has never happened before. Therefore, what is going on with data can be fully understood only against the background not just of capitalism but also the longer relations between capitalism *and* colonialism. As Couldry and Mejias forcefully argue: "The exploitation of human life for profit through data is the climax of five centuries worth of attempts to know, exploit, and rule the world from particular centers of power. We are entering the stage not so much of a new capitalism as of a new interlocking of capitalism's and colonialism's twinned histories, and the interlocking force is data."[60]

They emphasize that this is not just a narrative about the West, mentioning as a salient detail that the Chinese firm Tencent had already financed its fast-growing online chat platforms by a public offering in 2004, the year Zuckerberg launched Facebook. Neither is this a story about Western values with which capitalism's growth has long been associated. The emerging new order has important similarities whether one focuses on the United States, China, Europe or Latin America. The result is not a new type of capitalism per se but rather a new means and scale on which capitalism operates. In the long run it may provide the basis for an entirely new mode of production. What is at issue here is a new economic and social formation in which the orders of "liberal" democracies and "authoritarian" societies (including a state-led market society such as China) increasingly come to resemble one another.[61]

If historical colonialism involved the appropriation of land, human bodies (particularly slavery) and natural resources, data colonialism as understood here is a capturing of social resources, which represents both a progression and its return, potentially, to more brutal forms of exploitation. Couldry and Mejias acknowledge that, as data colonialism builds on the social relations of more than two centuries of capitalism, physical violence plays a lesser role in establishing it, but they do not deny the violence of data colonialism in other dimensions (symbolic, economic), or the fact that it can combine in forms of rule that amount to physical force.[62] Second, because of the globalization of science and economy, data colonialism is inevitably global, but it operates around two distinctive poles, China and the West, with other intermediate players. Third, data colonialism involves new forms of oppression. Historic colonialism was based on the violent seizure of land and physical resources, managed through the oppression of human bodies. Apparently, data colonialism may be less disruptive to existing ways of life, and unfold within very different timeframes, but nonetheless is expected to have fundamental long-term negative impacts.

This colonial reading of what is happening with data better addresses the scope of today's societal transformation through data extraction. Just as historical colonialism was an extractive model that reorganized life at every level, the turn toward data colonialism focuses not just on particular vectors, such as social media and search engines, but on wider habits of data collection all across economic life. This also means that solutions to the harm done by data colonialism requires fundamental societal change, and not merely the push for targeted reforms, as Zuboff seems to suggest.[63]

Couldry and Mejias claim that, since the dispossession of social resources operates in ways that replicate how the dispossession of natural resources once worked, data relations recreate a colonizing form of power. Data relations, as defined here, are ways of people interacting with each other and the world facilitated by digital tools. Through data relations, human life is not only annexed to capitalism but has also become subject to continuous monitoring and surveillance. These new types of human relations offer corporations a comprehensive view of people's sociality, enabling human life to become an input or resource for capitalism. The result of the incessant extraction of people's social interactions is to undermine the autonomy of human life in a fundamental way that threatens the very basis of freedom.[64]

In the new, datafied social world, earlier forms of social knowledge that had in various ways been publicly produced and controlled, become depreciated. A new social knowledge emerges that is entirely under corporate control, which Couldry and Mejias call "social caching."[65] It offers a virtually omniscient view of the social world, although parceled out among many corporations; segments which combine into a corporate social imaginary that justifies new ways of governing life for the benefit and convenience of corporate power. Thus, in the U.S. and other liberal market societies, the new data infrastructure involves market institutions for the first time in producing basic social knowledge. In a state-led market society such as the People's Republic of China, a similar process is called the "modernization of social governance." The state thereby acquires a significant new set of instruments to direct the production of social knowledge in its own interest.[66] It is now an explicit goal of the Chinese government to use artificial intelligence to "establish [an] intelligent monitoring platform for comprehensive community management." The result would be "a market improvement of the economic and social order," as the Chinese government put it.[67] Meanwhile in India—with its burgeoning IT sector's infrastructure being developed in close cooperation with the state in the form of public-private partnerships—a similar state-led production of social knowledge has developed since 2009 in the form of the Aadhaar identity-card system. It assigns a universal identification number to every citizen based on biometric data, supposedly to improve the distribution of social benefits but also to generate new commercial and security streams of data.[68] The government has made this a requirement for access to welfare services, tax dealings, and even the online bookings of train tickets. Interestingly, the system has faced greater civil-society opposition than the parallel developments in China.[69]

However that may be, the transformation of publicly-oriented government into "governance by proxy" has far-reaching consequences for societies' power relations. Data relations give corporations a privileged window into the world of social relations and a privileged grip on the levers of social differentiation. Nation-states have become increasingly dependent on access to what the corporate sector knows about the lives of their citizens, thus reversing the long-existing direction of knowledge transfer from states to corporations. The resulting relations have become highly controversial in the West—consider Apple's high-profile battle with the FBI over the encryption of its iPhones, and, more generally, opposition to U.S. security agencies (and their counterparts elsewhere) analyzing surveillance data about citizens' behaviors collected by platform companies. In other states such as China, however, the government has been heavily involved in encouraging platform development, in part because of their surveillance potential.[70]

This new social order relies increasingly on hidden forms of categorization and discrimination, implemented at many layers of an automated process. Two scholars in particular, Safiya Noble and Ruha Benjamin, focus on the workings of algorithmic power and code in their respective books *Algorithms of Oppression* (2018) and *Race after Technology* (2019), whereas race occurs only incidentally in other accounts of data power. Benjamin and Noble both argue that algorithms and code produced in the U.S. under conditions of deep and continuing racism, reproduce and amplify racial discrimination under a veneer of objectivity and scientific truth.[71] The result poorly serves "multiracial democracy"[72] and facilitates racial modes of social control. The implicit links to the colonial history which produced America's deeply racialized society and economy are obvious, including the ways knowledge, power and race have been intertwined since the colonial era.

The consequences of data colonialism for individual freedom and autonomy are just as serious, as Couldry and Mejias point out:

Whereas *specific* human subjects face danger from the new social hierarchies that are being built through managing extracted data, *all* human subjects face a threat to the minimal integrity of the self from data colonialism's reliance on ever-expanding mechanisms of surveillance and tracking. Lower-status work brings higher exposure to surveillance, with zero discretion, while higher-status work enjoys more protection from surveillance and greater discretion.[73]

It should be noted that while freedom is also a concern of Zuboff's "surveillance capitalism" thesis, her critique addresses first and foremost the citizens of the U.S. and Europe, and eschews the long histories of hierarchization between and among peoples that emerged through colonialism.[74] By emphasizing data extraction's continuity with colonialism's histories of ruling and dividing people on a global scale, the framework of data colonialism foregrounds continuities of today's quantification of the social (also known as datafication) with historic forms of inequality, rather than seeing it as an aberration of late modern Western democracies.[75]

As social relations are transformed this way, Couldry and Mejias also see the emergence of what they call the "Cloud Empire," "a totalizing vision and organizing of business in which the dispossession of data colonialism has been naturalized and extended across all social domains." This constellation of ideas and organizing structures is put into practice and extended by many players but primarily by "the social quantification sector," defined as "the industry devoted to the development of the infrastructure required for the extraction of profit from human life through data."[76]

This sector has been growing for a long time, in part through marketer accumulation of consumer data, such as credit card data, starting in the 1980s. Over the past twenty or so years, the social quantification sector achieved a new depth and complexity. As Couldry and Mejias explain, it currently includes the manufacturers of the digital devices through which people connect. This involves not only well-known media brands such as Apple, Microsoft and Samsung, but also the lesser-known manufacturers of "smart" (that is, internet-connected) fridges, heating systems and cars through which consumers never thought to communicate. Nor did people imagine that in the rapidly expanding Internet of Things, such devices would be able to communicate with other devices about their users. This sector further includes the builders of the computer-based environments, platforms and tools that enable people to connect with and make use of the online world—household names such as Google, Facebook and Amazon in America, and Alibaba, Baidu and WeChat in China. There is also the growing field of data brokers such as Acxiom, Equifax, Palantir, and TalkingData (China) that collect, aggregate, analyze, repackage, and sell all kinds of data while also supporting other organizations in their handling of data. Finally, the social quantification sector includes the vast domain of companies that increasingly depend on their basic functions to process data from social life, whether to customize their services (like Netflix and Spotify) or to link sellers and buyers (like Airbnb, Uber and Chinese ride-hailing platform Didi).[77]

Beyond the social quantification sector one finds the rest of the business sector, which has been transformed during this great data transition as well. Nowadays much of ordinary businesses' day-to-day operations involve crunching data from their internal processes and from the world around them, in which they depend increasingly on the work of the social quantification sector to target their ads and marketing. And then there is also the everyday context in which people are integrating the outputs of that sector into daily life.

In concluding this section on data colonialism it should be clear that the problems and challenges of current data practices are neither simply data nor the particular platforms that have emerged to exploit data, but the interlocking combination of the forces, as summarized by Couldry and Mejias:

> an *infrastructure* for data extraction (technological, still expanding), an *order* (social, still emerging) that binds humans into that infrastructure; a *system* (economically) built on that infrastructure and order; a model of *governance* (social) that benefits from that infrastructure, order, and system and works to bind humans ever further into them; a *rationality* (practical) that makes sense of each of the order levels; and, finally, a new *model of knowledge* that redefines the world as one in which these forces together encompass all there is to be known of human life. Data, in short, is the new means to remake the world in capital's image.[78]

Importantly, this analytical perspective does not assume that capitalism today is anything other than what it has always been. Contemporary societies are characterized by the ever-increasing importance of the circulation and processing of information. This has had profound impacts on the management of business, the organization of work, and the integration of social life into the economy. But that does not mean that the fundamental drivers of capitalism have changed; this is certainly not the case. When the term capitalism with a contemporary reference is used, this means capitalism as it is now evolving in societies in which the production, accumulation and processing of information is growing. Surveillance is certainly part of it, but not sufficient to label today's capitalism specifically as "surveillance capitalism," as within the longer history of colonialism and capitalism, surveillance has often accompanied the direct appropriation of laboring bodies for value extraction (think of the slave plantation in particular). "What is new today," Couldry and Mejias contend, "is not so much surveillance but rather the networks of social relations in which vast extended nodes of appropriating human life through data work to order economic and social life as a whole."[79] This then is the conceptual framework to be used in the following analyses of the various components of platform capitalism and the smart society.

2

The Internet

From a Government-Owned Network to a Privatized One

For a better understanding of the emergence of contemporary capitalism's relationship to data colonialism, it is necessary first to see how the internet was created, thereby laying the foundation for the later development of the social quantification sector. The internet began in the 1970s as an experimental technology created by U.S. military researchers, grew in the 1980s into a government-owned computer network used primarily by academics, followed by privatization beginning in the 1990s.

The Origins of the Internet

In the early 1960s, the Defense Advanced Research Projects Agency (DARPA), the R&D arm of the Defense Department, began investing heavily in computing, installing mainframes at universities and the other research sites where its community of contractors worked. But even for an agency so generously funded, this spending was not sustainable. In those days, a computer cost millions of dollars. So scientists at DARPA came up with a way to share its computing resources among its contractors more efficiently. This was to build a network, linking up Pentagon-sponsored computer mainframes belonging to government agencies, universities and defense contractors across the U.S. and NATO bloc.[1] The network was named ARPANET and it laid the foundation for the internet. It was designed as an "open and designable technology" through which scientists could communicate easily in a nonhierarchical environment. It was unlike the closed systems of corporate telecommunications in which private control over the bottleneck was the basis for profitability.[2]

ARPANET was set up with no central control, so that the system would be neutral or "dumb," leaving the power to develop specific applications to people on the periphery, who could participate as they wished.[3] This decentralized control meant that all machines on the network were peers, more or less; not one computer was in charge.[4] First connected in 1969, ARPANET linked computers through an experimental technology called packet switching which involved breaking messages down into small chunks, routing them through a maze of switches, and reassembling them at the other end. Today, this is the mechanism that moves data across the internet. At that time, the telecom industry considered it highly impractical; corporations had little interest in the internet during its formative decades. The following two examples are telling. In 1968, IBM declined to even bid on providing subnet computers, stating the venture was not sufficiently profitable. Four years later DARPA offered to let AT&T take control of the ARPANET. At that time DARPA would have preferred leasing lines; that is, to buy time on the internet, instead of managing it. Given the chance to acquire the most sophisticated computer network in the world, however, AT&T likewise declined the opportunity on the grounds that this would not be profitable.[5]

Under public ownership, ARPANET flourished. Government control gave the network two major advantages. The first was financial. DARPA could pour countless dollars into the system without having to worry about profitability. The agency commissioned research from the country's most talented computer scientists at a scale that would have been untenable for a private corporation. And, just as crucially, DARPA enforced an open-source ethic that encouraged collaboration and experimentation. The scientists who contributed to ARPANET had to share the source code of their creations. This fostered scientific creativity as researchers from a range of different institutions could refine and expand on each other's work without living in fear of intellectual property law.

As a result, the most important innovation was the internet protocol which first emerged in the mid–1970s. Initially, the protocol was a proposal for how computers should communicate. The proposal was subsequently implemented in software and refined through multiple experiments. This made it possible for ARPANET to evolve into the internet by providing a common language that let very different networks talk to one another. The language would be open and non-proprietary—a free and universal medium—rather than a patchwork of incompatible commercial dialects. Thus, it was the absence of the profit motive and the presence of public management that made the invention of the internet possible. Yet the internet would also reflect the institutional imperatives of the particular arm of the government that oversaw its creation: the military.[6]

In his book *Surveillance Valley: The Secret Military History of the Internet*, Yasha Levine posits that a primary purpose of conceiving the internet was the need for a computerized counterinsurgency tool that could predict and check the perceived global spread of communism and provide real-time surveillance of potential threat groups.

> The internet came out of this effort: an attempt to build computer systems that could collect and share intelligence, watch the world in real time, and study and analyze people and political movements with the ultimate goal of predicting and preventing social upheaval. Some even dreamed of creating a sort of early warning radar for human societies: a networked computer system that watched for social and political threats and intercepted them in much the same way that traditional radar did for hostile aircraft. In other words, the internet was hardwired to be a surveillance tool from the start. No matter what we use the network for today—dating, directions, encrypted chat, email, or just reading the news—it always had a dual-use nature rooted in intelligence gathering and war.[7]

The internet was first of all created to win wars.[8] DARPA had much leeway in selecting its projects, but it still had to develop technologies that might someday be useful for military ends. The internet was no exception. Its proponents within DARPA made the case that the internet was worth pursuing because it could give American forces an edge. This would come from taking computer power out of the lab and into the field. However, ARPANET had a significant limitation—it was not mobile. For ARPANET to be useful to forces in the field, it had to be accessible anywhere in the world. This required two things. The first was building a wireless network that could relay packets of data among the widely dispersed cogs of the U.S. war machine by radio or satellite. The second was connecting those wireless networks to ARPANET so that multi-million dollar mainframes could serve soldiers in combat, which the scientists called "internetworking."

Getting computers to talk to another—networking—had been challenging enough. But getting networks to communicate with one another—inter-networking—posed a whole new set of problems because the networks spoke different idioms. In response, the architects of the internet developed a common digital language that enabled data to travel across any network.[9] In 1974, two researchers, Vincent Cerf and Robert Kahn, published an early blueprint. Drawing on conversations happening throughout the international network community, they sketched a design for "a simple but very powerful and flexible protocol," a universal set of rules for how computers should communicate.[10]

These rules would make it possible to weave together a network of networks so versatile and so robust that, for instance, a soldier in the field or a pilot flying a war plane could connect to a mainframe computer halfway across the world. The experiments that DARPA conducted to test the new idiom of the internet in 1976 and 1977 were designed to do just that. They proved that the protocol developed by Cerf and Kahn had fulfilled its promise. Eventually, it would evolve into a whole suite of protocols called TCP/IP. Today, this is the *lingua franca* of the internet. Without TCP/IP's rules, the world's networks would remain a constellation of mutually unintelligible languages.[11]

In 1983, ARPANET switched over to TCP/IP, which allowed it to interconnect with other military and experimental networks. This new system became known as *the* internet, with ARPANET at its center. While the internet came into existence as a protocol, it would now become a place, one increasingly populated by civilian researchers—exchanging emails, accessing high-performance computers, collaborating, arguing and debating. The internet's apparent usefulness soon led scientists from outside DARPA's select circle of contractors to demand access. In response, the National Science Foundation (NSF)—a U.S. government agency tasked with supporting basic research—came up with a series of initiatives aimed at bringing more people online. These culminated in NSFNET, a program that oversaw the creation of a new national network. This network, which became operational in 1986, would be the new "backbone" of the internet, an assemblage of cables and computers forming the internet's main pathway. It resembled a river: data flowed from one end to another, feeding tributaries, which themselves branched into smaller and smaller streams. These streams served individual users, people who themselves never came directly into contact with the backbone. If they sent data to another part of the internet, it would travel up the chain of tributaries to the backbone, then down another chain, until it reached the stream that served the recipient.[12] In this model, the river is useless without the tributaries that extend its reach. That is why the NSF, to ensure the broadest possible connectivity, subsidized a number of regional networks that linked universities and other participating institutions to the NSFNET backbone. All this cost a lot of money, but it worked.[13] It has been estimated that the subsidies to the regional networks, together with the cost of running the NSFNET backbone, came to approximately $160 million. Other public sources, such as state governments and state-supported universities, contributed more than $1.6 billion towards the development of the internet during this period.[14] Thanks to all of this public largeness, the internet became widely available to American researchers by the end of the 1980s.

In retrospect, the internet's early success can well be explained in terms of its nonprofit origins and nonproprietary organizing principles; the principles of open cooperation that were to some degree built into its design and that encouraged its rapid global spread, arguably reflected the ethic of sharing and collective inquiry common to the research universities that fostered the internet's development. This is at right angles to the interpretation that the libertarian Electronic Frontier Foundation and similar organs of the computer culture gave at the time; that is, that the internet was a triumph, not of nonprofit principles or of cooperation between government and the private sector, but of what Thomas Streeter calls "a kind of romantic market entrepreneurialism"—a "frontier."[15] As this interpretation seeped into policymaking circles, eventually becoming the "common sense" of the day, any policy lessons that might have been learned from the internet's nonprofit origins thus were roundly ignored. Since the early 1990s, the only question has been how to completely commercialize the system, not whether or not to do this.

Encounters Between Bohemian San Francisco and Silicon Valley

The internet was noncommercial, even anticommercial during its first two decades, as Robert McChesney reaffirms. This is not only a matter of public subsidies with no return on investments, but it is also about a public ethos. Computers were regarded by many members of the 1960s and 1970s generation as harbingers of egalitarianism and cooperation, not competition and profits.[16] Apple's co-founder Steve Wozniak later recalled that everyone at his 1970s computer club "envisioned computers as a benefit to humanity—a tool that would lead to social justice."[17] As John Markoff has shown, in the computing world of the San Francisco Bay area in the late 1960s and early 1970s, industry engineers and computer hobbyists lived and worked side-by-side, and both were surrounded by countercultural activities and institutions.[18] In the Bay area in this period, the dynamic of personalization that had long been evident within some parts of the computer industry and the ideals of information sharing, individual empowerment, and collective growth that were alive within the counterculture and the hobbyist community, did not so much compete with as complete each other.

By the 1970s and 1980s, the computer professionals and students who comprised the internet community "deliberately cultivated an open, non-hierarchical culture that imposed few restrictions on how the network could be used."[19] The hacker culture that emerged at that time demonstrated a commitment to information being free and available, a marked hostility to centralized authority and secrecy, and a joy of learning and knowledge acquisition.[20] Here it is important to recognize the extraordinarily influential group of San Francisco Bay area journalists and entrepreneurs: Stewart Brand and the Whole Earth network. Between the late 1960s and the late 1990s, Brand assembled a network of people and publications that led to a series of encounters between bohemian San Francisco and the emerging technology hub of Silicon Valley. In 1968, Brand brought members of the two worlds together in the pages of one of the defining documents of the era, the *Whole Earth Catalog*. In 1985 he gathered them again on what would become perhaps the most influential computer conferencing system of the decade, the Whole Earth Electronic Link, the WELL. Throughout the late 1980s and early 1990s, Brand and other members of the network, including Kevin Kelly, Howard Rheingold, Esther Dyson, and John Perry Barlow, became some of the most quoted champions of a countercultural vision of the internet.[21] In 1993, they would help to create the magazine that, more than any other, depicted the emerging digital world in revolutionary terms: *Wired*.[22]

The history of this network of cultural entrepreneurs and journalists, which Fred Turner details in his book *From Counterculture to Cyberculture*, can help to explain the interweaving of two legacies: that of the American counterculture and the military-industrial research culture, which emerged during World War II and flourished across the Cold War era. Since the 1960s, both scholarly and popular accounts have described the counterculture in terms first expressed by its leading representatives—that is, as a culture antithetical to technologies and social structures underpinning the Cold War state and its defense industries. In this view, the 1940s and 1950s tend to be seen as a grey period characterized by rigid social norms, hierarchical institutions dominated by risk-adverse middle-management (the prototypical "organization man"), and the constant tensions of America's nuclear confrontation with the Soviet Union. The 1960s then appear to come onto the scene as a colorful whirl of personal exploration and political protest, much of it aimed at bringing down the military-industrial bureaucracy. Those who adopt this historical viewpoint tend to account for the persistence of the military-industrial complex decades later, as well as for the continuing growth of corporate capitalism and consumer culture, by arguing that the authentically revolutionary ideals of the generation of 1968 were somehow co-opted by the forces they opposed.[23] However, this version of the past obscures the fact that the same military-industrial research world that brought forth nuclear weapons—and computers—also gave rise to a free-wheeling, interdisciplinary and highly entrepreneurial style of work. In the research labs of World War II and after, and the massive military engineering projects of the Cold War, scientists, soldiers, technicians, and administrators dissembled traditional barriers of bureaucracy and collaborated as never before. In doing so, they embraced both computers and a new cybernetic discourse of systems and information. They began to envision institutions as living organisms, social networks as webs of information, and the collection and interpretation of information as keys to understanding not only the technical but also the natural and social worlds.

Turner points to parallels in the late 1960s, when substantial parts of the counterculture embraced rather similar views. Between 1967 and 1970, for instance, tens of thousands of young people set out to establish communes, many in the mountains and woods. It was for them that Brand first published the *Whole Earth Catalog*. For these back-to-the-landers (and many others who never actually established new communities), traditional political mechanisms for creating social change came to be seen as bankrupt. While their peers organized political parties and social movements and marched against the Vietnam War, this group, whom Turner calls the New Communalists, turned away from political action and toward technology and the transformation of consciousness as the primary sources of social change. If mainstream America had become a culture of conflict, with riots at home and war abroad, the commune world would be one of harmony. If the American state deployed massive weapons systems in order to annihilate faraway peoples, the New Communalists would deploy small-scale technologies—ranging from axes and hoes to amplifiers, strobe lights, slide projectors, and LSD—to bring people together and allow them to experience their common humanity. Finally, if the bureaucracies of industry and government demanded that men and women become psychologically fragmented specialists, the technology-induced experience of togetherness would allow them to become both self-sufficient and whole human beings once again.[24]

For this wing of the counterculture, the technological and intellectual output of American research culture held enormous appeal. Although they rejected the military-industrial complex as a whole, as well as the political process that brought it into being, many of them read Norbert Wiener, Buckminster Fuller, and Marshall McLuhan. Through their writings, young Americans encountered a cybernetic vision of the world, one in which material reality could be reimagined as an information system. To a generation that had grown up in a world dominated by massive armies and the threat of nuclear holocaust, the cybernetic notion of the globe as a single, interlinked pattern of information offered much comfort. In this play of information many thought they could envision the possibility of a global harmony.[25] Gregory Bateson's set of ideas about systems theory, ecology, and the human mind (involving an "everything is related" holism) especially caught their attention. Bateson's books from the late 1960s onward, most famously *Steps to an Ecology of Mind* (1972), were written in a highly accessible, engaging way that eschewed academic jargon and references.[26] These books could be read and understood by college students and literate members of the counterculture across the land, and even some precocious high-school students, who could make some sense of it without the guidance of professors or teachers. Stewart Brand elevated Bateson to the status of guru, particularly in the pages of *CoEvolution Quarterly*, a journal that Brand founded in 1974 using proceeds from the *Whole Earth Catalog*. In the early 1980s, *CoEvolution Quarterly* evolved into the *Whole Earth Software Review*; essays about solar power were replaced by reviews of the latest computer software, and *CoEvolution*'s non-profit egalitarian principles (e.g., all employees received the same pay) were replaced by a for-profit non-egalitarian salary structure.[27]

Cybernetics also presented a set of social and rhetorical resources for entrepreneurship to Stewart Brand and later, to other members of the Whole Earth group. In the early 1960s, not long after graduating from Stanford University, Brand immersed himself in the bohemian scenes of San Francisco and New York and quickly became what has sociologically been termed a "network entrepreneur."[28] He began to migrate from one intellectual community to another, serving as a link between formerly separate intellectual and social networks. In the *Whole Earth Catalog* era, these networks spanned the worlds of scientific research, hippie homesteading and organic farming, ecology, and mainstream consumer culture. By the 1990s, however, representatives of the Defense Department, the U.S. Congress, global corporations such as Royal Dutch Shell, as well as makers of all kinds of digital software and equipment were included.[29] Now there was a confluence of former counterculturalists, corporate executives, and right-wing politicians and pundits, which, at first glance, seems difficult to explain. But the history of the Whole Earth network offers significant clues. As those former members of the sixties counterculture turned toward technology consciousness and entrepreneurship as the guiding principles of a new society, they developed a utopian vision that was in many ways quite congenial to the insurgent Republicans of the 1990s (and like-minded corporate Democrats, one should add). Although Republican Speaker of the House Newt Gingrich and his circle (which included conservative author and media analyst George Gilder and futurist Alvin Toffler) loathed the hedonism of the 1960s counterculture, they shared its strong affinity for empowering technologically enabled elites, for building new businesses, and for rejecting traditional forms of governance.[30] Needless to say, the latter dovetailed with the economic libertarianism of those leading Republicans.

It is also noteworthy that, in the 1990s, both computer networks such as the internet and the social networks of the Whole Earth community became emblematic of what many claimed at the time was a new economic and political world. They began to imagine that the New Communalist dream of a nonhierarchical, interpersonally intimate society was on the verge of coming true. This was in large part thanks to the example of the Global Business Network and the writings of Kevin Kelly and Peter Schwartz, as well as to the work of *Wired* magazine.[31] The Global Business Network was founded in 1987 by a group of entrepreneurs including Peter Schwartz and Stewart Brand.[32] It was a consulting firm that specialized in helping organizations to adapt and grow in an uncertain and volatile world. It was particularly well known for using tools such as scenario planning and also offered experiential learning with the help of networks of experts and futurists.[33]

Despite their libertarian leanings, the writings of John Perry Barlow, Esther Dyson and Kevin Kelly in this period are imbued with a pining to return to a socially egalitarian world. For these influential writers, the early public internet seemed destined to model and help create a world in which each individual could act in his or her own best interest and at the same time produce a unified social sphere, a world in which we were "all one." In their view, people would reject the prevalent antagonistic politics, opting instead for the technologically mediated empowerment of the individual and the establishment of peer-to-peer gathering places. For these prophets of the internet, as for those who had headed back to the land some thirty years earlier, government was imagined as a bureaucratic behemoth that threatened to destroy the individual. And salvation, they claimed, lay in information, technology and the marketplace.

Stewart Brand and the editors, writers and entrepreneurs associated with the Whole Earth publications had thus completely reversed the political value of information and information technologies. The machines that once for them represented all social forces that threatened to end their lives and perhaps even to destroy the whole world, had now become portals to a way of living and working that promised to fulfill their youthful dreams of an egalitarian utopia, at least according to key members of the Global Business Network and the editors of *Wired*. Tied to the aspirations of the New Communalists, such heady assessments of computers and computer networks had become powerful ideological scaffolding for the techno-libertarianism of the 1990s and the internet bubble it helped generate.[34] Moreover, as Brand and the Whole Earth group realigned the cultural meanings of computing, they returned the technocentric, knowledge-oriented, collaborative social practices of the research world to the center of American culture at large.[35] Thus, the dominant computer culture came to consist of a deeply contradictory but politically potent fusion of sixties countercultural attitudes with a revived form of political libertarianism tied to economic neoliberalism.

The Transition to a Commercialized Internet

Before the early 1990s, the National Science Foundation Network (NSFNET) explicitly limited the network to noncommercial uses. NSFNET'S Acceptable Use Policy (AUP) banned commercial traffic, preserving the network for research and education only. The NSF deemed this politically necessary, since Congress might cut funding if taxpayer dollars were seen to be subsidizing industry. In practice, however, the AUP was unenforceable to a great extent as companies regularly used the NSFNET backbone. More broadly, the private sector had been making money off the internet for a long time, both as contractors and as beneficiaries of software, hardware, infrastructure and engineering expertise developed with public funds.

Yet the AUP did have an effect, but in another way. By formally excluding commercial activity, it spawned a parallel system of private networks. By the early 1990s, a variety of commercial providers had emerged across the nation, offering online services with no restrictions on the kind of traffic they would carry. Many of these networks traced their origins to government funding and recruited DARPA veterans for their technical expertise. But the fact that the commercial networks were prohibited from transmitting commercial content over NFSNET inevitably limited their value.[36]

At this juncture, the internet—which had thrived under public ownership—was reaching a breaking point. Massive demand from researchers strained the system, while the AUP prevented it from reaching an even wider audience. Meanwhile, the commercial networks were eager to expand (without restrictions) and capitalize on the growing enthusiasm for the internet. This demand was driven in part by the rise of the World Wide Web, which made being online much easier. The early internet was not very user-friendly, as it was dominated by text-heavy applications like email that generally necessitated some degree of technical skill. The web offered a new, more intuitive approach: a collection of hyperlinked "pages." What is a taken-for-granted practice now—clicking one's way through a chain of content—was revolutionary at the time.[37] The very first website was launched in August 1990; its single webpage contained information about the World Wide Web Project and instructions about how to use the new digital information management system.[38] The browser that popularized the web, Mosaic, appeared three years later. Once again, public money played a crucial role here. Tim Berners-Lee, the creator of the World Wide Web, worked as a scientist at CERN, the European research organization funded by nearly two dozen EU member states. Mosaic was developed at the University of Illinois's National Center for Supercomputing Applications, which had been created by the NSF in the 1980s. NSF next began to pursue privatization, taking the first step in 1991. A few years earlier, the NSF had awarded the contract for operating its network to a consortium of Michigan universities called Merit, in partnership with IBM and MCI. This group had significantly underbid, its members sensing a business that was realized in 1991. They then created a for-profit subsidiary that began selling commercial access to the MSFNET.

This move drew the ire of the nascent networking industry. Companies correctly accused NSF of cutting a backroom deal to grant its contractors a commercial monopoly and were successful in bringing about congressional hearings in 1992.[39] At these hearings, the desirability of privatization was not disputed, only its terms. Now that privatization had been set in motion, the other commercial providers simply wanted a piece of the action.

Previously the NSF had considered restructuring the NSFNET backbone to allow more creators to run it. By 1993, after the congressional hearings and extensive industry input, the NSF decided to take the more radical step of eliminating the MSFNET national backbone altogether. From now on there would be several backbones, all owned and operated by commercial providers. Supposedly the goal was to promote competition by creating a level playing field. But in reality the field remained tilted, open to only a few more players. Whereas the old architecture of the internet had favored monopoly, the new one would be tailor-made for oligopoly. There were not many companies that had consolidated enough infrastructure to operate a backbone—there were only five at the time. The NSF was not opening the internet to competition so much as handing it over to a small number of select companies waiting in the wings. Remarkably, this transfer of ownership came with no strings attached. There would be no federal oversight of the new internet backbones and no rules governing how the commercial providers should run their infrastructure. And subsidies for the nonprofit regional networks that had brought campuses and communities online in the NSFNET days, were stopped. Some had already created commercial spin-offs and those that had not done so were mostly acquired by bigger companies or went bankrupt.[40] Finally, on April 30, 1995, the NSF terminated the NSFNET altogether.

The swift changeover of the public infrastructure to fully private assets was facilitated by the fact that the government did not own it in the first place; the network had always operated on lines leased from contractors and had always relied on them to run the backbone on its behalf, just as ARPANET had done. Moreover, the TCP/IT protocols that held the manifold networks of the internet together, were non-proprietary, which means that anyone could use them and by the mid-1990s, many companies did so. Ben Tarnoff makes clear that the paths the internet took were not determined in advance; privatization was a political choice driven by commercial interests and pushed forward by the telecom companies' lobby. As he puts it, "The government had spent billions of dollars patiently developing a technology that had finally reached the point where it could serve as the basis for a business, and corporations were determined to reap the rewards."[41] However, as Robert McChesney points out—in comparison to the political debate that surrounded the emergence of radio broadcasting in the 1930s or the uprising against Western Union's telegraph monopoly in the late nineteenth century—there was hardly a murmur of public debate about whether this privatization and commercialization was appropriate and what its implications might be. As press coverage was nonexistent, the general public was uninformed, having no clue at all about what was happening.[42]

The Telecommunications Act of 1996 was a major legislative overhaul (with overwhelming bipartisan support) that abolished decades of public-interest regulation to aggressively implement a free market approach. Letting the private sector lead meant not only deregulation but also abandonment by the government of nearly all operational responsibility to oversee communication systems serving a public good. "The point of government deregulation, pure and simple, definitively became to help firms maximize their profits, and that was the new public interest," McChesney asserts. As he and like-minded critical media scholars have rightly argued, "deregulation" in communication is better understood, in effect, as "reregulation strictly to serve the largest corporate interests."[43] Rather than the rules being eliminated, they were revised to prioritize commercial interests.

All of this policy-making was the result of the neoliberal consensus, presided over by Bill Clinton and Al Gore and the New Democrats' third way policy program, that reflected a reorientation within the Democratic Party toward the interests of the technology and finance sectors and away from working-class politics.[44] Throughout this period, the free market orthodoxy applied not only to the internet, telecommunications, and media sectors, but also to banking, finance, and international trade agreements such as NAFTA. Despite turbulent partisanship and the impact of the 1994 Republican revolution (when Republicans won both the Senate and the House of Representatives), Democrats found common ground with the Republicans on a number of economic issues, particularly regarding telecommunications and media deregulation and the commercialization of the internet. The collapse of the Soviet Union strengthened the view of privatization as being both beneficial and inevitable, as the Cold War rationale for more robust public planning faded. There were significant structural dynamics at play as well, not least of all the overriding need for U.S.

capital to find new areas of profitable investment in the face of rising global competition. Neoliberal internet policy grew out of political and economic elites' response to the "long downturn" that began in the 1970s.[45] Domestic economic stagnation, a severe profitability crisis and increasing global competition incited elements of the private sector to drastically reorganize around computing, communication networks and global commodity chains.[46]

But despite its policy successes, the neoliberal consensus was not all-pervasive, impermeable or unchallenged. Throughout the early 1990s, there were contestations among government and private sector interests on various fronts, including battles over internet content deemed indecent (e.g., pornography) and encryption controls. However, these disputes were overshadowed by the broad agreement over privatization and commercialization. Due to the efficacy of business lobbying efforts, there was little room for alternative ideas regarding the fundamental state of affairs.[47] A telling example is the political trajectory of Al Gore. Prior to becoming a proponent of internet commercialization after he entered the White House as vice president in 1993, he had been a strong advocate of public investment in computing infrastructure development. In 1990 he argued that the foundation for the "information superhighway" should be a public network analogous to the interstate highway system. Commercial interests could use the network, much as commercial businesses used the highways. The telecommunication companies would play a role, get contracts and gradually increase their role, but the government would remain in charge, coordinate the system, and guarantee ubiquitous access and public interest standards.[48] Accordingly, then-Senator Gore introduced the High-Performance Computing and Communications Act of 1991. The bill was passed on December 9, 1991, which helped expand NSFNET. Gore saw the "information superhighway" envisioned by that bill as an investment that would stimulate economic growth and promote national competition—a politically significant issue at a time of widespread concern among U.S. policymakers over the rise of Japan. For the network to perform this function, Gore argued, the government had to ensure the benefits were broadly shared. His internet would therefore be a public-private partnership—operated by industry, but under federal oversight.[49] During the 1992 election campaign, and even into the first few months of governing, Gore argued that state-funded internet provision would be necessary to ensure that the information superhighway would be available to all, rather than a "private toll road open only to a business and scientific elite."[50]

The idea that the government would fund the new information infrastructure or reward the development of public interest applications was unacceptable to a broad swath of private sector interests. Once it came to Wall Street's attention, and under much pressure by lobbyists, Vice President Gore shifted his stance.[51] The public part of the public-private partnership gradually disappeared. The idea that the government would permanently play a significant role in the information superhighway receded, and the NSF pushed forward with full privatization.[52]

It should be clear that this privatization was a process, not an event. It did not involve a simple transfer of ownership from the public sector to the private but rather a more complex movement, whereby corporations inserted the profit motive into every level of the network. A system built by scientists and funded by public taxpayer dollars was refurbished for the purpose of profit maximization. This required the necessary hardware, software, legislation, and entrepreneurship, taking decades, and touching all of the network's many parts.[53]

3

Moving Towards Surveillance Advertising Led by Google

The political economy of surveillance advertising as this evolved later in the 1990s is a vital part of the larger history of the co-evolution of the internet and global neoliberal capitalism. After overcoming their initial reluctance to embark on surveillance advertising due to an anticipated pushback from the cyberspace community, marketers went rapidly digital. In 1993 the trade publication *Advertising Age* deplored how the internet was embedded in a culture that loathed advertising. Marketers feared that their efforts to use the web would be met by a tidal wave of opposition from a cyberspace community who regarded a commercial internet as "advertising hell."[1] But already by 1998, the effects of monetizing digital information access were obvious. Some proto-search engines sold "preferred listings" at the top of search results; others served advertisements outright. Whatever the path to commercialization, the end result was to manipulate what people saw on their monitors and steer people's clicks.[2]

In their milestone paper, "The Anatomy of a Large-Scale Hypertextual Web Search Engine," presented at the 1998 World Wide Web Conference, the founders of Google, Sergey Brin and Larry Page, addressed this troubling trend head-on:

> Currently, the predominant business model for commercial search engines is advertising. The goals of the advertising business model do not always correspond to providing quality search to users. We expect that advertising funded search engines will be inherently biased towards the advertisers and away from the needs of the consumers.
> This type of bias is very difficult to detect but could still have a significant effect on the market ... we believe the issue of advertising causes enough mixed incentives that it is crucial to have a competitive search engine that is transparent and in the academic realm.[3]

The original problem that Brin and Page were trying to solve was that of having too much data. "The number of documents in the [search engine] indices has been increasing by many orders of magnitude," they wrote in their paper, "but the user's ability to look at documents has not."[4] Their solution was an algorithmic method for ranking the quality and relevance of a webpage that relied primarily on counting how many other sites linked to it. This tended to produce more relevant results and, importantly, it scaled well. A vast and rapidly expanding patchwork would be made more intelligible through an automated analysis of how it was all tied together.

In introducing their search engine to the academic world, the founders of Google were remarkably prescient about the evolution of surveillance advertising as tied to search engines. They would soon make a complete turnaround, however. Abandoning their 1998 stance a few years later, Brin and Page would succumb to investor pressure to monetize their search engine via advertising and double down by serving targeted ads based on search queries.[5]

Key Moments in Internet Policy Making

For surveillance advertising, two moments of policy making are particularly important. The first was the overarching decision that the internet would be privatized and commercialized. Beginning in the late 1980s, federal policy makers worked closely with various commercial interests to establish what was framed as a "non-regulatory, market-oriented" approach to internet policy.[6] This meant that the private sector would lead internet system development and the government's primary role was to facilitate private profit-making. This left a regulatory gap regarding consumer data collection and gave the emerging online advertising industry free reign to build business models around hidden surveillance.

The second moment occurred at the end of the 1990s, when the predecessors of today's surveillance advertising giants faced the first public activism championing internet privacy. In response to increasingly invasive data collection, a coalition of advocacy groups campaigned to persuade legislators to reverse the laissez-faire approach to internet privacy. This public push notwithstanding, the U.S. Congress and the White House prioritized the growth of commercial internet over serious consideration of the implications of a surveillance-based digital economy. Though largely hidden from view by a mythos of friction-free markets and entrepreneurialism, the regulatory foundations of commercial internet surveillance were laid in this period through negotiations over privacy policies, user content, data merging, and industry "self-regulation."[7]

To the various private sector institutions that sought to mold the new interactive media of the 1990s into an efficient tool for marketing—what Matthew Crain calls the "marketing complex"—the World Wide Web was a double-edged sword: simultaneously a threat and an opportunity, the next frontier. The greatest threat was that interactivity would provide individuals with new kinds of media autonomy, perhaps even the ability to eliminate advertising altogether. To counter this threat, the marketing complex needed the support of the federal government as well as a backing from Silicon Valley investors. The politicians would make the rules that governed the web's commercialization while the venture capitalists financed the undertaking.

In the midst of the economic recession of the early 1990s, policy makers at the highest level of government were determined to stimulate growth through free trade agreements and deregulation of telecommunications and finance. In this context, the public sector and noncommercial origins of telecommunications were discarded so that private enterprise could take over to commercialize the digital revolution.[8]

Internet policy making under the first term of the Clinton administration took off with the formation of the Information Infrastructure Task Force (IITF), a group of federal officials charged with developing a master plan for the rollout of the National Information Infrastructure (NII). Even among high-level policy circles, there was still ambiguity about which technologies and services were about to be radically changed. But there was one common thought that whatever technological form it took, the communications revolution was destined to undermine individual privacy on a massive scale. A parallel study of the National Telecommunications and Information Administration (NTIA) affirmed the idea that the commercial internet would bring increased privacy risks and again mentioned consumer profiling as an area of particular concern.

Both the IIFT and NTIA warned that current regulations around electronic data collection were inadequate for the technological and market changes ahead. The analysts framed the problem in terms of balancing the social goods of individual privacy on the one hand and the economic benefits of "free flow of information" on the other. Building on a decades-old conception of "fair information practices," they outlined a privacy framework based on two basic principles: notice and choice. The idea was that people should be given notice about data collection practices so that they had the option whether to participate or not, but also that certain standards of privacy should apply universally. Warning about the historical lack of competition in the telecommunications sector, the NTIA advocated for a universal set of privacy safeguards. Without such standard protections, the agency argued, consumers would be at a considerable disadvantage in the coming privacy marketplace. A free market for privacy would mean that the most powerful, monopolistic internet companies would be allowed to dictate data practices without fear of competitive retaliation. Consumers would be forced to submit to unwanted surveillance or refrain altogether from using the new communications services.[9]

Ignoring the internet's potential regarding the dangers of commercializing personal information, decision makers at the highest levels made it a top priority to give business free rein to do just that. The fundamental organizing concept for federal internet policy in the 1990s was the maximization of private sector control. In matters of consumer data, this directive was put into practice via a sustained pattern of government inaction based on the simple belief that markets would solve any privacy problem that might arise. Time and again, calls for universal safety guards were ignored in favor of a patchwork of industry "self-regulation"—the very conditions that the NTIA had identified as detrimental to privacy protection.[10] It comes as no surprise then that this free-market approach was celebrated as an innovative policy that "worked so well" by economic analysts/ideologues at the time.[11]

Interactive Digital Advertising

Although there were strong social pressures to bring advertising to the internet, it was by no means a fait accompli that such efforts would be successful, particularly not on the World Wide Web. Early web technology was designed to be open-ended and flexible, but it was also anonymous and nonintuitive, therefore not optimally set up to serve the marketing needs of business. The formative business models and technologies of the internet advertising sector focused on the logistical problems of the advertising business. It was the intersection of data collection and networked ad distribution that enabled internet advertising to make its first generational leap. The prototype of this operational structure was introduced in the mid-1990s by a new kind of company that created ad networks to distribute banner ads to disparate sites across the web, that in turn provided unique opportunities to gather and combine user data. Among the first web advertisements, appearing in the fall of 1994, the most prevalent form was the banner ad, with a mix of text and graphical elements in a manner similar to print and outdoor media. But unlike other forms of such display advertising, banners offered a layer of interactivity. They could be configured as hyperlinks, enabling users to click through to visit a new website, dictated by the ad's sponsor.[12]

Combining long-established practices of ad sales outsourcing with the interactive properties of the new digital medium, ad networkers positioned themselves as intermediaries between web publishers looking to sell ad inventory and marketers looking to reach sizable audiences. Their point of intervention was logistical; utilizing the technical affordances of web communication to facilitate the distribution of advertising on a large scale, ad networks built an organizational and technical infrastructure to support subsequent innovations in collecting and deploying user data for targeted advertising.[13]

The most important ad network was DoubleClick, an early pioneer and enduring market leader that was among the first to incorporate surveillance into web advertising in a systematic way. DoubleClick exploited the flexible design of the web communications protocols in order to build a unique system in which every ad served was also an opportunity to gather data about internet users. This laid a foundation for the increasingly invasive forms of customer surveillance that became the center of the internet advertising industry. Ad networks drove the scale and precision of internet advertising far beyond what was available at the time, enabling the new industry to grow rapidly. This growth was dependent on a huge influx of capital that stemmed from the dot-com investment bubble in the second half of the 1990s. Many of the large-scale surveillance advertising operations conducted today by companies such as Google and Facebook (Meta) descend from this basic ad network model. In fact, in 2007 Google made a significant move to upgrade its surveillance advertising operations by acquiring DoubleClick, which now constitutes a major division of the advertising network.[14]

By systematically coupling banner ad delivery with data gathering, and by working to ensure that such practices were enabled as default web browser settings, ad networks set the stage for surveillance advertising's subsequent growth. But as banner ads spread, their novelty began to diminish significantly. Most of the corporate advertising done online in the late 1990s was ineffectual. Web publishers became well aware that the majority of users simply did not click on ads. And without robust interactivity, the low bandwidth at the time was a poor substitute for existing branding platforms like television. In the search to remedy the situation, the idea that became most prevalent among advertisers was that ads simply needed to be more "relevant" to consumers. This meant greater personalization of messaging that seemed to give consumers greater control over their media experiences. Personalization required greater knowledge about web users, which dovetailed neatly with evolving needs for data collection and user identification in other sectors such as internet retailing and banking.

In seeking to improve the relevance of ad targeting, internet advertisers found it necessary to develop the web's ability to recognize and remember individual users. Ad companies and publishers began to work on solutions that required users to register and log in. However, it was Netscape, maker of the then-leading web browser software, that developed a scalable fix for the problem in question. This was the HTTP cookie, an innovation that enabled tech servers to uniquely identify web browsers. Passing a cookie back and forth created a continuous communication state between browsers and servers, which facilitated a new range of data collection practices on the web. Released as an open technical standard, cookie functionality was quickly integrated into all major browsers and put to a variety of uses encompassing e-commerce, credentialization and personalization. The widespread implementation of cookies altered the web's trajectory by introducing a new capability for surveillance into what had formerly been an anonymous communications environment.[15]

The early success of the ad network business model, based on the integration of tracking technologies like the HTTP cookies, thus brought about an important shift in the web's configuration of user privacy. Before cookies, no company was able to reliably identify activities of individual users without forcing people to register and log into websites. Cookies offered any business on the web an inconspicuous means to link online actions with individual web browsers. Because they were built into the web's standard communication protocols, cookies were automatically passed back and forth behind the scenes during the browsing experience. Voluntary site registration and user questionnaires became outmoded by the new methods of data collection that need not deal with the problem of getting the user's permission.[16]

Ad networks had the ability to reach a more holistic understanding of the situation as users moved from site to site. The deployment of cookies gave ad networks a way to develop improved user tracking and profiling capabilities and to make a renewed business case to publishers, marketers, and investors by showing the capabilities of targeted banner advertising. Cookies added an unprecedented level of granularity, generating early explorations of what would later be called behavioral profiling. The regularly updated information stored in cookie files provided the crucial material used when building databases of user profiles, enabling surveillance of individual web browsers as a proxy for individual consumers.

A very popular innovation was called clickstream analysis, which involved tracking the movements of individual web browsers over time. This made it possible for ad networks to compile browsing histories that included information about the time, duration and order of the sites users visited.

From the start, cookies were infrastructural, a background technical feature to support a variety of second-order practices. There were no options for users to manage cookies, nor was there any indication that they were being placed on users' machines as they surfed the web. It was only several years later that browser makers did add user controls over cookies. (This occurred in the midst of mounting public scrutiny and debates regarding a revised specification that disabled third-party cookies by default proposed by the Internet Engineering Task Force [IETF], a volunteer standard-setting community that developed internet protocols.) But even then, browsers were configured to accept all cookies by default, including those from third-party ad networks—undeniably a significant victory for ad networks and the surveillance advertising business model at large.[17] It must be emphasized that the proposed IETF cookie specification was not against third-party advertising when conducted on a voluntary, opt-in basis. It was only opposed to hidden third-party surveillance as a default setting on the web. The proposal was specifically aimed at the lack of transparency involved in third-party cookies and the absence of consumer control over their use.[18]

The Dot-Com Bubble and Increased Demand for Web Advertising

Between 1995 and 2000, the U.S. economy was dominated by a financial market boom and bust that revolved around the commercialization of the internet: the dot-com bubble. This bubble played a significant role in generating surveillance advertising through a marketing/finance feedback loop in which the most important business competence was the ability to attract investment capital. This was achieved to a significant degree through advertising and public relations, whereby companies sought to demonstrate their potential to become dominant in a given online market. Dot-coms with a strong market position and favorable media profile found it much easier to attract investors, while securing risk capital through IPOs (Initial Public Offerings) and other means was deployed as a public relations event in its own right. At the same time, companies that had secured investment funding spent heavily on advertising to further build market share and enhance brand image, which in turn fueled more fundraising.

Internet advertising companies deployed venture capital to roll out new services, acquire competitors, and invest in infrastructure, all the while operating at enormous losses. Tech startups were among the web's biggest ad spenders, legitimizing the commercial internet at a time when many traditional marketers were still ambivalent about its prospects as a sales channel. Large internet advertising outlays were rationalized through a "new economy" discourse in which conventional metrics of economic valuation based on objective indicators of business performance and underlying market fundamentals were superseded by measures of publicity such as brand recognition and public image (as assessed through behavioral analysis). Advertising thus became an important dot-com business strategy, necessary not only to acquire customers but also to attract the next round of vital investment capital.[19]

Many traditional marketers remained doubtful about the efficacy of online advertising campaigns since the prevailing banner format was not producing adequate proof of return on investment (ROI). Although data collection techniques developed by ad networks such as DoubleClick provided new opportunities for ad targeting, marketers pointed out that, by 1998, average click-through rates had plummeted to below one percent for most campaigns. The fact that ads were bought and sold according to numbers of impressions, not clicks, made the ROI problem worse. Today, many internet ads are effectively free if no one clicks on them, as marketers only pay for click-throughs. This kind of performance-based pricing was far less common in the 1990s, when almost all ads were sold on the basis of cost per thousand ad impressions (CPM), a standard pricing system for print and broadcast media. Marketers began to question why they should pay for impressions on an interactive medium that seemed to produce hardly any such interaction. Marketers stressed the need for better data to justify spending real money online. The web's user base was growing fast, but it was still far away from achieving the kind of scale and saturation readily available via broadcasting. This was also the pre-broadband era; limited bandwidth and network latency issues (regarding the time it took for data to go from the source to the destination) made it virtually impossible to deliver quality video and audio content. To prove its worth, the internet advertising industry needed to improve its capacity to target specific groups of consumers and demonstrate that online ads could indeed change consumer behavior as marketers intended.[20]

As the financial bubble intensified in the late 1990s, leading ad networks like DoubleClick and CMGI began to reconfigure their business around two overlapping objectives. The first was to get big fast in order to squeeze out competition and secure a dominant market position in the web advertising sector. The second objective was to transition from a business model that primarily served web publishers (the supply side of the advertising market) to one that also provided services to marketers directly (the demand side). Until then, much of the industry had focused on helping publishers sell ads, leaving traditional ad agencies to help marketers buy them. Now major ad networks attempted to facilitate both supply and demand, providing new products and services geared toward marketers and their ad agency proxies.

These efforts joined together to form a business strategy that prefigured an approach aimed at what Nick Srnicek calls "platform monopoly."[21] With this turn to the advent of platform capitalism, the trend towards a wholesale commodification of social relations and a universalization of the neoliberal corporate paradigm only accelerated.[22] In the case of a platform monopoly, the goal is not just to successfully compete within a market but to arrange the market itself as an essential intermediary. Using dot-com financial capital, ad networks began to create services not only for web publishers who were paid to host ads on their sites, but also for the ad agencies, marketers and retailers who bought the ads and increasingly demanded proof of their results. Crucially, ad networks thus aimed to deliver on their promises to provide marketers with enhanced ROI. In a move that anticipated the later platform monopolies of Google and Facebook, ad networks began a process of what has been called proto-platformization, constructing a sociotechnical infrastructure of business relationships and technical capacities to facilitate the capture and exchange of consumer information at an accelerating scale and pace. The trend of proto-platformization expanded surveillance capacities and generated strong market pressures for all commercial entities to participate in consumer data collection. It also offered leeway for increasingly invasive, manipulative and discriminatory practices of behavioral profiling and ad targeting.[23]

One of the most worrying consequences of surveillance advertising's expansion in the late 1990s was the increased use of behavioral profiling to categorize people according to determinations of their social worth.[24] The information technology-based sorting in question accelerated forms of discrimination in accordance with institutionalized biases regarding race, gender, class, age, culture and so forth. Of course, the practice of tailoring ad messages to specific audiences is not new. But online behavioral profiling is of a qualitatively different character than the probability-based methods traditionally used to analyze and target mass media audiences. Consumer classification is nowadays greatly enhanced by surveillance practices that combine past purchasing records with online behavioral data and demographic information.[25]

Digital Enclosure of the Internet

The trajectory of proto-platformization in the late 1990s was a key moment in the internet's digital enclosure—the creation of an interactive realm wherein every action and transaction generates information about itself.[26] As noted earlier, this constitutes the core of data colonialism, which has certain similarities and continuities with historical colonialism, including both colonialism abroad and internal colonialism as epitomized by the enclosure movements in Britain. The original theory of enclosure is relevant here because it highlights the deep connections between capital accumulation and social division. It implies a Marxist understanding of the land enclosure movements associated with the historical transition from feudalism to capitalism, the process whereby communal land over time was subjected to private control, allowing private landowners to set the conditions for its use.

The enclosures (particularly those between 1760 and 1820 in England) were fundamentally about bringing public domains (that had up until then been exempted from private ownership) into the new and expanding commercial relations that marked the growth of capitalism. Former ways of providing food and sustenance—strip farming; labor relationships based on obligation and deference; widespread access to, and availability of, common land for grazing, hunting and collection of fuel—were abandoned in the name of efficiency, progress and private property rights. In the process, agriculture was radically marketized; the farm laborer lost access rights to land as common land was privatized. A large component of the new industrial labor force was created by these developments as dispossessed land laborers were forced (out of bitter necessity) to seek employment in the growing towns and cities.[27]

Land enclosures solidified social cleavages around property ownership and the means of production, in which those who did not own those means had to sell their labor for access to arable lands or factories. Likewise, digital enclosures create class-like divisions between those who own and control privatized interactive spaces (virtual or otherwise) and those who submit to particular forms of monitoring in order to gain access to goods, services and conveniences.[28] The latter involves a deliberate engineering to facilitate what David Harvey calls "accumulation by dispossession." This is an updated and more fine-tuned version of what Marx called "primitive accumulation"[29]—with privatization and commodification, and more extreme forms of financialization, as key components of the neoliberal formation. These were taken to a whole new level during the formative years of internet policy development.[30]

The term "digital enclosure" is used not only to invoke the notion of a space—virtual or otherwise—that is made interactive, but also to highlight the *process* of enclosure, whereby places and activities become captured by the monitoring embrace of an interactive (virtual) space. This movement was accompanied by a significant shift in social relations; entry into the digital enclosure involved, in most cases, the condition of surveillance. This then normalized the default mode of media engagement in which participation in interactive spaces demanded submission to surveillance, a stark reversal of the celebrated emancipatory potential of the internet's interactivity.

Consumers are free not to interact, of course, but they may find themselves compelled to engage in such interactive exchanges (and to go online) by what has been described as "the tyranny of convenience."[31] Indeed, interactivity is not necessarily a two-way street; often it amounts to the offer of convenience in exchange for submitting (either knowingly or unknowingly) to increasingly detailed forms of information gathering.[32] But it is not always just a matter of convenience; it has become harder to opt out because many businesses now demand opting in from their customers. This may even be a necessity when items that are unavailable in brick-and-mortar stores can only be purchased in online shops such as Amazon. Moreover, governmental agencies may store citizens' data they collect on major tech companies' servers, often with the help of private contractors. Currently, in the case of the Centers for Medicare and Medicaid Services (CMS), for example, the U.S. government uses Amazon Web Services, purportedly "to deliver streamlined customer experience" through Healthcare.gov, the portal where "consumers" can find information and sign up for insurance plans under the Affordable Care Act.[33] Obviously, the result is that it has become increasingly difficult for citizens to voluntarily refrain from using the internet.

The digital platforms that came to dominate life in this interactive realm are arguably best understood as digital shopping malls, as Jathan Sadowski suggests.[34] The shopping malls of the internet are nothing if not privately owned public spaces. They are corporate enclosures with a wide range of interactions taking place within them. Just as with a physical mall, some of these transactions are commercial, such as buying clothes from a merchant, while others are social, such as hanging out with friends. But what distinguishes the online mall from the physical mall is that within the former, everything one does generates data: every move, however small, leaves a digital trace. And these traces present an opportunity to create a completely new set of arrangements. Brick-and-mortar malls are in the rental business; the owner charges tenants rents, essentially taking a portion of their revenues. Online malls can make money more or less the same way, by taking a cut of the transactions they facilitate. But online malls are also able to capture another kind of rent: data rent. In other words, their owners can collect digital traces generated by the activities that occur within them. Since they control every part of the enclosure—and modifying the enclosure is simply a matter of deploying new code—they can introduce architectural changes as needed in order to cause these activities to generate more or different kinds of traces.[35]

What the digital enclosures and land enclosures have in common is the purpose of squeezing profits from as yet unexploited social domains, and thus expand the sphere of commodification. In the end, platformization makes digital media engagement more productive for capital by creating new opportunities for information commodification, centralizing control over interactive spaces and making it increasingly difficult for people to opt out of digital surveillance.[36]

Consumer Feedback versus Meaningful Participation

With the support of the White House, Congress and the Federal Trade Commission (FTC), the emergent internet advertising sector gave a positive spin to the policy vacuum regarding public-interest funding and regulation of the internet, framing it as victory for "consumer empowerment." Foundational U.S. internet policies that were heavily skewed toward corporate interests were misrepresented to the public as a kind of cyber-enhanced exercise of participatory democracy.[37] Individuals were deemed "free" to bargain with companies over their data collection practices—"empowered" to take their privacy into their own hands.[38]

In these juxtapositions of empowerment and participatory democracy versus centralized control, the very meaning of the term *democratic participation* is at stake. If participation simply means the ability to provide increasingly detailed information about ourselves, then the offer of participation is only the perfection of marketing strategies—both commercial and political. If, however, participation means a conscious, considered, informed and meaningful contribution to the governing process, it is important to distinguish this from the advertising sector's rhetorical view of participation, which equates submission to detailed monitoring with participation.

Here one hits upon the crucial difference between meaningful participation and consumer feedback. Marketers have a bottom-line goal that remains inaccessible to consumers, even though information about those consumers' needs and desires is crucial to attaining this goal. It is necessary to differentiate between two layers of feedback in its broadest sense: the first allows for the adjustment of strategies to achieve a given end (e.g., boosting sales of a product or service); the second influences the goal-setting decisions themselves (e.g., whether profits are more important than other aims). The market-based model of interactivity promises shared control of the second level but it delivers only on the first. This model aims at generating cybernetic feedback about the transactions themselves, which becomes the property of private companies that can store, aggregate, sort and in many cases, sell the information to others in the form of a database or some other digital commodity.[39]

Democratic politics, on the other hand, promises public participation all the way up to the goal-setting process itself. It is this level of participation that might be considered one element of the definition of meaningful participation. A second element is the creation of optimal conditions for public deliberation of shared goals. As constitutional scholar Cass Sunstein suggests, the adoption of marketing and advertising techniques by political campaigns ignores an important difference between consumer decisions and political decisions. While the former relate to individual preferences and only indirectly influence society as a whole, the latter are explicitly about collective decisions that directly influence the societal whole. Moreover, as Sunstein argues, political participation envisions a decision-making process that "does not take individual tastes as fixed or given. It prizes self-government, understood as a requirement of 'government by discussion,' accompanied by reason-giving in the public domain."[40]

The database-informed campaign model of political marketing transposes the workings of a consumerist model onto the political process. Far from contributing to democratic participation and deliberation, the version of interactivity envisioned by the consultants and target marketers concerned offers to perfect a cybernetic form of public relations: the customization of marketing appeals that are based on detailed profiles of individual voters. The consequences of this model of interactivity are destructive to democracy as they entail the further disaggregation of the citizenry as a public, the facilitation of sorting and exclusion when it comes to information access and the further normalization of surveillance as a legitimate political tool.[41]

The danger of this model of interactivity as cybernetic feedback is that it propagates a form of participation that amounts to actively setting the stage for citizens' submission by helping marketers—both political and commercial—increase their leverage over the former. By contrast, the unrealized promise of genuine interactivity poses a challenge to the forms of asymmetrical top-down control that its deployment often facilitates. As Mark Andrejevic points out,

> The technological capacity of interactivity will not, on its own ... dismantle social, political, and economic hierarchies. It will not on its own foster a version of democracy based on collective control over the shaping of political goals. Only the deliberately political use of new media technologies can have such results—and only a political struggle for control over the means of interactivity and the databases can enable such usage.[42]

It should become clear that much of what passes for interactivity in the digital economy is better understood as techniques for facilitating the scientific management of consumption. This entails indeed a reliance on detailed forms of continuous data collection to help allocate resources more effectively not only in the realm of production but also in those of marketing and advertising. As media theorists Frank Webster and Kevin Robins put it in their insightful history of information technology, "The contemporary ideologies of the information society have their tap-roots in that philosophy of productivism and progress propounded by Frederick Winslow Taylor and the ideologues of Scientific Management."[43]

The deployment of the promise of interactivity in the current era harkens back to early twentieth-century techniques of management of production and consumption. In those decades one can identify the origins of organized and systematic information gathering, analysis and distribution in the workplace, in the organization of consumption, and in political relations. This period marked the real take off of what James Beniger considers the "control revolution" according to his seminal study of the origins of the information society, which offers a detailed overview of the rise of technologies and techniques of communication and information processing, and their use for controlling social and economic processes.[44]

Social Taylorism in the Spheres of Marketing and Advertising

Even if the workplace is subject to forms of direct control quite different from those associated with the market place, there is a complementary relationship between the two that allows strategies associated with the one to be applied to the management of the other. As Webster and Robins suggest, this means that the surveillance-based rationalization of both got under way not at the end of the twentieth century, but from its inception with the emergence of social Taylorism.[45] They contend that "in so far as they represent endeavors to better manage the affairs of the corporation beyond the workplace, then the development of market research, mail order, credit agencies, annual product models, public relations, advertising and so forth can be interpreted as an extension of Taylorism from the factory throughout society."[46] So if watching workers helped make them more efficient, monitoring consumers became an integral component of managing distribution.[47] The complex and expanding system of mass production and mass consumption could only be coordinated and regulated if the criteria of efficiency and cost-optimization were extended from the factory to society as a whole. The system of consumption in particular needed to be brought within the purview of scientific management. It became increasingly apparent that both economic and social stability depended upon continuous and regular consumption along with matching demand to cycles and patterns of production.[48]

Whereas in premodern society producers benefited from a direct relationship with consumers that enabled them to gauge their needs and desires directly and personally, in the era of mass production, these consumers became part of an imagined consumption community whose needs and desires needed to be both studied and, most importantly, managed. Ultimately what was required was the scientific management of need, desire and fantasy and their reengineering in terms of the commodity form.[49] Thus, Taylorist principles and practices were extended to the marketing sphere.

As nascent businesses, the marketing and advertising industries drew on the management techniques of their era, including the monitoring-based approaches of scientific management. The lesson of the rise of social science's "instrumental association" with social management in America—which began in the late nineteenth century—was that effective forms of control and guidance relied on knowledge derived from close observation and detailed research.[50] This shared understanding provides a layer of continuity between scientific management, marketing, and corporate and political public relations. The affinity between the last three domains has been well documented, but their connection to the first one—workplace management—tends to be downplayed by separating the analysis of consumption from that of production. The notion of "social Taylorism" reconnects these domains and identifies their affinities. Cultural historian Stuart Ewen succinctly argues: "As Ford's assembly line utilized 'expensive single-purpose machinery' to produce automobiles inexpensively and at a rate that dwarfed traditional methods, the costly machinery of advertising … set out to produce consumers, likewise inexpensively and at a rate that dwarfed traditional methods."[51] A further link between the spheres of production and consumption is epitomized by the fact that increased worker productivity was a precondition for two things: the production of consumer goods and the formation of what can be described (through analogy to the "workforce") as a "consumerforce" with the time and resources to afford a consumption-oriented lifestyle.[52]

The fields of marketing and advertising also have an intriguing relationship with journalism that should be taken into consideration. Seen from the perspective of career training and specialization, those fields remain distinct forms of social practice. But viewed as business concerns, they are thoroughly intertwined. As media historian Robert McChesney notes, "Those media that depend upon advertising for the lion's share of their income … are, in effect, part of the advertising industry."[53] The implication here is that the mass media play a particular role in the feedback loop whereby marketing messages are targeted to consumers. It can be argued that, in important respects, the commercial mass media—despite being described by the promoters of the new media "revolution" as being too one-directional and top-down—have been interactive from their inception. In other words, the version of interactivity on offer by commercial media in the digital era represents not a decisive break with previous forms of mass media, but rather an extension and intensification of the forms of monitoring that their development helped pioneer.

The slippery nature of the term *interactive* pops up again here. Accounts of early newspapers conclude, for example, that their product was truly participatory, in the sense that the boundary between readers and writers—the consumers and producers of the public debate—remained blurred. All this changed with the increasing commodification of information and the birth of commercial mass media. As they became more commercial, increasingly professionalized and less participatory (in the sense of allowing readers to share in the production process), mass media simultaneously engaged in more systematic forms of consumer monitoring and information gathering, largely driven by their dependence on advertising. As readers came to have less input into content, they made a greater indirect contribution in the form of the data they produced about themselves to pollsters, the ratings industry and market researchers.[54] The formation of consumer management industries in the early twentieth century—marketing, advertising and the mass media they supported—was characterized by the development of a cybernetic interactivity, an increasing reliance on feedback-based forms of social control. Here "consumption engineers" were particularly important; they used strategies and interventions to regulate economic transactions and consumer behavior.[55] These advocates of big business were the first who turned to the "rational" and "scientific" exploitation of information in the wider society. And it is their descendants—the advertisers, market researchers, opinion pollers, data brokers and so forth—who are at the heart of information politics in the current era.[56]

The portrayal of the mass media by promoters of the "interactive revolution" brought about by digital media and the internet tends to downplay or overlook this historical fact, perhaps in part for the purpose of highlighting the contrast between the interactive media and their mass media precursors. Relegating the forms of interactivity that previously characterized the mass media industries to the background, makes it easier to depict the interactivity—even in its cybernetic form—as inherently democratizing and subversive of centralized forms of command and control. By contrast, recognizing the central role that feedback played in the development of the mass media and their supporting industries, highlights the continuity between the old and the new media, as well as between allegedly novel forms of interactivity and the strategies of cybernetic management they extend and perfect.[57]

In sum, far from representing a radically new economic model, the current ("total information awareness") society promises to realize the dream of information-based management conceived at the dawn of industrialized mass production. The goal of interactive digital technology represents not the end of mass-consumer society but its customized culmination: the prospect of its perfection in the form of markets that extract from each worker *and* every consumer the maximum productivity and profit. While Taylor relied on the collection of detailed information about workers to customize workplace tasks and payment schemes, marketers rely on detailed forms of surveillance in order to customize advertising appeals, products, and pricing schemes that extract the maximum value from consumers. Taylor's influence is ubiquitous in the formation of the marketing industry and the rise of the managerial side of business education and training. However, the drive toward customization, differentiation and specialization characteristic of the system was limited for many years by the costs of information gathering and by the technological infrastructure needed for customization.

These limits dissolved with the arrival of networked, interactive communication devices and programmable manufacturing equipment. The interactive era made the comprehensive forms of data collection envisioned by Taylor both feasible and relatively inexpensive. Rather than countering the forms of managerial control associated with industrial society, these changes allow for its increased penetration into the fabric of everyday life. Along with it come the features of scientific management developed a century earlier: meticulous monitoring, differentiation of tasks and customization of payment as a means of disaggregating the workforce, as well as the formation of increasingly detailed databases and a focus on individual acts of production and consumption. And this all takes place against a largely hidden background of unequal control over the means of surveillance. Crucially, the process of disaggregation remains a basic component of managerial control and resurfaces with a vengeance in the form of surveillance-based mass customization in the spheres of marketing, advertising and public relations.[58]

Online Industry's Self-Regulation in Dispute

The late 1990s saw a popular backlash against the merging of offline and online data that created a potential crisis for the marketing complex. A fledgling advocacy community pressured the U.S. Congress to consider opt-in legislation mandating that companies obtain prior consent from users before collecting their data. The pivotal moment was the controversy surrounding ad network DoubleClick's acquisition of data broker Abacus Direct. DoubleClick sought to merge its web-based consumer profiles with Abacus' off-line purchasing records, which contained names, addresses and other personal identification information. Mixing online and off-line records was a boundary-pushing move at the time, but it was merely the highest-profile case in a more protracted battle over whether federal regulation should establish guidelines for internet privacy or whether it should be left to industry self-regulation, as had become the norm since the Clinton administration had taken an early position that the private sector should lead internet system development in the United States. This position was apparently supported by the majority of federal lawmakers between 1998 and 2000.[59]

Privacy issues resonated with the American public and across the political spectrum. At the height of the debate, Congress considered adopting legislation that would require companies to obtain affirmative consent to "opt in" from web users prior to collecting their personal data. Facing negative publicity and potential legislation, a coalition of marketing trade associations and newly-formed internet advertising groups lobbied to consolidate an "opt-out" status quo in which surveillance, not privacy, was the default setting.

The politicians' commitment to industry self-regulation was put to the test as activists began to agitate against the resulting wave of unchecked consumer surveillance. Public awareness of online data collection was enhanced to a great extent by the efforts of a coalition of civil society groups that began to challenge surveillance advertising. Organizations including the Center for Media Education (CME), Electronic Privacy Information Center (EPIC) and Center for Democracy and Technology (CDT) formed a privacy advocacy network to generate publicity and policy proposals around internet privacy issues.[60] These groups became the mouthpiece of a broader advocacy network that worked against industry opposition, pressuring the federal government to act.

Yet, importantly, the range of policy discussion during this period remained largely confined to a neoliberal framework that took the validity of the market as the internet's core organizing principle for granted. Although public concern over internet data collection grew quickly, there were no mass protests or organized events of civil disobedience in this regard. Privacy activists stayed within the world of Beltway activism, using the tools of mainstream politics and public opinion making:

conducting research, writing reports, filing lawsuits and bringing complaints to federal agencies like the Federal Trade Commission.[61]

Only one effort to federally legislate online privacy protection proved to be successful, although its protections were limited. At the end of 1998, Congress passed the Children's Online Privacy Protection Act (COPPA) with bipartisan support. The Clinton administration's tentative support of this legislation to safeguard children's internet privacy (amid their otherwise staunch defense of self-regulation) seems to have been a political expedient meant to deflate the pressure stemming from the White House's trade talks with the EU over the harmonization of international standards for data collection and privacy protection.[62] In 1995, the EU enacted a Data Protection Directive containing a series of regulatory measures set to take effect in October 1998. Among the Directive's stipulations were guidelines for EU member states that disallowed "data transfers" to countries that failed to provide an "adequate level of protection" for consumer information. The internet advertising sector in the U.S. at that time was a far cry from meeting these requirements. The EU directive therefore posed an obstacle to intercontinental trade for the United States (threatening to impede billions of dollars' worth of trade involving "personal information"), which challenged the Clinton administration's hands-off approach to advertising regulation. Consequently, Clinton deployed the Department of Commerce and the FTC to oversee the development of more robust self-regulation in order to appease EU officials concerned about U.S. data practices.[63]

COPPA created standards for the collection and use of children's data to be enforced by the FTC and specifically required that websites obtain parental consent before collecting personal information from children under the age of thirteen. But the law's opt-in provisions were undermined by vague language and poorly designed enforcement mechanisms. Privacy activists who initially supported the legislation later criticized its implementation.[64] Moreover, the bill's passage also put the marketing complex on the defensive. Its members began to view the threat of formal privacy rules as a major obstacle to the continued growth of e-commerce. Companies and trade associations formed new partnerships to coordinate self-regulatory efforts and lobby government officials. Among the most prominent was the Online Privacy Alliance (OPA), a coalition of marketing trade associations and newly formed online ad industry groups with the express purpose of securing industry self-regulation on internet data issues.[65]

The years of activism culminating in the passage of COPPA involved a political struggle that tested the Clinton administration's commitment to self-regulation. COPPA seemed to create an impasse, as did the Commerce Department's ongoing negotiations to carve out an agreement to exempt U.S. companies from EU data oversight. To cap it off, DoubleClick then announced its intention to merge with the data broker Abacus Direct. This represented the most sweeping attempt to date to link online profile information with personal identification information gathered from everyday off-line consumption. Privacy activists were fervently opposed to this strategy and organized to block the merger. Their efforts reinvigorated the privacy debate about COPPA and prompted a formal FTC investigation into DoubleClick's data collection practices.

Both companies' stock prices rose on news of the deal, which was completed in late November 1999, when the dot-com bubble ballooned, sharply increasing DoubleClick's market capitalization to nearly $9 billion. A few months later the company quietly modified its privacy policy, removing its pledge to keep consumer profiles anonymous.[66] This indirectly revealed DoubleClick's plans to merge Abacus' data with its own in order to build profiles that each would include name and address; retail, catalog and online purchase history; and demographic data.[67]

The rhetoric of notice and choice—of consumer empowerment—deployed by DoubleClick in its PR counter-offensive against the wave of privacy activism, was an all too familiar refrain among defenders of online profiling. In practice, DoubleClick's implementation of these pro-consumer ideals was far less empowering than the company claimed. DoubleClick's default practice was to link online profile data with off-line identifiable information whenever possible. And the company off-loaded the burden of disclosing its data collection practices to its huge network of publisher and marketer affiliates and clients. The problem was that DoubleClick did not actually hold its network partners accountable for following its guidelines. DoubleClick provided an opt-out tool on its own website; however, the average user was unaware of DoubleClick's existence, let alone its opt-out mechanism. As a third-party platform, DoubleClick's widespread internet presence was largely behind the scenes.[68]

There was a temporary change of privacy policy as the dot-com bubble began to burst in March 2000. DoubleClick had been under public scrutiny in the media for months and its stock price was falling fast. This forced the company to halt its plans for expansion and DoubleClick announced it would suspend its data merging program, giving privacy activists momentum. A few months later, these activists achieved what appeared to be yet another important victory. After four years of waiting in vain for industry to implement self-regulatory privacy protections on its own accord, the Federal Trade Commission reversed its opposition to federal privacy legislation. The commissioners recommended in a 3–2 vote that Congress enact legislation that, together with continuing self-regulation programs, would ensure adequate protection of consumer privacy online.

Following up on this, both parties in Congress increased their focus on internet data collection. The majority of congressional proposals closely adhered to the FTC's longstanding recommendations around the "fair information practices" of notice and choice. Most of these legislative proposals such as that of Senator John McCain (R-AZ) supported opt out. They required companies to post privacy policies that left data collection as the default practice from which consumers could opt out. The opposite was the case for the Consumer Privacy Protection Act proposed by Senator Fritz Hollings (D-SC). This act was one of several opt-in bills that would have required all websites and ad platforms to obtain affirmative consent in advance from consumers before personal data could be collected or shared.[69]

However, aggressive industry lobbying simply overpowered privacy activists with their meager resources. Congress held as many as ten hearings on online privacy issues between 1996 and 2000. Legislators introduced dozens of bills containing various degrees of consumer data protections. But the only bill to make it out of committee, let alone be passed into law, was the Children's Online Privacy

Protection Act (COPPA). The marketing complex was successful not only in defeating opt-in measures but also in preventing any privacy legislation outside of the narrowly targeted COPPA. By 2001, the issue of internet privacy had been substantively abandoned by all branches of federal government.

In the absence of government privacy regulation, commercial entities remained free to conduct covert consumer surveillance on an increasing scale. Under the auspices of industry self-regulation, the "fair information principles" of notice and choice were implemented in such a manner that they served the exact opposite purposes for which they were designed.[70] Rather than providing genuine notice, privacy policies informed users "as little as possible about data collection activities, in as polite but complex a fashion as possible that they wouldn't understand what was really going on but could feel good about them."[71]

In hindsight, it should be noted that what has been called "the doctrinal fix of conspicuous notice" as the supposed solution to the privacy issue, falls short of what consumers might actually need. This kind of legal reform of electronic contracting meant to protect consumers more generally from surreptitious contracts, provides the appearance of protection through notice but actually only reinforces the existing regime of capitalism tied to data colonialism. That is why designers arrange the digital contracting environment to create a practically seamless user experience with minimal transaction costs. Rather than requiring people who intend to use online services to read lengthy pages filled with boilerplate legal jargon—which they cannot reasonably be expected to understand and will not be able to negotiate with—a simple mouse click in response to mere "conspicuous notice" of the existence of terms, suffices as consent for entering legally binding contract relationships.[72] As legal expert Nancy Kim describes the heart of the matter in more down-to-earth terms:

> Notice triggers the duty to read. The duty to read is onerous when the terms are convoluted and voluminous. It is frustrating when there is no ability to negotiate for different terms. It is troubling when the duty is triggered by notice that terms exist rather than by the terms themselves. But it is unrealistic and maybe sadistic given the burdensome nature of some standard wrap contract terms.[73]

Not only does this type of intervention fail to address the fundamental problem, but in practice, the conspicuous notice requirement may even contribute to engineered complacency by creating a false sense of security, autonomy and participation.[74] It is also true more generally that neither a lack of conspicuous notice nor ineffective notice is what causes people to act like "simple stimulus-response machines." They are likely as susceptible—if not more susceptible—to techno-social engineering when they are put on notice and then prompted to click-to-contract automatically.[75]

The Rise of Google as Trendsetting Model of Online Surveillance

The financial craze of the dot-com bubble did not last, but the business practices, technical infrastructures and political framework for surveillance advertising persisted. Moreover, as Mark Andrejevic suggested in 2007,

...the ongoing attempt to equate new media technologies with the promise of empowerment, individuation, and creative control remains alive and well even in the post-bubble tech economy.... The important ideological role that this equation plays in legitimating the ongoing rationalization of economic and political control suggests it's not going away any time soon.[76]

After the bubble burst, the broader enterprise of internet advertising quickly rebounded beyond pre-crash heights. By 2003, the sector's resurgence was evident, both in terms of aggregate revenues and as a percentage of total ad spending across all media. While weaker online advertising went out of business, many of the biggest players survived, with surveillance advertising infrastructures readily at hand. Further, the collapse of the new economy did not deter marketers, as increasing numbers began to include the web in their advertising mix. Another contributing factor to internet advertising's revival was an expanding market of a different kind of ad product: keywords on web search engines. Although early web portals as Yahoo, AOL and AltaVista had run display ads in preceding years, companies like GoTo developed paid advertising as an alternative to the prevailing ad network model of targeted banners. This kind of advertising did not prove to be very useful until the early 2000s, when Google entered the market.

Founded in 1998, Google developed a search engine that was powerful, fast and free. When the company's founders and financiers made the decision to monetize the business through advertising, Google's large and expanding user base instantly made it a major player. It also made Google a formidable competitor to ad networks like DoubleClick. Google's search advertising business was, in significant ways, an antithesis to DoubleClick's advertising approach.[77] DoubleClick pioneered the ad network model of targeted web advertising. At first focused on providing data-driven ad services to web publishers, the company then extended its reach to marketer clients in adopting a strategy of platformization. By contrast, Google's original base consisted of marketers; its ad inventory was not distributed across a network of partners but rather aggregated around the traffic of its own search engine.

The biggest difference between Google and DoubleClick was their approach to consumer surveillance. DoubleClick leadership's overarching principle was that consumer data was the keystone for closing the loop between ads and sales, and in their bid for platform monopoly, they expanded surveillance capacities to all of the company's market and publisher clients. Google's founders did exactly the opposite.[78] Rather than collecting consumer information to target ads, Google relied instead on users' key words to display "contextual advertising."[79] This meant that the ad targeting was based on the context of the user's activity rather than a composite of their IP address, browsing history and department store purchases. Contextual advertising was a new twist on a classic ad-targeting strategy used in print and broadcast media for many decades. The premise here is that content—whether a web search, a TV show, or a magazine spread—can serve as a reasonable proxy for consumer interest. Google was neither the first nor the only internet advertising company that moved in this direction, but it quickly emerged as the market leader in this domain. As DoubleClick's advertising model grew more complex and multivariate, Google's search advertising was characterized by simplicity and speed. Most importantly, DoubleClick stuck to a pricing structure based on impressions, while Google developed a system to sell ads via auctions on a cost per click basis. This implied that marketers bid on advertising opportunities and only paid when an ad was clicked, which proved to be attractive to marketers bent on improving efficiency and return on investment.[80]

Google had been selling ads since 1999, though with meager results, which was largely self-inflicted. Its founders hated online advertising, sharing a distaste for the tacky commercialization of the dot-com era. Moreover, advertising was not central to Google's initial business model; revenue was mostly expected to come from licensing Google's search technology to other sites.[81] But this expectation did not come true, and the dot-com implosion of 2000–2001 meant that Google, like every other unprofitable internet company, was under pressure from investors to figure out a more appropriate business model.

The breakthrough was in 2002, the year that Google became profitable. It involved AdWords, a system for selling and displaying ads. First, advertisers submitted bids on generic search terms, like "clothes," "cars," or something more specific. The system automatically selected the winner and placed its ad in the choicest position on the search results page—the one the user was most likely to see—the ad of the runner-up in the second-best position, and so on. But the bids were not the only factor that determined ad placement. The other factor was a quality score computed for each ad, which was so important that it could take an advertiser to the top spot even if it did not have the highest bid. For calculating the quality score the servers' logs became useful. User data provided information about the ad's relevance; the more users clicked the ad, the more relevant the ad was assumed to be. The click-through rate thus became the initial measure of ad quality. Over time, the ad quality score would become a much more complex calculation, involving a number of different metrics and employing advanced statistical techniques known as "machine learning."[82]

By the mid-2000s, contextual advertising and targeted display were the two main prongs of web advertising. Together these two accounted for three quarters of industry revenues, but financially the archetypical companies of Google and DoubleClick were moving in opposite directions. DoubleClick left the stock market after being acquired by a private equity firm for $1.1 billion, no more than about one-tenth of its market capitalization at the height of the dot-com bubble. Google, on the other hand, having just successfully gone public in 2004, was valued at over $100 billion, clear proof that web advertising could flourish without relying on consumer surveillance. Unlike the dot-coms of the 1990s, Google had achieved profitability soon after launching its ad business. And this was accomplished without stealthily collecting consumer data or building privacy-invasive profiles for targeted advertising.[83] However, Google was facing a saturated domestic search advertising market that accounted for nearly all of its revenue. Now subject to the harsh scrutiny of Wall Street, Google needed to maintain a forward momentum, for which maintaining a successful search advertising business was not enough.[84] The company needed to move into new markets, with the adjacent market of targeted banner ads as one obvious place to grow.

Google had already created a program called AdSense (launched in 2003) that enabled any web publisher to host Google's text ads. It extended the AdWords model into the wider web by enabling site owners to sell space to advertisers and split the revenue with Google.[85] AdSense was a distributed service that turned the web into a giant Google billboard. Still, Google remained distinct from surveillance advertising, as these ads were targeted on the context provided by the surrounding web page rather than consumer data. Google began to move into display advertising, adding some graphical options to its text ads and entering into partnerships with the likes of AOL to utilize its sales force to sell ad banners on Google's AdSense partner sites.

As broadband internet spread, online video emerged as a promising new format. In 2006, Google acquired YouTube, the enormously popular video streaming platform. However, to effectively exploit networked display and video formats, Google needed to get into the businesses of consumer surveillance and to do so quickly, as competition was closing in. Yahoo and Microsoft were also signaling their intentions to move forcefully into targeted ads.[86]

To make things worse for Google, social networks including MySpace, Twitter and Facebook came on the scene in the mid–2000s and began to attract large numbers of users. Fortunately for Google, the leading purveyor of surveillance advertising came up for sale about that time. Outflanked by search advertising and badly missing the dot-com era market valuation, DoubleClick was downsized in its private equity takeover. Still, the company's strong market position and technological proficiency in targeted display put it in a better position than most to survive the bubble's collapse. In early 2007, Google went into a bidding war with Microsoft and Yahoo to buy DoubleClick from the private equity owners. Google emerged as the victor; the final cost was $3.1 billion (in stock), nearly double what Google had paid for YouTube the year before. The acquisition gave Google a major market position in display ads and provided an extensive register of DoubleClick clients from both the supply side (web publishers) and demand side (marketers and ad agencies) of the online advertising market.[87] In addition, the deal gave Google control over the surveillance advertising platform that DoubleClick had been building for over ten years. Google's overarching goal then was to integrate the companies' operations to offer an expansive range of advertising services to publishers and marketers alike.[88]

Even before Google's DoubleClick purchase was finalized on March 11, 2008, Google's biggest competitors went on a shopping spree to avoid begin crowded out. Within a few months, Microsoft, AOL, Yahoo, and the advertising holding giant WPP each announced plans to acquire one or more internet advertising companies with core competencies in consumer monitoring and networked ad distribution. The cluster of mergers marked an inflection point where the differences among internet advertising forms and their respective data practices began to dissipate. Driven by market forces, the formerly distinct practices of contextual search advertising and targeted display converged on consumer surveillance. In March 2009, Google announced a new service called interest-based advertising that would allow advertisers for the first time to target consumers on the basis of user profiles populated with behavioral data. The new service would disseminate targeted display ads across Google's publisher networks, thereby revitalizing the surveillance advertising campaigns pioneered by DoubleClick in the dot-com heyday. Important in light of the privacy issue, the new "interest-based ads" would be enabled for everyone by default. Over the course of the next decade, Google systematically incorporated surveillance advertising into its operations, profiling billions of people worldwide using data collected from its own services, as well as a network of millions of publishers and mobile partners.

In 2012 Google began tracking users universally across its many services in order to further refine its ad-targeting capabilities. Up until then, Google's data collection was more like a patchwork operation: the information was kept in silos not connected to composite profiles. As critical observers have argued, the move toward universal profiles undermined the ad industry's long-standing claim that online surveillance was anonymous. From this point on, personally identifiable information gathered from Gmail, Maps and the ill-fated Google Plus social network,[89] would be combined with Google's extensive histories of search, web browsing and YouTube viewing habits. Although Google allowed users to modify certain settings via a renovated privacy manager and promised to keep targeting data anonymous, there was no option to opt out of the broader centralization of data.[90]

Google's appropriation of surveillance advertising laid out the historical trajectory of the entire online advertising sector, and more broadly the commercial internet writ large. By the end of the 2000s, the five largest U.S. advertising companies—Google, Facebook, Microsoft, AOL and Yahoo—all provided profile-based targeted advertising and collective consumer data across expansive networks and applications. Each also offered ad exchanges—software-powered markets for buying and selling access to individual consumers in real time, a process dependent on continual transfer of identification data among any number of buyers and sellers.

This configuration is the realization of the platform monopoly that the likes of DoubleClick sought to build in the dot-com era. By merging consumer monitoring and ad distribution, the first generation of ad platforms created a technical prototype and provisional organizational model for web advertising that placed surveillance front and center. Once surveillance advertising gained a foothold, competitive pressures made it increasingly difficult for companies to abstain. This is the discipline of the market in action. As an executive at the advertising giant Interpublic argued in 2007, adopting data-driven advertising was then "less about competitive advantage and more about survival."[91] Conducting consumer surveillance was not optional for marketers and ad agencies; it was simply the price of entry. Surveillance advertising stems from the instrumental desires of marketers and ad platforms to grow their businesses. It is also deeply rooted in structural factors; the competitive pressures and growth imperatives of the global capitalist economy. Today there are companies like search engine DuckDuckGo that have managed to build a business purely on contextual advertising and collect no consumer data. This is proof that the discipline of the market is not totalizing, but DuckDuckGo only exists at the margins of the advertising industry.[92] It is tiny compared to Google, operating entirely in the tech giant's shadow.

4

Identifying the Distinctive Features of Various Types of Online Platforms

The platform economy teems with tensions and contradictions. This is partly due to the fact that there are so many types of platforms using a variety of business models.[1] From advertising platforms such as Google and Facebook to brokerage (Uber, Deliveroo), e-commerce (Amazon, Alibaba), streaming (Spotify, Netflix) and cloud platforms (Amazon Web Services, Microsoft Azure), platform companies generate revenue in different ways.[2] There are three distinctive features of platforms that may help to explain their emerging role, with the caveat that all three features do not apply equally to each of the different types. The first concerns how platform owners and investors benefit from rentier relations. These result from the platform's role as a digital intermediary that facilitates the value-creating activity of others. A rentier is defined as someone who earns income from capital without working. This is generally done through ownership of assets that generate cash such as rental properties, shares in dividend paying companies or bonds that pay interest.

Digital platform owners can be described as rentierist because many of the respective platforms generate revenue through the ownership and control of the platform under conditions of limited competition that allows them to undertake traditional gatekeeper activities such as charging transaction and usage fees.[3] Data collected on platforms also enables profit generation through advertising revenue even if the users themselves do not have to pay for the service. Platforms are designed primarily for value capture rather than facilitating efficient market outcomes. As digital intermediaries, platforms provide value by maintaining the platform and lowering transaction costs for participants, but their primary source of profit comes from capturing value generated by others. This form of profit making differs from traditional models of goods and services delivery and enables platform owners to profit from "having" rather than "doing."[4] Importantly, this form of rentierism has always been a significant element of the capitalist economy, that includes many other forms of rentierism such as exploiting land, natural resources, infrastructure and intellectual property.[5] Secondly, platforms generate two opposing movements: the concentration of profit and the spread of risk and responsibility. At the same time as they centralize profits for shareholders—and much of the generated wealth ends up in the pockets of a few billionaire founders—platforms outsource labor to "independent" contractors and service providers to externalize the traditional costs and risks of doing business. This goes together with a less direct form of management, which keeps parties at a distance to avoid the costs associated with either maintaining a permanent labor force or fixed assets such as rental homes or a fleet of cars. By acting primarily as intermediaries and denying greater levels of responsibility for the services provided on the platform, these businesses have found new ways to benefit from social and commercial activities without bearing the burden of direct control.[6]

The independence certain labor platforms grant to workers has a Janusian face. Workers may find the flexibility of platform work to be a genuine benefit. But this flexibility is designed to enable the companies to avoid longstanding labor protections such as ensuring a minimum wage, health insurance, holiday and sick pay, and retirement income, as well as bearing the full costs of employment.[7] Platforms are able to externalize the risks associated with service delivery onto poorly paid and precarious workers. Issues such as long wait time between jobs, damage to assets or bad behavior from clients must all be borne by the workers in question.[8]

Despite this evasion of responsibility by platform owners, a wide network of service users and workers became dependent on the platform for their livelihoods and social interactions. However, there is no universal experience for platform workers, since platform work tends to be hierarchically organized in ways that intersect with workers' existing positions within the labor market. For example, compared to the hosts at Airbnb, the marketplace for short- and long-term home stays, drivers in the ride hail sector (Uber and the like) tend to be more dependent on the platform as their sole source of income and are therefore more vulnerable and precariously situated. Airbnb hosts, on the other hand, can be property owners leveraging an existing asset to earn a supplemental income with a good effort-to-earnings ratio. But despite their different experiences, a unifying theme is having to bear the main risks and responsibilities of operating on the platform.[9]

Thirdly, many platforms benefit from network effects that enable them to scale rapidly, dominate markets and gain a monopoly over their sector. Network effects result from platforms acquiring large numbers of users that help them improve the service and increase the value of the platform. Platforms also are able to grow exponentially due to the reduced demand for infrastructure, stock and personnel. Google, Facebook and Amazon have all achieved an effective monopoly that enables them to dictate the terms of exchange to relevant parties and charge high fees or unfavorable terms of service regarding ownership of data.[10]

A closer look at the leading types of platforms reveals major differences. Google, whose development and distinctive features were already discussed in the previous chapter, remains (to date) at the top of the digital capitalist ecosystem. The next biggest player is Facebook, founded shortly after Google began its ads-driven ascent.

Facebook, a Social Media Pioneer

Facebook was part of a wave of startups that together came to be defined as "social media." Facebook appeared in 2004, YouTube in 2005 (subsequently purchased by Google the following year), while Twitter followed in 2006. What characterizes these sites is a reliance on user-generated content. Social networks give people the ability to create their own unique profiles, share content and connect with others. At the time, this aspect was hailed as one of the pillars of what tech guru Tim O'Reilly called "Web 2.0": a new, more participatory web that emerged from the wreckage of the dot-com era.[11] These enclosures would be particularly well-suited to the business model popularized by Google, because their social nature encouraged people to provide more data about themselves, data which could be applied to the task of ad targeting. The fast growing social networks began to capture firsthand a range of personal information that was otherwise difficult to obtain. Soon these companies began using data about user demographics, interests and social connections to inform ad campaigns.[12]

Facebook's success as a business owes much to the way in which it established proprietary rights over a social commons embedded in online communities. The business model could be understood as an enclosure of this commons based on an exploitative relationship in which Facebook's owners capture value created by its users. Facebook both sells itself as a space for communities and in turn, relies on these communities as one of its primary sources of value.[13] Facebook claims ownership over the data produced from the daily interactions of people's social lives. The data, which is co-created by communities of individuals on the social network, becomes the exclusive property of Facebook to be analyzed and sold as advertising commodities. Facebook takes unfair advantage of their position by capturing this resource from a digital commons. At the beginning of the digital enclosure, the loss of public goods that was going to take place (about which more later) was not immediately obvious. But as these networks have grown within a privatized sphere of commodification, the extent of the robbery has increasingly become more apparent.[14]

At Facebook's founding, CEO Mark Zuckerberg began with an idea still prevalent today, which could be called "social network as database." The platform is envisioned as a network connecting different "nodes" (technically defined as points or areas where two lines, paths, or parts intersect or branch off), while people's posts and interactions provide the input for the collection of data. In 2005, Zuckerberg thought of Facebook as an online directory that could be used to look up individuals and find information about them.[15] In the early years, Facebook was (for him) an information bank that contained publicly available and searchable data on people's lives. This concept utilized the social graph, a diagrammatic representation of a social network in which people are depicted as nodes and their relationships as lines called "edges." It was a way of codifying relations, creating a global map of everybody and how they were related. In the beginning, when Zuckerman was still in college and discussed this with his Harvard roommates, the social mission was about transparency, as he later recollected. "We thought that the added transparency in the world, all the added access to information and sharing, would inevitably change big-world things. But we had no idea we would play a part in it.... We were just a group of college kids."[16] The engine of this system was information freely provided by people about their lives that fueled the platform's growth. The social graph, it was hoped, would make access to information easier and provide a community asset through which data could be used for social good. However, the problem with this model was that the graph was two dimensional and only offered a superficial understanding of the network. It lacked depth, content and most of all, meaning. The model of the database required a sociologically richer account of why people used the network and what value they received from it.[17] In 2016, Facebook's PR messaging addressed this problem through a new framing of the model, when an important new phrase entered Zuckerberg's vocabulary: "meaningful communities"—groups that, upon joining, quickly became the most important part of people's network experience and an important part of their physical support structure.[18] In the updated view, Facebook was not a simple database but a tool for people to share experiences, support each other and solve collective problems. It was pictured as users woven into a rich tapestry of a global community consisting of millions of subcommunities with a wide variety of specific interests and objectives.[19]

Following the political upheavals of 2016 in the U.S., Zuckerberg began to define Facebook's own purpose in terms of epochal change. It was now an essential player in a world-historical process of building global community through its infrastructure and tools. Facebook had morphed into a key battleground for electoral parties and a source of political information for citizens. Zuckerberg imagined it playing a decisive political role, in which the "forces of freedom, openness, and global community" were up against "the forces of authoritarianism, isolationism, and nationalism."[20] It was Facebook's stated mission to connect people through the platform and give them tools to join meaningful communities.

As political scientist James Muldoon points out, this rhetoric about the new paradigm draws on a familiar theme of a longing for lost community that appears time and again in political writing throughout the modern period. A sense of malaise began to be felt more acutely in the 1980s and 1990s than in previous decades due to the effects of neoliberal economic policies that eroded community cohesion and membership of social organizations.[21] The political consensus of the 1980s favored economic liberalization, deregulation, privatization and the expansion of "free markets."[22] These policies were also instrumental in disrupting social collectivities like trade unions and depoliticizing political struggles. In their stead, neoliberals promoted an ideology of personal responsibility and a series of market-based solutions to individuals' problems.[23]

In a 2017 open letter (and manifesto) "Building Global Communities," Zuckerberg offered his view on this theme, positioning Facebook as offering a potential solution to these social problems. Facebook's mission was to provide its own digital infrastructure and tools to support online communities, which have the potential "to strengthen existing physical communities" and "enable completely new ones to form."[24] In the wake of a decline of social capital and meaningful relationships, Facebook could step in to fill the gap.

It is deeply ironic therefore that one of the world's most successful capitalist entrepreneurs portrayed his company as a champion of grassroots community. Throughout its long history, the growth of capitalism has led to the disintegration of community across the globe and the destruction of traditional ways of life. Apparently the time had come at which the very forces responsible for the erosion of community, were attempting to sell it back in a monetized form to the people concerned.[25]

Notions of community are at the heart of the practices as envisioned in the rhetoric of many of the latest generation of platform companies. Tech companies have come to see community as a new domain of profitable extraction. Community is such a powerful marketing tool because it taps into one of people's deepest sources of meaning. Facebook draws on the fundamental human motivation of seeking a sense of belonging through the promotion of its groups and communities as a way of attracting and retaining users. Belonging is more than just connecting, it entails a deeper sense of significance and meaning. People desire not simply to share information but to be part of a collective, participating in something bigger than themselves. Social media might fulfill this purpose imperfectly, but following the decline of real-world communities in the U.S., it is increasingly a place where people seek out these relationships. Other companies as diverse as Airbnb, eBay and TaskRabbit have also deployed a sense of community among users to motivate people to use their online platforms and maintain an emotional bond with the service.

Paradoxically, this utilization of virtual community has negative ramifications for real-world community life itself. Whereas in industrial capitalism, human labor power is exploited to generate surplus value through wage labor, in the platform economy this exploitation is combined with new sociotechnical systems that capture and control the bonds of community itself and extract informational resources from them.[26] This leads to a loss of community as a space in which people can come together for free and open dialogue and experience collective self-determination. In its place a new kind of enclosed digital community emerges whereby connection with others is mediated through a digital architecture.[27]

The digital enclosures of social media look somewhat different than the online mall of search exemplified by Google. The goal is to keep users locked inside of it as long as possible, to maximize "engagement," as executives have instructed their software engineers in increasingly aggressive business schemes. Thus, the social media apps are deliberately designed to hook users, because that is how these companies make their money. These apps are part of what is known as the "attention economy," in which users' attention is attracted (and data about what they are likely to pay attention to extracted) rather than any goods or services being sold. In this economy, users of the apps are not the customers—advertisers are. Users are essentially the product, manipulated into giving their most valuable asset—their attention—away for free.

App makers aim to hook users by mimicking techniques used by slot machines, considered to be some of the most addictive machines ever invented. This is because slot machines are designed to trigger the release of dopamine, a neurotransmitter that (among other things) helps people's brains record behavior worth repeating—and then motivates them to repeat it. As in all addictions, the biological substrate is the brain's reward circuit. The tricky thing about people's dopamine systems implicated in these reward circuits is that they are nondiscriminatory; if a behavior triggers the release of dopamine, people are destined to repeat that behavior, regardless of whether it is good for them or not. This means that if one wants to create a product (or algorithm) that hooks people, it is quite simple; incorporate as many dopamine triggers as one can into the product's design. Smartphones and apps are packed with so many dopamine triggers that Tristan Harris, co-founder and executive director of the Center for Humane Technology, has referred to these phones as slot machines that people keep in their pockets. For example, flashy colors are dopamine triggers. So are novelty, unpredictability and anticipation—all of which users experience almost every time they check their phones. Rewards are also huge triggers; on smartphones, some of the most common rewards come in the form of social affirmation, such as a like or a comment on a post. This is why apps such as social media, news notifications, email, games and shopping are so easy to lose time on; they are the ones with the most dopamine triggers.[28]

Engagement on social media may take many forms: a like, a retweet, a view, a share, a commitment, a post. These posts need to be flexible enough to accommodate a satisfying range of expression—for social media to work, it must feel genuinely social—but also structured enough to be easily interpretable by software. The latter requires that computers impose a "grammar" on human activity to make it intelligible, just as the grammar of a language makes it intelligible to its speakers, as the theorist Philip E. Agre observed.[29]

Interestingly, a grammar is not a straitjacket; it is a remarkably elastic system. "Just as the speakers of English can produce a potentially infinite variety of grammatical sequences from the finite means of English vocabulary and grammar," Agre wrote in 1994, "people engaged in captured activity can engage in an infinite variety of actions, provided these sequences are composed of the unitary elements and means of combination prescribed by the grammar of action."[30] Thus, a crucial virtue of the system of rules that grammar imposes on its users is, oddly enough, the sense of freedom it allows. This sense of freedom helps explain why users tend to find social media pleasurable. Even as their interactions are being structured by the user interface and the code underneath, they experience a feeling of autonomy, a feeling of being free to express themselves.

The power of social media thus rests on a strange kind of sovereignty, one that pretends it does not exist. This denial has its legal basis in Section 230, passed into law as part of the Communications Decency Act of 1996. This Section shields online services of all kinds, as well as internet service providers (ISPs), from legal liability for the speech they circulate. The protection it affords is especially vital for the owners of social media, who can disclaim responsibility for the activities of their users even as they are covertly involved in shaping those very activities.

Only in exceptional circumstances—e.g., a high-profile user is banned or an intrusive new feature is introduced—does the sovereignty of social media become manifest. Most of the time, the "community" allegedly organizes itself. It is essential to the social media's functioning as an online mall that users behave as if their behavior is entirely their own, while being induced to behave in ways that are maximally readable to the automated systems that track and analyze them, ultimately for the purpose of selling ads.[31]

Amazon's Online Marketplace for Goods with a Universal Logistics System and Cloud Services

Advertising plays a central role in the development of the internet privatization. But creating a marketplace for attention is not the whole story. Tech companies also needed to find out how to create a more old-fashioned entity—a marketplace for goods. This was easier said than done. E-commerce had raised high expectations during the dot-com boom, but plummeted in the crash. It was not hard to put a product catalog online, but the logistics was an enormous challenge. Warehousing and delivery were complicated and expensive. It was Amazon that ultimately managed to solve the problems in question. In doing so, it replaced eBay, the favorite of the dot-coms, and joined Google and Facebook as one of the behemoths of the modern internet. Amazon's business model would differ from that of the search and social media giants but it resembled them in one key aspect—its evolution into a platform, that is, an online mall.[32]

Amazon opened its online bookstore in July 1995, a couple of months before the website that would become eBay appeared. eBay's setup was simple; it charged sellers a fee while saddling them with the fulfillment of their own orders. Amazon, by contrast, shipped goods to consumers. This meant large fixed costs and immense efforts in making its sprawling network more efficient. Amazon's margins were further minimized by the fact that it undercut competitors to capture market share. Low prices helped the company grow but further deferred profitability. This hardly mattered during the dot-com craze, but when the bubble burst in 2000–2001, Amazon began to look like any other dot-com with a cash flow problem.[33]

Amazon recovered again after the crash by a number of policy changes, probably the most important of which was to become more of a middleman—that is, more like eBay. It made its first move in the fall of 2000, when the company unveiled a new feature called Marketplace. From then on, third-party sellers would be able to list their goods on Amazon's product pages and, most significantly, compete against Amazon on price. This move paid off. In January 2002, Amazon announced its first profitable quarter, with third-party sales already accounting for 15 percent of all orders.[34]

In the following years, as Marketplace grew, Amazon began aggressively poaching sellers from eBay. Perplexed by these attacks and weakened by a series of bad business decisions, eBay faltered. By 2007, Amazon had gained the upper hand. That year, the company reported $14.8 billion in sales, more than eBay and bookseller Barnes & Noble combined.

Amazon's streamlined website and rigorously optimized logistics network made it well positioned to collect a large share of Americans' spending on online purchases.[35]

Building a third-party marketplace enabled Amazon to meet more of those needs by opening its site to millions of small and mid-sized businesses. By 2017, third-party sellers accounted for most of the sales on the site. The revenue that flowed from them to Amazon was substantial. They paid the company fees and, if they chose to have their orders fulfilled by Amazon, purchased warehousing and delivery services. They could also obtain financing, as Amazon began providing business loans. Finally, they could be induced to buy ads that made their products more visible to customers, whether in searches or elsewhere on the site—a miniature marketplace for attention embedded within the marketplace for goods.[36]

Importantly, third-party sellers not only paid in money, but also in data. Their activities were surveilled no less thoroughly than those of Amazon's customers. This is when Amazon came to resemble Google and Facebook more closely. Its rise, like theirs, rested on the manufacture and processing of large quantities of information. This business identity had already started to take shape in the late 1990s when CEO Jeff Bezos began placing greater emphasis on technology and hiring more tech people. The priority became personalization, or individualized customer targeting. Software engineers created a system that used data about customers—what they bought, what they looked at—to generate automated product recommendations. Just like at Google at about the same time, user behavior would be monetized to make automated predictions about what someone might buy. These predictions also informed the quantity of each of the items the company should stock in its warehouses and determined their optimal selling prices. It was data that helped Amazon solve the logistics challenges that had led to the demise of many dot-coms.[37]

Adding the third-party marketplace greatly expanded the scope of Amazon's operations. The company could now observe not only buyers, but also sellers; in fact, it could monitor entire markets. It would become a proper online mall, facilitating interactions in order to create data about them. The data would prove invaluable (especially as a market research tool) for Amazon's own product lines.

When businesses sell through Marketplace, Amazon watches them closely and records detailed information about sales, shipping and marketing. Then it uses this information to come up with its own copycat products. The third-party marketplace thus serves as "a petri dish," where "independent firms undertake the initial risks of bringing products to market and Amazon gets to reap from their insights, often at their own expense."[38] This means that Amazon not only competes with sellers, it also uses their data against them. And because it controls the online mall, it can promote its own merchandise over those of its rivals.[39]

Nowadays, Amazon is less the "everything store" and more a universal logistics system. The vast warehouses, the delivery vans, the brick-and-mortar Amazon stores, are all physical manifestations of a computerized logistical system that distributes labor, goods and information. And every aspect of Amazon's business model is geared toward augmenting its computational power. Amazon Prime, for example, allegedly loses money on each order and primarily exists to attract customers onto the platform who, in turn, leave the data required to power its logistics and cloud services. In the process of becoming a logistical giant, the company developed Amazon Web Services (AWS), initially an internal service for data storage, software applications and computational power, which has since come to provide the majority of Amazon's operational income.[40] Through AWS, Amazon—valued at $1.6 trillion on the New York Stock Exchange in February 2024—leads the lucrative cloud computing market, though Microsoft—then worth $3 trillion—has encroached on its lead with more extensive integrations of artificial intelligence.[41]

Thus, as Amazon created a marketplace for consumer goods, it also created a marketplace for capital goods or the means of production. The ones that Amazon sold would be the machinery needed to make software for a commercializing internet, as Ben Tarnoff points out.[42] In the post-dot-com era, as firms began to find more promising paths to profitability, they also made the internet more complex. The simple static web page faded from view and was replaced by the dynamic and interactive web application (linked to elaborate underlying systems of data collection and analysis), designed to grab a user's attention and stimulate their engagement. The online mall then would be a computationally intensive entity. Just as the capitalist transformation of manufacturing meant replacing the workshop with the factory, the capitalist transformation of the internet would bring on "factories" of its own. Importantly, they are not generally considered factories because they came to be known by a name that obscured their fundamentally industrial character: the cloud. Materially, the cloud is a globally distributed set of climate-controlled buildings—"data centers"—filled with servers. These servers supply the storage and perform the computation for the software running on the internet. At its most basic level, the cloud is a computer that is used by another computer through a network.

This is a surprisingly old idea. In the 1960s, the idea of "time-sharing" systems offered an early way to stretch computation over a network. Such systems enabled multiple users to run programs on a single mainframe simultaneously from individual terminals. This made computing more accessible at a time when computers were large and expensive. The desire to share computing resources through a network inspired both the making of ARPANET in the 1960s and the making of the internet in the 1970s.[43] While the modern cloud has several points of origin, Amazon is the company most deeply involved in its creation. It was within Amazon that the idea of constructing an on-demand computing service took root, leading to a very profitable line of business.[44]

In 2006 Elastic Compute Cloud (EC2) was launched, which was named "elastic" because the infrastructure could quickly stretch to meet demand. This feature also made it easier to sell computing on a metered basis, like electricity, water or gas. Along with Simple Storage Service (S3), an online storage system released the same year, EC2 would become the company's flagship offering—Amazon Web Services (AWS). AWS created, and still dominates, the market for cloud infrastructure services. Its portfolio, which expanded beyond EC2 and S3 to include dozens of different services, would enable smaller, newer or less technical companies to create online malls and to develop some of the same capacities as firms such as Google without having to build their own infrastructure or recruit top engineering talent.[45]

These dynamics speeded up with the arrival of the "big data" era in the 2010s. During this period there were rapid advances in machine learning, a set of techniques for pattern recognition that could be applied to a number of tasks with increasing precision, from predicting consumer preferences to understanding human speech to recognizing human faces. Because machine learning takes place by training on data, the abundant amount of data being generated through the internet greatly accelerated the development of the technology. Online data-making became the indispensable precondition for the growth of automated systems that came to be associated with "artificial intelligence."[46] The sophistication of these systems and the wealth and power they appeared to bestow on major tech firms, stimulated an even stronger and more generalized desire for data. What has been called the "data imperative" settled itself throughout the corporate world and in many government agencies.[47] Organizations also began to store as much data as possible in attempts to secure some artificial intelligence for themselves. Here, too, AWS could furnish the necessary machinery by selling machine learning services in the cloud.[48] Now the global leader in cloud computing, AWS provides governments with algorithmic power and other companies with logistics solutions and machine learning. More and more businesses and governments rely on Amazon to organize and store their data, which takes an infrastructure just as large as the one needed for physical entities, comprised of huge data centers expanding in number and size every year.[49]

Algorithmic Management by Uber and Its Imitators

The elasticity of the current internet, its capacity to transmit data across heterogeneous networks over large distances, makes it also a powerful tool for what is called "algorithmic management." This concept can be broadly defined as the delegation of managerial functions to algorithmic and automated systems.[50] What distinguishes algorithmic regulatory systems from pre-digital regulation, is that the former is highly adaptive and opaque. The ultimate goal is to obtain the consent of the labor force and to regulate the platform's unstable labor supply in a cost-effective manner. It is ride-hailing company Uber that perhaps best represents the features of algorithmic management. Founded in 2009, it became an online mall in the mold of Google, Facebook and Amazon, though it did not make a market in attention or in goods but rather in labor, matching customers who wanted a service performed with the workers who could carry it out, on demand.

Uber facilitates interactions (connecting passengers with drivers) while exercising meticulous control over the terms of those interactions, control that reaches far beyond that of a market maker. In addition to setting the fares for trips—how much a ride costs in each city based on distance and supply and demand—the platform plays a crucial role in determining how services are performed. Software running in Uber's cloud directs the driver through a smartphone app. Sometimes this direction is quite direct. The app guides the driver along the optimal route for picking up and dropping of a passenger or a delivery, computed from factors like current and historical traffic patterns. Sometimes the direction is less direct; the app uses "surge pricing" to lure drivers into specific zones with the promise of higher rates, or displays messages that encourage drivers to keep driving when they are about to log off.[51]

4. Identifying the Distinctive Features of Various Types of Online Platforms

In 2018, Uber released a "completely overhauled" mobile application for its "independent contractors" workforce, which the company marketed as a panacea for driver complaints about the lack of transparency, autonomy and flexibility while working on the platform. (Uber did not address the broader issues about fair pay and algorithmic management, however.) Based on the gamification of work, which has become a pervasive trend in algorithmic governance and management, the application linked individualized "rewards" with Uber's need to maintain a "frictionless marketplace." More generally, gamification frames mundane work tasks as "puzzles" and "challenges" that offer workers the opportunity to earn "points," "badges," and virtual goods through an adaptive representation, direction and intervention of the on-demand workplace. Representation entails the creation of social knowledge through algorithmic instruments. Direction refers to the regulators' desired standards and mental states that are prescribed in algorithms. Intervention encompasses the techniques of modifying the governed subjects' behaviors to achieve the desired states. In other words, within the context of Uber—and more broadly, the platform-mediated gig economy—gamification is a key management technique for reorienting how workers should pay attention to work-related information on the app and then work accordingly. This technique does so by re-organizing work-related information (e.g., representation), normalizing scores and rewards (e.g., direction), and nudging drivers to increase their productivity (e.g., intervention). It is underpinned by the same psychological and design insights that are used in video gambling machines to condition addictive behaviors. More specifically, these software programs employ variable reward structures and multi-sensorial, positive feedback that are designed to place players in a state of "habituation" or "the zone," and keep them there as long as possible.[52] As such, gamification can be considered a form of "soft control" consisting of veiled but effective articulations of managerial authority, while enabling platforms to manage their volatile labor supply.[53]

The deployment of metrics can be seen as another key technique of gamification whereby workers are categorized into different levels based on their metrics. As metrics become a form of capital and are often used as markers of visibility, they stimulate workers to modify their behavior. Meanwhile, workers' metrics are essential to platform-based rewards programs and loyalty schemes. As these metrics are not transferable, workers have to stay on the platform if they want to keep their reward status.[54]

These techniques are only possible thanks to the abundant quantities of data that Uber manufactures about its drivers. They are observed just as closely as users within the digital enclosures of Google, Facebook and Amazon. When they drive, how often their rides last, how fast they are going, how hard they hit the brakes—the app records all these data points (among many others) and transmits them to the cloud for analysis, which further improves the algorithms. The routes become more efficient and the nudges to persuade drivers to keep driving become more personalized.[55]

Algorithmic management thus makes it possible for Uber and its many "gig economy" imitators to coordinate the labor of millions of workers without the need for middle managers, and with more technical sophistication than attainable by middle managers. It should be clear, however, that this does not mean that Uber has entirely replaced managers with algorithms, as is a common criticism of the company. In reality, much of the managerial role is split between algorithms and a crowd of workers on platforms like microwork site Appen, who check, for example, whether Uber's facial recognition software (which is prone to errors) has made correct decisions about a driver's identity during their daily photo authentication. The microworker who accepts this task, gets thirty seconds to validate whether the driver is who they claim they are. If this worker decides in the affirmative, the ride goes ahead; if not, the ride is canceled and the driver is locked out of his/her account. This means that for less than a minute, the microworker indirectly takes the role of Uber's manager, in effect supervising the algorithm that oversees the labor process and makes decisions about the company's workflow. This example shows that automation of management is actually the breakdown of a unified job into a variety of tasks—some done by machines, others by (micro)workers.[56]

There is another important advantage of algorithmic management from the corporate standpoint. By having software (with microworkers operating out of sight in the background) rather than humans instructing workers what to do, and having the software implement techniques like nudges and gamification, gig companies can pretend that nobody is telling their workers what to do and that therefore they are not really workers at all. Uber refers accordingly to its drivers as "partners" or "entrepreneurs," and classifies them as "self-employed." Legally, they are considered "independent contractors" and attempts at challenging this legal status in court have been unsuccessful so far, at least in the United States.[57] This classification is a mainstay of the gig-economy business model, since it keeps labor costs low by absolving firms from having to pay a minimum wage or to comply with the other legal protections afforded to direct employees.[58] This is the primary reason why all sorts of companies have turned to outsourcing wherever possible during the past few decades. It has created what has been called the "fissured workplace"; rather than hiring workers directly, firms increasingly parcel out work to contractors.[59] Uber is certainly not the first to embrace this model, let alone the first to combine it with the internet. In fact, the internet is closely associated with the rise of this type of contracting in general.

Networks offer good opportunities to move work around and exercise managerial control from a distance. As soon as corporate offices became networked in the 1990s, executives began using those systems to reorganize their workforce. Wide area networks (WANs) spanning multiple sites let firms relocate "back-office" functions like data processing to places with lower wage and real estate costs. Then, as these networks became connected to the wider internet, such functions could be pushed even farther out and, increasingly, abroad. In addition, certain categories of work were displaced from the corporate core and also outsourced. As the best-known example initially, the directly employed customer service representative became a subcontracted call-center operator in another country.[60] The internet thus has enabled the creation of a massive global reservoir of human labor power for companies to tap into, according to their specific and varying needs.

Low Interest Rates and the Role of Venture Capital Investments

Uber is an internet company; the internet provides the connective framework that links the smartphone apps of its drivers with those of its riders, and both to the software of the cloud.

But technology plays a smaller role in Uber's success than is often thought. It is politics that has been crucial here, particularly the pressure the company has brought to bear on regulators and policymakers.[61] In city after city, this pressure has enabled Uber to dodge long-standing municipal rules regarding fares and the number of vehicles allowed on the road.[62] Equally important has been the limitless supply of investors' money which was used to bankroll lobbying, as well as subsidize the cheap fares that undercut regular taxi drivers. Finally, the legal classification of ride-hail drivers as independent contractors is a pillar of Uber's business model since it keeps labor costs low.[63]

Intriguingly, unlike the other major online malls, Uber has not been profitable, losing billions of dollars every year. It was only in August 2023 that Uber, for the first time, announced it had made a profit—a modest amount of $326 million.[64] Over 2023, the company would make a profit of $1.1 billion, compared to a loss of $1.8 billion the year before, Uber reported in February 2024.[65] Up until then, the company had recurring yearly financial losses yet still managed to raise capital and increased its stock value. This bizarre state of affairs reflects the influence of venture capital, a popular form of financing that gives money to startups in exchange for equity. It is indicative of a time when declining *average* labor productivity rates (obscuring important variation across states, cities, sectors and firms), frequent financial crises and low traditional asset profitability, incentivize investors to take more risks for extended periods. Backers are willing to bet that a data company will eventually convert this data into profit, as was the case earlier with Google, Facebook, and Amazon.

In recent decades, macroeconomic conditions made it even easier to source capital for such ventures. The responses of Western governments and central banks to the financial crisis of 2007–2008 released a tidal wave of cash, brought about by the monetary policy action of quantitative easing.[66] Permanent low interest rates—the legacy of the Federal Reserve's response to that crisis and its long aftermath—reduced the returns on various financial assets. This period lasted from 2009 until December 2022 when, after a series of federal interest hikes that set in earlier that year, the federal interest rate increased to 4.25 percent–4.5 percent, approximating the level prior to the start of the Fed's crisis interventions.[67] The result of low interest rates was that investors seeking higher yields felt compelled to take more risks by investing, for example, in unprofitable and unproven tech companies.[68] More generally, business models that promised world-changing outcomes were encouraged, even if they were highly unrealistic and/or hostile to the public interest (e.g., the gig economy, self-driving cars, cryptocurrency, and Zuckerberg's short-lived "metaverse" dream of the digital future).[69] This came at a time of little or no regulation of Big Tech and an accepted business culture (of shareholder capitalism) that contended that executives should maximize shareholder value at the expense of everything else (e.g., democracy, public health, public safety, labor's working conditions, wages and benefits). Thus, Uber losses could coexist with profit generation through stock valuation and speculation over data expectations. What Uber and other comparable companies offered (especially in their early stages) were exceptionally lucrative ties between financialization and the internet.

Venture capital is only one financing source for tech companies, however; a larger share is made up from hedge funds, mutual funds, investment banks, and the bond and equity markets. Yet venture capital exerts considerable influence because of its close historical ties with the tech industry. As venture funds are designed by definition to demand large returns, this pressure towards hyperscale is put on start-ups as an imperative to grow at any cost. Once the company has acquired enough market share, the expectation is that it can use its dominant position to achieve the profits that its investors demand.

But what exactly makes investors believe this is going to happen? In the case of Uber, how can one explain the investors' faith that the company, *despite losing billions*, is worth tens of billions of dollars? Understanding how such a paradox is possible acknowledges the nature of Uber's role in the privatization of the internet, and more generally, the cutting edge of networked moneymaking it represents. As an online mall, Uber collects both monetary rents and data rents. The monetary rents are the commissions that the company charges on each ride. They do not come close to covering its costs, however, which is why it loses money. The data rents are the manifold streams of information that Uber draws from its drivers, which are used to develop and refine the algorithms that manage their labor. But all this data also serves another important purpose; it helps persuade investors that Uber is worth a lot of money despite being so unprofitable, especially in its core ride-hailing business (at least until very recently). In other words, the data has both an operational value and a speculative financial value, and the two are closely connected.[70]

The reason that this data harvesting "attracts venture capital and grows financial valuations," according to Niels van Doorn and Adam Badger, is because "investors expect data-rich platform companies to achieve competitive advantages by creating data-driven cost efficiencies, cross-industry synergies, and new markets."[71] The hope is that manufacturing large quantities of data through the internet will boost revenue through the workings of machine learning and other modes of analysis, leading to optimization that produces value. In van Doorn and Badger's estimation, data manufacturing is at the heart of Uber's business model. Its human cloud is a factory for the production of financial value. The labor of drivers is channeled to create "data assets" that help secure the capital that allows the company to keep growing. In other words, data is converted into money through its interaction with the psychology of financial markets. In this phase of the internet's privatization, capital is so abundant and the potential returns are so excessive that investors can live on hope alone.

The fact that Uber was unprofitable for all those years does not mean that certain well-positioned persons did not profit from it. But these profits are derived from speculation, not production. In contemporary capitalism, profits from speculation are a primary means by which capital accumulation occurs. What Uber and other comparable companies offer, particularly in their early stages, are exceptionally lucrative linkages between financialization and the internet. Investors can ride a rising tide of paper wealth as the valuation of the firm grows, and then convert that paper wealth into real wealth during a "liquidity event," such as when the company goes public (through an IPO on the stock exchange) or is acquired by a larger company, regardless of whether the firm ever turns a profit.[72]

It should be noted that before its highly anticipated IPO in 2019, Uber was valued at one point as high as $120 billion by Wall Street analysts, which would have made it the largest company ever to debut on the stock market. But after going public on May 9, 2019, it made history with the biggest first day dollar loss in U.S. history when it was valued at only about $69 billion. Since then, Uber has worked on becoming profitable, partly through investments in services other than its basic ride-hailing app and acquisitions of other companies.[73]

Airbnb's Online Platform for Short-Term Rental Accommodations

After moving to San Francisco in October 2007, roommates Brian Chesky and Joe Gebbia came up with the idea of putting an air mattress in their living room and turning that space into a bed and breakfast. In February 2008, Nathan Blecharczyk joined as the Chief Technology Officer and the third co-founder of the new venture, which they named AirBed & Breakfast (later to be changed into Airbnb). They put together a website that offered short-term living quarters and breakfast for those who were unable to book a hotel in the saturated market. The site was officially launched on August 11, 2008.[74] Within a few years of its inception, Airbnb had become one of the most successful "sharing economy" platforms. As of April 2019, Airbnb was available in more than 1,000 cities across the world and was expected to have hosted more than 500 million guests worldwide.[75] The company ended 2020 with a highly successful public offering in which its share price more than doubled on the first day of trading at the New York Stock Exchange, with a valuation of just over $100 billion. It was a stunning recovery, as lockdowns eased after the firm's business was heavily damaged by the social effects of the COVID-19 pandemic earlier that year.[76] However, despite the company having a "solid business model, an impressive growth rate, and a large addressable market," according to a business analyst, it failed to turn a profit until 2022, when the company made $1.9 billion—its first profitable full year.[77]

The company developed its business model based on a compelling value proposition; it integrates economic benefits for travelers and residents of tourist areas via a trusted marketplace that enables the platform to scale up and leverage its assets through network utilization. According to a systematic review of the mainstream research literature, Airbnb offers many benefits to its stakeholders. For customers, Airbnb accommodation is typically cheaper than traditional hotels. Moreover, Airbnb offers local "authenticity," giving customers the opportunity to live like locals in a listed apartment, house or private room. For property owners, Airbnb enables them to maximize the utilization of their underutilized assets. For other stakeholders, such as the community, Airbnb may in some instances increase community economic and business opportunities.

However, on the flipside, some have been on the receiving end of the negative effects that Airbnb's growth has brought about. For example, some hotels have experienced revenue losses in response to increases in Airbnb property listings. Tourist sites have been negatively impacted by the increased concentration of tourists in particular places, inviting environmental problems such as water scarcity, waste management and carbon emission issues, as well as noise pollution. The problems created by Airbnb as a "shared economy" accommodation have also generated challenges for the government as a regulator because Airbnb's disruption of the accommodation industry has changed the tourism landscape, creating taxation issues and discrimination problems on the platform. Additionally, Airbnb has negatively affected availability and affordability of housing for permanent residents by taking potential properties off the housing market, driving up housing prices. The platform has also contributed to gentrification and more generally the destruction of the character of local communities.[78] These problems have led to social movement opposition and local attempts to create and impose restrictions in various locations, including San Francisco, New York City, Portland, Maine, Berlin, Barcelona, Paris, Amsterdam and dozens of other cities worldwide.

Airbnb's social mission is reflected in its slogan, "our product is our community." To understand its origins one needs to go back to 2012. From 2012 to 2017, Airbnb underwent a similar positive image transformation as Facebook's. Back in 2012, its Twitter bio read: "Airbnb connects travelers seeking authentic experiences with hosts offering unique, inspiring spaces around the world." But since "connecting people" belonged so much to the 2000s, Airbnb soon changed it to "the world's largest community-driven hospitality company," one which enabled you to #BelongAnywhere (as expressed in this Twitter hashtag). To belong anywhere may have been a comforting slogan, but it was also somewhat ominously prescient of the company's growing omnipresence. For to belong *anywhere* in the way Airbnb management envisioned this, Airbnb properties had to exist *everywhere*—in every country, in thousands of cities and in millions of neighborhoods across the world. Paradoxically, this meant that the growth of the Airbnb "community" has often been at the expense of the very communities it purports to serve.[79]

In developing this "product identity" Airbnb's co-founder and CEO, Brian Chesky, was heavily influenced by the work of brand guru Douglas Atkin. During the 1990s, Atkin worked at a number of ad agencies and pioneered techniques in community building and brand storytelling that he would take into the tech startup world. Atkin's book, *The Culting of Brands: Turn Your Customers into True Believers*, examined how and why people join cults in order to gain insight into how communities could foster brand loyalty and commitment.[80] Atkin claimed that the lessons he learned were applicable to any kind of community, whether one was building a cult, a religion, or a company. He moved from the world of marketing to digital community building and started a consultancy service called Purpose. Along the way he acquainted himself with the language of social movements finessed in generic catchphrases that aligned a sense of higher purpose with corporations. Atkin's twenty-first-century corporate mysticism basically repackages insights borrowed from social movement activists using the language of empowerment and community building. The result found a welcoming audience with Silicon Valley executives. Atkin went on to become the global head of community and mobilization at Airbnb and played a key role in developing its mission, proclaimed values and culture.[81]

Airbnb's vision of its newly defined role was to facilitate meaningful relationships in an increasingly alienated and disconnected world. The company's professed purpose was to create a safer, more open and more tolerant world in which everybody could experience a sense of community and belonging. It spread this message by sharing stories of its hosts and other members of the community through its social media accounts and the pages of *Airbnb Magazine.*

There one can find hundreds of stories of unconventional superhosts, unique homes and authentic travel experiences. It is all part of a carefully crafted PR exercise in brand storytelling designed to associate the company with experiences of belonging and community.

Since 2013, Airbnb has re-imagined itself not only as a community but also as a grassroots organization and social movement. This kind of discourse has profoundly shaped how the company engages with the wider world. Airbnb has developed a global, corporate-driven social movement by rallying its hosts to oppose regulators who stand in its way. It has borrowed from the toolbox of community organizing and social movement history to develop new strategies to advance its corporate goals. According to Atkin's vision, Airbnb grew from a short-term rental matchmaker into a global force for democratization, economic empowerment and community building.[82] This is, of course, a distortion of how things have actually evolved.

From 2014, a key part of Airbnb's political response has been the use of grassroots lobbying, a phenomenon whereby businesses influence societies by creating "independent" social movements to act on their behalf.[83] Airbnb presents carefully curated and extensively coordinated groups of landlords who rent out a single room or property as "people's power," an Airbnb "community" or movement. This offers the company legitimacy and additional political clout to protect a business model that is increasingly dominated by professional accommodation providers, *not* "home sharers."

One of the more innovative aspects of Airbnb's strategy is its combination of lobbying efforts combined with hosts' grassroots-style organizing. In 2015, the company announced that it would support the creation of 100 independent Home Sharing Clubs in 100 cities around the world. This was an attempt to foster the development of civic society organizations that would function as what Airbnb defined as "a powerful people-to-people based political advocacy bloc."[84] To help create these "independent" organizations, the company began hiring "community organizers"—paid professionals who would embed themselves in local Airbnb communities and encourage hosts to take action. It was an approach to community mobilization that drew on the "snowflake" model of the Obama election campaigns of 2008 and 2012, in which paid staff trained community leaders to recruit their own volunteers.[85] In this endeavor, new recruits would be given ever larger roles as they moved up the commitment curve and became more involved in the campaign. When Obama's campaign team adopted certain aspects of the rhetoric and tactics of community organizing, they created, in fact, a fundamental contradiction: were communities organizing for their own power or to get Obama elected?[86] Although the objectives were not mutually exclusive, in the heat of an election campaign it is very likely that voter identification and turnout are prioritized. There is something deeply disturbing—adverse to the spirit of community organizing—in the practices of mobilizing citizens for the instrumental goal of electing a politician or securing a policy victory for a company.[87]

The Home Sharing Clubs are associations of selected Airbnb landlords who are resourced, mobilized, and coordinated by Airbnb public policy teams to advocate for favorable regulation. The support, resources, and influence offered by Airbnb is extensive. These include numerous forms of support and influence: protesting alongside landlords, organizing many aspects of protests, political education and training, editing and rehearsing curated "stories," and suggesting the policies that the company wants. These Home Sharing Clubs are made up of an unrepresentative segment of Airbnb landlords—mostly those who share their own homes or rent them out for a short term. Professional landlords on the platform—the most controversial segment and accounting for a majority of listings—are excluded. This is apparently done for the purpose of presenting a benign image of the company. After an intensive search for suitable recruits, Airbnb staff hold a series of meet-ups with those who have "good stories," building trust and increasingly asking them to take on tasks that become ever more political and demand more responsibility. Specific landlords' personal biographies or curated "stories" are subsequently used in marketing and for court hearings and campaigns to lobby key decision makers.

The Home Sharing Clubs are created predominantly and in disproportionate numbers in cities where the effects of Airbnb are leading to calls for stricter regulation. Like more traditional lobbying and PR practices, they target public officials and public opinion. These associations have been deployed in hundreds of cities and towns globally through the hiring of "community organizers" whose major role is to create host campaigns that are politically useful to the company. Clubs hold meetings, attend and give evidence in legislative hearings, lobby officials by phone-banking, letter writing, in-person appointments or open petitions, and link up with media as well as convene protests. By 2021, the number of clubs had grown to an estimated 350–400 globally; 40 percent were outside of the United States. Airbnb then entailed the most extensive and sustained platform-sponsored grassroots lobbying strategy in the world.[88]

The tradition of community organizing constitutes an ideal resource for the company, because uncompromising local councils have typically been the target of community organizers. The tradition is also malleable, as the pragmatic and non-ideological approach of community organizers leaves it without an overarching vision of an ideal society. The framework in question encourages building people power and focusing on immediate, specific and winnable issues for the purpose of growing the organization and building the collective power of its members. The crucial difference is that in traditional community organizing, it is people versus power rather than the powerful using people as a tool to achieve their policy objective, as in Airbnb's case. It should also be noted that platform-sponsored grassroots organizing, as a key tactic for disruptive new businesses facing regulation, has become more widespread across the new digital platform economy—from ride-hailing companies such as Uber and Lyft, to delivery companies such as DoorDash, to electric scooter companies such as Getaround, Lime, Scoot, Spin, Bird and Lyft Scooters.[89]

Importantly, one must keep in mind that many of the Airbnb hosts are platform entrepreneurs themselves. Today, the majority of hosts, in fact, have multiple listings and their enterprises are run by professional rental companies. Since their business is affected by tighter regulations on the short-term rental market, they are staunchly committed to helping Airbnb fight regulators because the company's lobbying efforts and continued legality enable their own profitable enterprises. This reflects the changing nature of Airbnb and the short-term rental industry, which is shifting from a person-to-person model of individual hosts on one platform to a business-to-consumer model of property management companies with multiple listings on multiple platforms simultaneously. Needless to say, the idea of professional managers is at odds with Airbnb's self-portrayal as a company that simply wants to help middle-class families rent out their spare rooms.[90]

Indeed, meticulous research by the Economic Policy Institute (EPI) has shown that Airbnb bookings are increasingly concentrated on a relatively small number of professional landlords who act essentially like "miniature hotel companies."[91] The highly informative data and advocacy site Inside Airbnb says on its website: "Airbnb claims to be part of the 'sharing economy' and disrupting the hotel industry. However, data shows that the majority of Airbnb listings in most cities are entire homes, many of which are rented all year round—disrupting housing and communities."[92] The site has collected data on dozens of cities and countries around the world and is constantly updating. It provides a breakdown of the top hosts with multiple listings in cities around the world.

A pivotal development took place in September 2023, when New York City began to crack down on Airbnb (and platform Vrbo, an online marketplace for vacation rentals owned by Media Group) with tough new restrictions, designed to bring back thousands of rental properties to the housing market for residents to live in. With Airbnb calling it a "de facto ban," it might become the death knell for the company if this policy spreads further to other cities. New York's Local Law 18 requires hosts to register with the mayor's office and prove that they will live in the home they are renting out for the duration of the renter's stay. More than two guests at a time are not allowed—effectively banning families—and hosts in violation of the ban can be fined up to $5,000. The effects will likely be closely monitored by policymakers in cities worldwide who are hoping that local government interventions like in New York City will show them a way to take back cities for people who actually live in them.

By operating in 220 countries, with over six million active listings in 100,000 cities rented out through Airbnb (as of December 31, 2021), the company exacerbates a growing global crisis in the housing market. The huge number of homes lost to short-term rentals is inextricably linked in many places to a local housing crisis. It is further pushing up already unaffordable rents for people living in cities and tourist areas with large numbers of second homes that are rented out. It should be apparent, however, that this is only one major contributing factor; crucial is the lack of affordable public housing in many places. The main causes of the global housing crisis can be found in the financialization of housing, which means it is seen primarily as a financial asset rather than a social good, disregarding the need for investments in public housing. Combined with the consequences of monetary policy over almost fifteen years—from the inflationary impact of quantitative easing to low interest rates facilitating a credit boom, followed by rising rates pushing up rents—controlling the growth of Airbnb can only be part of a much broader solution to the housing crisis.[93]

5

The Wider Array of Smart Tech Devices and Systems

In addition to the birth of the online mall, the rise of the cloud, and the spread of the data imperative, the internet's diffusion was another change that profoundly altered its nature. During the 2010s, the internet went mobile thanks to the smartphone, which would become ubiquitous. At the same time, more objects became connected to a network, ranging from cars to thermostats to security cameras to industrial equipment. This "Internet of Things" (IoT) would account for an increasingly greater part of all networked devices. From then on, the internet was no longer something people logged onto but morphed into an entity that was always on—fastened to one's hand or wrist or in one's pocket—and woven through homes, workplaces and cities. "Smartness" came to inundate the spaces of everyday life. The Internet of Things provides the perfect cover for attempting to convert all aspects of human life into raw material for capitalism, in the process, capitalizing everything and everyone. The proliferation of smartness is aimed at making digital surveillance as deeply integrated into people's physical world as it is in their virtual one. By putting internet-connected devices in more places, companies can put more of people's lives online, which means more data about their lives can be manufactured.[1]

This growth of the Internet of Things is incorporating consumption—what people do with products after they purchase them—into an extended chain of profit extraction through the processing of data. The bigger vision of the IoT has been well articulated by IBM, a company normally seen as the enemy of social media platforms' entrepreneurs. IBM has suggested that by turning the human environment into a network of listening devices that capture data about all activities, they can "liquify" areas previously inaccessible to capital. The company stated it this way: "Just as large financial marketplaces create liquidity in securities, currencies and cash, the IoT can liquefy whole industries, squeezing greater productivity and profitability out of them than anyone ever imagined possible."[2] In this view, according to Couldry and Mejias, "every layer of human life, whether on social media platforms or not, *must* become a resource from which economic value can be extracted and profit generated. The processing power of artificial intelligence is a key tool in all these developments."[3] They point out that the essence of the Internet of Things can be well captured through the lens of colonialism: "Whereas the extractivism of historical colonialism 'relat[ed] to the world as a frontier of conquest—rather than as home,'[4] data colonialism brings extraction home, literally into the home and the farthest recesses of everyday life."[5]

Thus, the smartphone and the Internet of Things have opened great new opportunities for profit-making. It is the major online malls that have been central participants in, and beneficiaries of, this diffusion of the internet. For example, smartphones' geolocation data allows Google and Facebook to deliver more accurate forms of targeting to advertisers. Likewise, the Echo and Alexa "smart speakers" let Amazon learn more about customers by placing listening devices in their living rooms.[6]

At this point it is necessary to broaden the theoretical perspective on digital capitalism to the wider range of data-driven, network-connected and automated systems. Building on theorizing about the "control revolution," scientific management and social Taylorism, a beneficial approach is to look through the lens of technopolitics, focusing on politics, power, and profit and the way they are channeled through and changed by technology.

The framework of technopolitics recognizes that far from being objective or neutral, technology is permeated by values and intentions. Technology is the result of decisions and actions made by humans, driven by their motivations and goals. Importantly, no technology's existence is inevitable, and all technologies are socially shaped or constructed. Behind every technology lies a number of human choices about what problems should be solved, how resources should be spent, why people should use a particular device or system, what trade-offs should be made, and many other choices that boil down to doing X instead of Y or Z. Politics comes into play when one begins to focus on questions such as: Whose interests are represented, who is included and who benefits? On the flip side: Whose interests are erased, who is excluded, who loses?[7]

In his book *Too Smart*, Jathan Sadowski analyzes the technopolitics of the emerging "smart society" in terms of three broad focal points, and then considers how they play out in different ways and places across society:

- Smart tech advances the *interests* of corporate technocratic power over values like human autonomy, social goals and human rights;
- Smart tech is driven by the *imperatives* of digital capitalism, extracting data from, and expanding control over, potentially everything and everybody;
- Smart tech's *impacts* are a bargain of convenience and connection (as exemplified by Facebook), in exchange for a wide range of (un)intended and (un)known consequences.

While *interests* are about whose values and voices are included, *imperatives* concern the overarching principles and goals that have deeper influence and wider reach. The profit motive is an imperative of capitalism, which drives firms to maximize profits, usually as the primary or only imperative. Similarly, the imperatives of smart tech are core parts of digital capitalism. Two major imperatives drive the design, development and use of smart tech: *collection* and *control*.

The imperative of collection is about extracting all data, from all sources, by any means possible. It compels businesses and governments to collect as much data as they can, wherever they can. That is why so much smart tech is designed to absorb data. For many industries, data is a new form of capital, and thus they are always seeking and exploiting new ways to accumulate data. It should be clear that the imperative of collection involves more than just passively gathering data; it means actively creating it. Data is not simply waiting "out there" to be discovered as if it already exists in the world like crude oil or raw ore. Data is produced by people using technical processes.[8]

The imperative of control is about creating systems that monitor, manage and manipulate the world and the people within it. It is represented by the ubiquitous surveillance systems that help corporations and governments preside over people, regulate access, and modify behaviors. This imperative leads to sensors embedded everywhere, connecting everything to the internet, and a reliance on automation to oversee it all. Smart tech is built to expand and enhance powers of control, whether that is remote control over objects via software applications or social control over populations via algorithmic analysis.

Needless to say, the imperatives of collection and control are deeply interdependent. Harvesting data requires the technical ability and social authority to probe things, people and places. Control systems are fed by data, which allows for more granular, more effective, and more instantaneous command over those same things, people and places. Smart tech is the offspring of both imperatives. As imperatives, collection and control are also not something new; they have been integral components of capitalism from the start. Capitalism continuously innovates, coming up with new ways to extract profit and exercise power over everything—society and nature, human and nonhuman, mind and body.[9]

The *impacts* of smart tech having gone viral, are monumental—spreading, infecting, reproducing, disrupting and thriving in nearly all spheres of society and aspects of people's lives. Even though they may share the same label, different types of smart tech (e.g., various gadgets involving the smart self; devices and sensors of the smart home; and technologies of the smart city) are often treated as if they were separate entities, independent from one another. For example, the smart watch one wears, the smart home one lives in, and the smart city one inhabits are rarely scrutinized together. This is ironic since the explicit goal of major tech companies like Google and Cisco is to plug everything and everybody into a single mega network—a "system of systems"—which is constructed and controlled by them. Against this kind of disjointed analysis, Sadowski argues, there are shared interests and imperatives that influence the design as well as the use of smart tech across different types, scales and spaces. In other words, instead of seeing these technologies as discrete and unrelated, they should be seen as parts of a powerful, yet still emerging technopolitical regime of digital capitalism.[10]

Data and Varieties of Value Generation

As mentioned earlier, Zuboff's analysis of surveillance capitalism lacks an account of how all of the parts of value—not just those accruing to behavioral surplus—are produced in the digital economy. Sadowski's description of the different ways in which data is used to generate value can be seen as a worthy attempt to fill this gap:

- Data is used to *profile and target people*, thus generating value in building up behavioral surplus. Many business models and services in digital capitalism are based on the value proposition that knowing more about people will, in some ways, translate to more profit. This is how internet-based companies often make their revenue by serving personalized advertisements, and it is the central focus of Zuboff's analysis. Through this kind of data usage, retailers can charge different prices based on the customer's characteristics. Political consultants analyze data to decide who is susceptible to certain kinds of messaging and influence.
- Data is deployed to *build digital systems and services*, as these require data to operate, utilizing existing data and collecting new data. This is what Zuboff recognizes in Google's early phase when the collected data was primarily used for service improvement. As devices become "smarter," they also become more data driven and internet connected in order to facilitate the flow of data to and from the device. Advances in artificial intelligence and machine learning are predicated on the input of myriads of data used to train these systems.[11] This development requires further explanation.

Though the terms "AI" and "machine learning" are often used interchangeably, machine learning is actually a particular line of AI development. It relies on large data sets to train models which are then used to make further predictions. Integrated into this process are algorithms that analyze data to extract patterns and make predictions, and then apply those predictions to generate further algorithms. In learning and creating new rules, these products develop in ways that supposedly resemble human intelligence. Of those technologies currently available, artificial neural networks (ANNs), closely modeled on the brain's neuron connections, are the most sophisticated and widely used. In a process known as "training," the neural network is repeatedly exposed to instances of a specific data object, an image of a dog or an audioclip of a melody, for example, and then the weighed interplay of the network's various layers is manipulated by an algorithm until the network is able to recognize the object. This new data then feeds back automatically into the network, creating a more sophisticated algorithm.[12]

- Data is used to *optimize systems*, making machines more efficient, workers more productive and platforms more seamless. By showing how to eliminate waste and do more with less, data helps to generate big savings. This might mean an industrial manufacturer installing sensors and actuators on mechanical systems or a management consultant using algorithmic analysis to assess how a city government should run its programs.
- Data is used to *manage things (and the people associated with those)*. This essentially boils down to the interdependency between knowledge and power; they are in a mutual relationship and each requires the other. In

this respect, data is a digital, formal, mobile, and machine-readable form of knowledge. The idea is that amassing data about an object or people's behavior enhances the ability to control it. This might be as mundane as a person keeping track of their diet and exercise in attempting to improve their health. It might be more complex as in the case of management digitally overseeing the activities of workers in a company so that it can control their behavior more thoroughly,[13] or even more complicated as in the case of an engineer overseeing the traffic pattern of a big city in order to manage how millions of people move through the space.

- Data is used to *model probabilities*. With enough data covering a wide range of variables over time—fed to the right algorithms and data analysts—many highly valued companies promise that they are able to predict the future, as stated earlier. In reality, however, these "predictions" are better thought of as probabilities instead of crystal balls as they are frequently treated. Some areas where data-driven predictions go far beyond forecasting the weather, include "predictive policing" systems that create "heat lists" and "hot spots" that signal the people and places having a high likelihood of criminal activity.
- Data is used to *increase the value of assets*. Things like buildings, infrastructure, vehicles and machinery tend to depreciate in value over time as wear and tear take their toll. (However, due to their location, some properties may increase in value, even if falling apart.) Upgrading assets with smart tech that collects data about their use may help combat the normal cycle of deterioration. Ideally, the usage of these assets becomes more adaptive and responsive, thus extending their useful lives. Rather than depreciating, smartified assets can, in principle, maintain and even gain value. Or, if they do not grow value, at least data can help slow their decay.[14]

Control Systems and the Colonialization of Everyday Life

Today's society is filled with digital control systems that have colonialized everyday life. These systems watch and track people by capturing data about specific activities and actions. They use various metrics to judge, rate and rank people, and they establish checkpoints that regulate access and enforce exclusion. A focus on technopolitics involves the analysis of the power of technology and the technology of power. Here it is helpful to tease out three major forms of power. The first form is conceptualized in terms of force, that is, the ability to make people do what you want—to obey a command, follow a rule and/or change their behavior—usually by the threat of punishment. This kind of power is called "sovereign power" because it is how monarchs and lords historically exercised power over their subjects. Today it is better to conceive it in terms of fear and force, backed up by authority and assault. The police and parents both wield this type of power. For much of human history, this is the main way power has been exercised in society and it is undoubtedly still a common feature of life for many people.[15]

The second form of power is what social theorist Michel Foucault called "biopower," but can be indicated by a simpler name: discipline.[16] Power works in this way by inculcating people with certain ways of thinking and behaving. People internalize these beliefs, habits and norms through which they are then shaped into certain kinds of people (or "subjects") by institutions like schools, workplaces and prisons. In this form, power is the ability to determine the boundaries of what is good, normal and just, and establish institutions and methods to discipline people accordingly.

Foucault described the panopticon as an exemplary form of modern disciplinary power—a structure built in such a way that the subject knows one could always be watched, but one never knows when or if one is being watched. Consequently one always behaves as if one is monitored. Its archetype is the prison with the architectural plan originally conceived by the English philosopher and social theorist Jeremy Bentham in the late eighteenth century.[17] The ultimate goal of this monitoring system is for the subject to internalize the "disciplinary gaze" of the authorities.[18] For power to be exercised in this way, the inmate does not have to believe that s/he is under constant observation, but only that the possibility of being observed is constantly present. Discipline leverages in equal parts the subject's paranoia, guilt and shame.[19]

To recap, sovereign power is wielded by giving commands, threatening force and handing out punishments (or rewards). In contrast, discipline operates by fostering ways of living, rejecting "deviant" lifestyles, and watching people to make sure they behave appropriately. As good, disciplined subjects, individuals police themselves and those around them by making sure everyone follows social norms. If someone deviates from those norms, s/he is gossiped about and ostracized from social life. If the offense is serious enough, the community resorts to the time-honored ways of removing deviants from society via prisons and asylums.

Discipline loathes too much individuality; people need to conform to the model of a "good student," "good worker," "good citizen," and so forth. There is no tolerance for deviance and abnormality, whereas control, the third form of power, operates on different principles. Control works by setting parameters for what is allowed or not, and establishing checkpoints that regulate actions. People can act freely within those parameters as long as they allow themselves to be tracked all the time.[20]

The preeminent example of control is the computer network that invisibly and continuously records every activity, rejecting any action that does not fit with its code. Control systems do not rely on mere threats of surveillance. They follow through on monitoring, judging, and inhibiting one's freedom if one deviates from their parameters. Unlike the panopticon, it does not matter if one knows about the system, but it is likely better for the system to function optimally if one is unaware of its operations.

Obedience to authority in the data colonialism at issue here is promoted through a self-monitoring, delusional feeling that "we live public," but that no one needs to bother too much about it (at least as long as one has not done anything wrong) because everyone else is also being surveilled. While the panopticon effect relied on one's inability to tell exactly when one was being watched, so that one behaved all the time as though one was, what is called the "inverse panopticon effect" of pervasive surveillance rests upon people *knowing* that they are being watched all the time, but habituating to behaving as if they are not.[21] This means "naturalizing" acceptance of a world in which surveillance and continuous tracking operate unnoticed in the background.[22]

In practice, these three forms of power are not totally separate from each other, rather they merge together and coexist. Control is the most relevant one for the analysis here, as it is a crucial component of digital capitalism. But it must be recognized, too, that there are many cases where force and discipline are also channeled through smart tech.[23]

The arrival of the flexible network of the digital enclosure may well mark a rearrangement of the boundaries of what Michel Foucault described as the great enclosed institutions of the modern era: hospitals, prisons, factories and schools. In each case, a widening digital enclosure allows for the spatial dispersion of institutional control; the dissolution of the walls of enclosure. Prison officials are increasingly reliant on digital tethers for nonviolent offenders (think electronic ankle bracelets). The work space has expanded beyond the office and factory walls, while schools embrace distance learning. Even the medical profession in some places is taking recourse to digital outpatient monitoring as a low-cost alternative to hospitalization.[24]

However, an individual's activities may be only partially captured by the digital enclosure, depending on the limitations of the interactive capability of both the space and the individual moving through it. For example, cash transactions may fall completely below the network's radar, but, if cash is totally replaced in the future by electronic money transfers via smartphone or some other device, all economic transactions may fall within the purview of the enclosure.[25]

With respect to the fate of labor in the digital enclosure, media critics Kevin Robins and Frank Webster have argued that networks provide the "technological means to break the times of working, consumption, and recreation into 'pellets' of any duration, which may then be arranged in complex, individualized configurations and shifted to any part of the day or night,"[26] a phenomenon extended to the hotel room, airport, home or café. However, such pellets of consumption and production are also internally de-differentiated; an act of consumption doubles as an act of production to the extent that it generates feedback commodities. Thanks to de-differentiation, "'free' time becomes increasingly subordinated to the 'labor' of consumption."[27]

Importantly, checkpoints regarding digital enclosures mediate access and restriction, freedom and constraint. Each checkpoint requires a "password" (understood in a more general sense than usual) to move through it. In addition to the code for unlocking a smartphone or logging onto a computer, other common "passwords" can be the keycard needed to open an electronic door, a personal identification number used to purchase things, or a visa that allows one to enter a country legally. Thus, possessing the right "passwords" is necessary to navigate a smart society.[28] In short, control in the digital world works through various sprawling, connected, hidden systems, which monitor people by breaking them down into data points that can be recorded, analyzed and assessed in real time, so that their freedom of access, action and so on, can be regulated via checkpoints and passwords.

The following area of digital behavioral control deserves special attention here. The prevailing online contracting regime revolving around boilerplate clauses discussed earlier is a compelling example of how the legal rules in question, coupled with a specific technological environment, can lead people to behave like simple stimulus-response machines—"rational," but also predictable and ultimately programmable. The boilerplate "contracts" in question are standardized form contracts, also known as contracts of adhesion or take-it-or-leave-it contracts. This type of contract is between two parties, in which the terms and conditions are set by one of the parties (in this case the service provider). The other party (the user) has little or no ability to negotiate more favorable terms, and is therefore placed in a take-it-or-leave-it position. The environment disciplines users to go on auto-pilot and arguably, helps to create or reinforce dispositions that will affect users in other walks of life involving similar technological environments. Such environments might become commonplace given the direction of innovation implicated in the further development of capitalism interlocked with data colonialism. Law professor Margaret Jane Radin notes how status quo bias can lead us down a familiar path,

> Given our tendency to stick with what we've done before, it is hardly surprising that after we've received boilerplate many times without having any negative repercussions, we will persist in our acceptance of it. Once we are used to clicking "I agree," we'll keep clicking "I agree." It would take some extraordinary event, some real change in context, to make us stop doing what we're used to doing when it seems to work.[29]

The effect may be more powerful than Radin suggests, however. A variety of heuristics (that is, "rules of thumb") and cognitive biases might reinforce people's behavior in this instance.[30] Decision fatigue may also play an important role, with the opportunity costs of slowing down and deliberating becoming increasingly untenable. And ingrained habits have in themselves immense impact of course.

Radin examined how the use of boilerplate has expanded significantly in legal contracts. She explains how "boilerplate creep" gradually erodes public ordering (e.g., law of the people, political and social institutions, government) and replaces it with private ordering (e.g., law of the firm, market). Not surprisingly, boilerplate creep only exacerbates the effects of the abovementioned heuristics and biases as users become more comfortable, complacent, and easier to manipulate.[31]

At this point it must be emphasized that the electronic contracting environment that prevails in the U.S. (and extends to countries within its digital orbit) is contingent—a product of, and fully dependent upon, evolved contract law and practice, which could both be different. Contract law has permitted and encouraged the development of a contracting environment in which it would be unfeasible for users to read the terms of the contract. The contract law could have accommodated changes in economic, social and technological systems differently. The current state of affairs is neither necessary nor inevitable; contract law could still change, although this seems highly unlikely as of now. The technological systems through which people interact, communicate, transact and form relationships also could be different. They reflect a series of design choices in their architecture and deployment.

These systems are set up and optimized to obtain predetermined results given the current legal and technological constraints along with the predictable behavior of users/consumers.[32]

Smart Spaces

The three major spaces where the regime of digital capitalism is being built are the smart self, smart home and smart city. Each space has been subjected to a rapid, widespread, large-scale wave of smartification. By uncovering the technopolitics of smart tech in each of these spaces, it becomes clear how their impacts go far beyond the usual set of concerns about privacy intrusions and cybersecurity breaches (that dominate the criticism of, and opposition to, surveillance capitalism).

There are three formidable efforts behind the growing scale, scope, influence and power of the engineering involved in all of this smartification that Brett Frischmann and Evan Selinger highlight in their widely acclaimed book, *Re-Engineering Humanity* (2019). First, instrumental reason is valued to such a degree that it has become fetishized. This concerns the fundamental problem first identified by the pioneer computer scientist Joseph Weizenbaum (who became later in his life one of artificial intelligence's leading critics). He argued that the way computers (and other tools) have usually been deployed have ultimately "reified complex systems that have no authors, about which we know only they were somehow given to us by science and that they speak with its authority, permit no questions of truth or justice to be asked."[33] The "science" Weizenbaum refers to is the type of rationalism and instrumental reason that can be boiled down to "computability and logicality."[34]

Second, the scientific management of human beings in general and data-driven efficiency management in particular, are rapidly spreading, which is best understood as the extension of Taylorism from the workplace context to nearly every setting within which we develop and lead our lives (an issue addressed earlier in this book). Lastly, in Frischmann and Selinger's words, "it's rapidly becoming easier to design technologies that nudge us to go on auto-pilot and accept the cheap pleasure that comes from minimal thinking; smart environments are poised to significantly exacerbate the situation."[35] This refers to "the social costs associated with rampant techno-social engineering that devalues and diminishes human autonomy and sociality as we become adjusted to being nudged, conditioned, and more broadly, engineered to behave like simple stimulus-response machines."[36] These observations should be kept in mind when considering the various smartification trends outlined next.

Smart Self

Smart tech (in principle) is used to measure, monitor, manage and monetize all aspects of people's lives. But the rise of the smart self is based on much more than simply people choosing to track themselves. The most important consequences of the smart self arise from what others do with our data and how they direct or otherwise influence our behavior, whether we want them to do so or not.

A good example to begin with is the starter interrupt device that allows auto lenders to track the location of cars, both in real time and over time, and remotely shut off vehicles if the borrower falls behind on payments or drives outside an approved area. It is customary for many lenders who lease cars to subprime[37] borrowers to require the installation of starter interruption devices in their vehicles. Research has shown that the techniques used by subprime lenders are less about extending credit to risky borrowers than they are about maximizing profit margins.[38] The starter interrupt device is an extreme example of how disadvantaged consumers are now exploited in new ways.

The starter interrupt device is not an isolated case of smart tech run amok. It is just a more blatant example of the numerous ways in which smart tech is used to monitor and manage people. Everything from location tracking and remote control to value extraction and behavior modification is part of digital capitalism's basic modus operandi.[39] This example could at some time become even more onerous and also include automatized repossession. At the time of writing, automaker Ford had submitted a patent application for a system that would enable a computer to remotely disable a vehicle's functionality (radio, air conditioner, cruise control, door locking system, etc.) over delinquent car payments, or place the vehicle in a lockout condition. Eventually this could lead to cars self-driving to repossession lots. When asked, however, a spokesperson for Ford said there were no plans to deploy this system, adding that Ford submits patents on new inventions as a normal course of business, but they are not necessarily an indication of new business or production plans.[40]

As Virginia Eubanks, a scholar of technology and justice, has pointed out, a good way to predict the future of data collection and social control is to ask poor and marginalized groups, as they are typically the test subjects for surveillance technologies.[41] Smart tech, if not being tested and trained on vulnerable communities, may first appear on the market as high-end consumer goods purchased by early adopters and people with disposable incomes. In each case, if the technology proves to be effective and lucrative, the smart features are integrated into products as standard, rolled out to the rest of the population and spread throughout society until it becomes a normal part of everyday life. This is what happened with many early developments in electronics and digital computing. For instance, the Progressive insurance company began to give customers a device called Snapshot to install in their cars. This device records where, when and how one drives, and then streams that information back to the company. Snapshot is a launch pad for personalized insurance pricing based on real-time assessment of individual driving behavior.[42] Many other insurers are now using similar devices. These devices and analytics can be seen as the middle-class version of the starter-interrupt device. Those who are able to avoid hard control by auto lenders are enticed to submit to soft power by auto insurers.[43]

Smart tech further plays a pivotal role in the Quantified Self movement, which involves communities of users and makers of self-tracking digital tools, who share an interest in self-knowledge through numbers.⁴⁴ Quantified self practices overlap with the practice of "lifelogging" and other trends that incorporate technology and data harvesting into daily life, often with the goal of improving physical, mental and emotional performance. The widespread adoption in recent years of wearable devices such as Fitbit, Jawbone Up, Misfit, and Apple Watch, have been designed to feature self-tracking sensors and software to monitor and measure health and bodily movements. Combined with the increased presence of the Internet of Things in health care and in exercise equipment, these devices have made self-tracking possible for a large segment of the population.

The Quantified Self movement has attracted some serious criticism related to the limitations it contains or might pose for other domains. Most criticisms target the issue of data exploitation and data privacy but also the lack of health literacy skills in self-tracking. While most of the users engaging in self-tracking practices are using the gathered data for self-knowledge and self-improvement, in some cases, people are prodded and forced into self-tracking by others. Children may be required to participate in behavior monitoring at school using apps or software. Doctors may demand that their patients with chronic health conditions engage in the prescribed self-monitoring program at home. Employers expect their employees to sign up for workplace "wellness" programs requiring health and fitness self-tracking. Health and life insurers invite clients to upload their medical and exercise data to receive rewards or lower premiums. In other cases, self-tracking is imposed on people by a court of law as part of substance addiction programs (alcohol and other drug monitoring). They may not have any choice but to comply. Needless to say, the data gathered from many of these activities are exploited by many different actors and agencies for commercial, managerial, governmental or research purposes.⁴⁵

Another issue associated with self-tracking, particularly in health and wellness programs, is what is called "data fetishism." This phenomenon occurs when active users of self-tracking devices become enticed by the satisfactions and sense of achievement that the collection of the quantitative data offer. Proponents of this line of criticism claim that data in this sense becomes simplistic whereby complex phenomena are translated into reductionist data. This criticism generally incorporates fears and concerns about the ways ideas about health are redefined, as well as doctor-patient dynamics and the experience of selfhood among self-trackers. For these reasons, the Quantified Self movement has been criticized for providing predetermined ideals of health, well-being and self-awareness. Rather than increasing the personal skills of self-knowledge, the quantified self distances the user from the self by offering an inbuilt normative and reductionist framework.⁴⁶

However, the most important issue here is the illusion of autonomy through self-measurement that plays a quintessential role. Continuous automated data collection can be justified for the social benefits that data relations supposedly involve. When self-trackers celebrate their community, they do not mean the "community" of platform owners who profit from the data streams they generate, but the unseen assemblage of users encouraged by those corporations to openly share their self-tracking experiences. Self-tracking is a social process, and that social aspect may be meaningful. There is a clear link here to the wider value of "sharing" in contemporary culture. What is striking about this sharing is the unlimited extent to which data are to be shared: from thoughts and pictures to information sources and life histories to everything and anything else. Sharing our lives has come to imply a broader value of "openness and mutuality" at the risk of mystifying the communities we supposedly form online. In a world in which we are constantly encouraged to share personal information, data collection becomes part of a social life within which, somewhere, that autonomous subject allegedly still exists.

In other words, the Quantified Self movement has been all too ready to cede individual property rights regarding their own data. What Quantified Self adherents and corporate "data colonialists" alike see as a benefit (that our moment-to-moment existence is understood better by external data processing systems than by ourselves) actually collapses the space of subjectivity in which the self enjoys freedom. The problem is not that we have relations with external systems or that we delegate certain actions to those systems. These features have been an integral part of people's life throughout modernity, and with increasing regularity and intensity, such projects have been the object of the self's reflection. The problem is that contemporary data relations involve more than just relations; they involve systems that insert themselves within the self's needs, desires and choices.[47] Most importantly, smart self data is not stored in our own private vaults, but resides in databases owned by others. The insights offered by our self-tracking devices and profiles generated by data brokers represent two sides of the smart self but they are unequal in power. The issue is not just about transparency. Even if people had access to all the data collected about themselves, what individuals can do with their data in isolation differs enormously from what various data collectors are able to do with this same data in the context of everyone else's data.[48]

With the current state of affairs, consumers have little influence over how, why, or for whose benefit most of the data about themselves is used. And they lack the power to understand and derive value from that data in the same ways. As Jathan Sadowski adroitly puts it, "Beyond *observing* and *understanding* the world, data-driven systems are central to ways of *making* and *managing* the world. If knowledge is power, then a massive database of personal information updated in real time is like steroids injected into already-muscular corporations and governments."[49] This interpretation stands in sharp contrast to most accounts of the smart self that focus on how digital tech leads to self-knowledge and self-empowerment. Contrary to that hyperbolic narrative, close attention should be paid to the fact that more often than not, the most important impacts of the smart self are the result of how others use our data and how they influence our behavior.

The smart self involves much more than the abovementioned forms of digital tracking, data collection and analysis. An intriguing current feature of digital self-tracking is "function creep"—the spread of the mentalities, motivations and technologies of the smart self beyond the personal, domestic or medical sphere into other social domains.[50] The impacts of the smart self that result from what powerful people and institutions do with the data, and how they direct the individual's behavior, manifest themselves particularly in two other areas: scoring systems and managerial control of workers systems (the latter to be discussed in the next chapter).

Scoring Systems

Data brokering has become a massive industry, estimated at $200 billion worth in 2017. Over 4,000 companies, such as Acxiom, Cambridge Analytica, Epsilon, CoreLogic, Datalogixx, PeekYou, LexisNexis, Accurint, Spokeo, Zabasearch, and Thomson Reuters CLEAR, collect and aggregate information from public records and private sources, then make these data available to whoever is able and willing to pay the access fee. Moreover, a wide range of companies and agencies generate profits not only from their primary business, but also from monetizing their customers' information by selling it to data brokers. The list is long and growing: pizza chains, consumer health websites, repossession agents, payday lenders, liens, online surveys, warranty registration, rebates, internet sweepstakes, loyalty-card data from retailers, social media, mortgages, bankruptcies, drivers licenses, professional credentials, charities' donor lists, magazine subscription lists, public records, Social Security death master files, and credit headers (the identifying information at the top of a credit score report that includes name, spouse's name, address, former address, Social Security number, and name of employer). All these and more are involved in selling people's data.[51]

Data brokers use all this information to slice society into market segments. The categories used can be callous and nasty, such as specific ones building on racial, ethnic, gender or social class stereotypes, or psychographic segmentation that is problematic. In the mission to squeeze value from data, everything is allowed if it boosts the industry's ability to target and exploit people. But data brokers also sort people by more neutral categories like demographics, consumer choices and political views. If that data is not readily available, data brokers claim they can then use other information to infer people's identities and predict their preferences. (Even when data is supposedly anonymous, many studies show that it is not too difficult for those with the right tools and knowledge to figure out somebody's identity and other intimate details by using just a handful of data points about them.) Information can then be merged from multiple databases to obtain an even more complete profile, which can sometimes be used in unexpected and disconcerting ways. For example, some hospitals and health insurers have used data brokers to buy people's credit card purchase history and then adjust the insurance premiums according to the presumed healthy or unhealthy purchases those people made.

In addition to building profiles, many data brokers provide credit scoring services, which means that on top of categorizing people, they judge and rank them. The profiles and scores they create are then used to make decisions/choices that directly impact many facets of people's lives, such as obtaining a loan, renting a home or finding a job. Even if the data brokers' profiles in question are totally wrong or incomplete, and even if they are based on dubious assumptions, it is hard to correct the record. That is, if one is aware of them. But meanwhile these inaccurate assessments do still have real consequences.[52] These companies uphold the notion that their data-driven systems are objective and neutral. They depict their outputs as accurate reflections of the world, which should allow them to escape blame for any harmful, unjust outcomes. But in reality, their methods of analysis smuggle in a host of stereotypes, biases and errors. They both reflect and reproduce long-standing inequalities.[53]

Extensive profiles, compiled from disparate off-line and online sources, have become the basis for the commercial messages people encounter on the internet, but these processes also dictate the availability of broader social opportunities and material provisions. By 2000, the term "weblining" came to be used by critics to describe the practices of denying people opportunities on the internet based on their marketing profile.[54] The concept derives from "redlining," a description of earlier discrimination practices in which the boundaries of neighborhoods in urban areas where typically poor, often black, Latino, or other marginalized people lived, were outlined in red to indicate that services such as banking or telecommunications should not be offered. It should not be forgotten, however, that the legacy of redlining still impacts communities across the United States today.

With weblining there is likewise a hardening of social inequalities in that surveillance advertising practices reproduce social discrimination and economic exploitation on the internet and beyond. Big data technologies are then used to "digitally redline" unwanted groups, either as customers, employees, tenants or recipients of credit. Rather than relying on overt discrimination, however, these companies can rely on proxies for race, gender and other legally "protected categories." Then they calculate correlations to draw conclusions that have discriminatory effects, while maintaining enough plausible deniability to avoid regulatory intervention. Thus data brokers (in tandem with marketers) have created a system that, when functioning optimally, classifies individuals as either valuable or worthless via a range of processes focused merely on profit maximization.[55]

Credit Scores Chinese Style

A good way to see how these data-driven systems are further developing is to look at China, a country that has appropriated the kind of systems developed in the United States and Europe, and taken those to further extremes, driven by its government's ideological tenets and policy goals. In China these systems are integrated even closer with financial and banking operations, so corporations are better equipped to track credit scores and other bank-like services; these include micropayments featuring facial ID and various forms of platform-based commerce innovations that the West has begun to replicate. China is also much more interested in integrating the rural economy into the whole. For instance, e-commerce giant Alibaba (comparable to a mixture of Amazon, eBay and PayPal) has an extensive network of rural service centers that help farmers and small producers. JD.com (the third largest tech company in the world, after Amazon and Alphabet) has developed a system that allows people even in remote rural villages to buy products online and have them delivered quickly and efficiently through a system relying on high-tech methods such as drones as well as traditional means such as delivery by workers recruited from the regional population.[56] There is also a more concerted effort to invest in artificial intelligence on the scale of the economy and society as a whole.

5. The Wider Array of Smart Tech Devices and Systems

Although Silicon Valley is still leading the way, China may soon overtake it in terms of resources invested in AI-fueled education, research and development.[57]

At this point, the development of the Chinese social credit system deserves further scrutiny, beginning with the mobile payment app Alipay (enormously popular in China) and developed by Ant Financial, an affiliate of Alibaba. In many ways, Alipay is more like a bank combined with a large mobile ecosystem. People use it to buy groceries, pay bills, buy insurance, invest money, order dinner, call an Uber, book flights and much more. Alipay has built its reputation on being convenient and reliable. Its slogan is "Trust makes it simple." Yet sometime in 2015 users of Alipay were confronted with a new icon on the app's home screen for a service called Zhima Credit, which takes trust to a new level.

Zhima Credit (or Sesame Credit) analyzes the wealth of data collected by Alibaba about every user of Alipay plus data acquired from partnerships with a long list of other companies, in order to assign each person a social credit score. This score is an assessment of an individual's worth, reputation and status wrapped up as a three-digit number. The aim is to leverage the explosion of personal data in order to improve citizens' behavior. This scoring system has rapidly spread through China and it exerts real influence over how individuals are treated by other people, businesses and even government agencies. At first glance it appears to resemble the FICO credit score, widely used in the United States for decades, but there are differences. This is because the Chinese social credit score system skipped the long course of development that credit scores took in the West, emerging in a society that is ubiquitously equipped with surveillance systems and digitally connected lives.[58]

Just as data brokers in the U.S. now integrate a wide variety of data into their profiles and scores, the Zhima Credit score is based on much more than just financial data. Behaviors such as cheating on a college exam, neglecting a traffic fine, having friends with low scores, playing videogames too often, or posting criticism online against the Chinese Communist Party, could all impact your credit score.[59] Alibaba keeps its complex algorithm secret, but it has disclosed five broad factors that are taken into account: credit histories, personal characteristics, interpersonal relationships, the fulfillment of contractual obligations, and behaviors and preferences.[60] The score, ranging from 350 to 950 in this nationwide system, dictates your standing in society and how others treat you. The benefits of having a high score also go well beyond low interest rates; they include, for example, jumping the waiting list for health care or skipping the security line at airports. A low score may lead, for instance, to being denied access to good job opportunities or securing a foreign travel visa.[61]

In 2014, the Chinese State Council released the "Guidelines of Social Credit System Construction (2014–2020)," outlining the goal of establishing a basic social credit system by 2020. It is a national credit rating and blacklist being developed by the government of the People's Republic of China. The social credit initiative calls for the establishment of a record system so that businesses, individuals and government institutions can be tracked and evaluated for trustworthiness. The national regulatory system is based on blacklisting and "whitelisting." It is a mandatory government-run version of Zhima Credit, built with the help of Baidu, one of the world's largest tech companies.[62] If people commit certain "social misdeeds," like failing to pay fines, they will be blacklisted and banned from traveling on planes and trains for up to a year. In due course, the Social Credit System (SCS) will take over the other scores in China. It is planned to be the ultimate, unified smart system, and aims to be even more extensive, invasive, and consequential, composed as it is of over four hundred data sets, five hundred variables, and fifty central government agencies.[63]

In this system individuals and enterprises are scored on various aspects of their conduct: where one goes, what one buys and who one knows. These scores are incorporated into a comprehensive database that links to government information as well as to data collected by private businesses. The system tracks "good" and "bad" behavior across a variety of financial and social activities, automatically assigning punishments and rewards to decisively shape behavior toward "building sincerity" in economic, social and political life—a contemporary version of the old communist goal of the "engineering of human souls."

As of 2022, more than sixty different Social Credit System pilot programs had been implemented by local governments. The national government oversaw the creation and development of these governmental pilots by requesting they each publish a regular "interdepartmental agreement on joint enforcement of rewards and punishments for 'trustworthy' and 'untrustworthy' conduct."[64] A year later, however, most private social credit initiatives had been shut down by the People's Bank of China and government regulations had cracked down on most local scoring pilot programs.[65] These are clear signs of a strengthening of the national government's grip on, and centralization of, the Social Credit System. This coincides with the Chinese government's tightening of internet access (since 2020).[66]

The existing trend of social engineering and nudging individuals toward "better" behavior is also part of Silicon Valley's approach that holds that human problems can be solved once and for all through the disruptive power of technology. In this sense, perhaps the most disturbing element of the China story is not the Chinese government's agenda, but how similar it is to the path Big Tech companies are taking in the U.S. and the many regions within its digital orbit.[67]

The social credit score systems in China are only more extreme versions of what already exists in the United States and Europe. People in the West nowadays tend to rate and rank everything, and in return, they themselves are rated and ranked. They are assigned an untold number of scores created in hidden ways by secretive organizations and used for "unintended" purposes such as intelligence gathering by security agencies, or perhaps for deciding which immigrants are allowed to settle in the United States by the Department of Homeland Security.

The existing systems in China do not actually predict the future in the West. They reflect China's social and political context and will look differently in the United States or Europe, even if their processes may lead to many of the same results. One should look at them as examples of how these smart systems can develop and then consider how they are likely to continue advancing under a deeply stratified, massively surveilled, highly corporatized version of platform capitalism, the kind that is being perfected in the United States and other countries within its political-economic orbit.[68]

Along with the Social Credit System, one finds in China, too, the "regular" corporate surveillance as an integral part of platform capitalism. But mass surveillance and repression of political dissidents or minorities (most notably the Uyghurs and other mostly–Muslim ethnic groups in the north-western region of Xinjiang) have been left to other, more invasive initiatives, such as the *Golden Shield* and *Sharp Eyes* projects launched by the central government.[69] It is important to recognize here that the Chinese firm Hikvision, one of the world's largest makers of video surveillance equipment (blacklisted by the U.S.), is linked to the repression of Uyghurs, while its cameras also blanket Israel's occupied West Bank.[70]

In China, mass surveillance involves a vast network of surveillance systems used by the government to monitor citizens through the internet, surveillance cameras (enabled by facial recognition software), surveillance drones, robot police, alongside big data collection that targets online social media platforms. Mass surveillance has significantly expanded under China's Cyber Security Law (2016) with the help of a large number of local tech companies. In 2019, Comparitech reported that eight out of the ten most monitored cities in the world were in China, with Chongqing, Shenzhen and Shanghai being the world's top three. In addition, the Chinese government encouraged the use of various mobile phone apps as part of a broader surveillance initiative. Local regulators launched mobile apps for national security purposes and to allow citizens (spying on others) to report violations.[71]

In 2017, Human Rights Watch signaled that public security bureaus (PSBs) across the country had begun rolling out "Police Cloud" systems in order to aggregate data from health care, social media activity and internet browsing activity, purportedly to track and predict the activities of activists, dissidents and ethnic minorities, including those alleged to hold "extreme e thoughts." The Central Political and Legal Affairs Commission was planning to construct a network of police clouds in every provincial and municipal public security bureau, eventually interlinking them into one unified national police cloud system. Reporting by Human Rights Watch likewise revealed that PSBs also intended to purchase data such as navigating data on the internet (browsing histories) and the logistical purchase and transaction records of major e-commerce companies from third-party brokers in order to more effectively predict crime while targeting and cracking down on any potential dissent.[72]

On closer inspection, China's social credit and mass surveillance systems are problematic not so much because they are uniquely antidemocratic (the West has its own attacks on civil rights in this regard) but because they "perfectly illustrate" where datafication might be leading eventually to everywhere, as Couldry and Mejias rightly argue. They suggest that, "Ultimately, the close collaboration between a government that grants economic and legal advantages to tech companies and companies that help the government to conduct its surveillance is a global feature of data colonialism, not one restricted to China, Russia or the United States."[73]

The final chapter will further explore the basic features of China's state-driven regulatory model regarding datafication and the digital society. It will do so in a comparison with the two other major approaches that exist: the U.S.'s market-driven regulatory model and the EU's rights-driven one.

6

The Smart Self and Algorithmic Management Expanded and Intensified

An earlier chapter already discussed the rigorous "soft control" of management that Uber and its many gig-economy imitators exercise over workers legally defined as "independent contractors." Similar algorithmic management techniques also have been deployed regarding directly employed workers in many other companies. Over the past four decades, just-in-time techniques have come to dominate the big box retail stores, with Walmart leading the way, optimally taking advantage of manufacturing and logistics brought about by intensified industrial rationalization methods known as "lean production." The system consists of a tightly controlled chain of suppliers, just-in-time movements of goods, and the logical and telecommunications technology that pulls it all together. Lean production originally stemmed from Toyota and General Motors (GM). The Taylorized work patterns, coupled with extreme uniformity ("one best way") of doing tasks, and the use of the term "associates" instead of employees or workers, are also familiar components of lean production. Walmart has become a dominant retailer by strategically combining just-in-time delivery techniques and corresponding advanced logistics, with an advanced IT inventory system, control over its suppliers, and globalization—meaning above all, outsourcing to Chinese producers and suppliers. To this Walmart has added a level of central control and managerial authoritarianism that goes far beyond what most corporations achieve, according to labor historian Nelson Lichtenstein (writing at the turn of the 2010s).[1] This assessment needs further qualification, however, in light of similar developments at several other big companies.

In the past decade, Walmart has added digital control technologies that rely on factory discipline to extract higher output from its already overstretched shopfloor workforce. Foremost among these technologies is "Task Manager," a targeting and monitoring system that Walmart introduced in its stores from 2010 onward. The system tells employees what to do, the maximum time the task should take, and whether they have met their target. Workers sign on to the system by swiping their identity cards on a terminal and then the system delivers its instructions. Employees who fail to meet the target times mandated by Task Managers are then punished, using an elaborate system of penalties overseen by management. These penalties include:

6. The Smart Self and Algorithmic Management Expanded and Intensified

written reprimands in the form of Walmart's own "pink slips"; spoken reprimands in the guise of "coachings"; "decision-making days," when an employee must explain why s/he should not be fired; and then dismissal itself.[2] Amazon equals Walmart in its use of monitoring techniques to track the minute-by-minute movements and performance of employees and in settings that go beyond the assembly line to include their movements between loading/unloading docks, and between packing/unpacking stations. Workers are closely monitored, too, in their movements to and from the miles of shelves of what Amazon calls its "fulfillment centers"—gigantic warehouses where goods ordered by Amazon's online customers are sent by manufacturers and wholesalers, to be shelved, packaged and shipped to Amazon customers.

Thousands of workers dash around the warehouse, never pausing for a moment, always in motion. Some pick items off shelves, while others pack things into boxes. The pickers, packers and stockers robotically complete their tasks. All the pickers are equipped with a handheld computer, which issues commands to the worker, telling him/her what product must be retrieved and where it is located in the warehouse.[3] The device then counts down the seconds left for the picker to find and scan the correct item. If the item is not found in time, the picker's rate of success falls. And if the rate dips low enough, the worker is fired and replaced by a new hire—and the cycle begins again. Productivity quotas are valued so highly and humans so lowly that the tracking system even automatically terminates workers based on their performance without input from supervisors.[4]

The smart tech deployed in the warehouses is not designed to make the job easier or complement people's skills. It is more like a handheld overseer that barks orders, tracks productivity, cracks the whip and terminates "slackers." Every second is monetized, every movement is monitored and optimized. Accruing mere minutes of "unproductive" activity is something that must be weeded out. Walmart calls these infractions "time theft," because when employees are on the clock, the company owns every second of their lives.[5] Even going to the toilet too often, by the company's standards, is grounds for disciplinary action. On the other hand, however, companies never discuss the many instances of "wage theft," like having to wait in security lines before and after clocking out.[6]

There is no room for error and no time for less than peak performance in this rigid work regime. Amazon calculates the minimum number of bodies working at maximum productivity needed to meet its demanding targets—and that is the number of people hired. Workers are typically expected to pick over a thousand items in a ten-hour shift, which means walking (or jogging) an estimated twelve to fifteen miles while constantly crouching, standing and reaching on tiptoes.[7]

Meanwhile Amazon has been granted patents for an ultrasonic wristband that can track the hand movements of warehouse employees in relation to inventory tasks in real time. The wristband can also provide "haptic feedback" to alert workers when they are about to put an item in the wrong place, or pick up a wrong one. In reality, it is a smart version of classical time-motion study. This idea of wearing a monitoring bracelet obviously seems rather creepy, not unlike the idea of wearing an ankle bracelet during house arrest. However, this is only at the patent stage and not a guarantee that such a product will ever be created or put into use. Nevertheless, it points again to the fact that workers are being treated more like automatons controlled by Amazon, that is, until Amazon can develop actual robots advanced enough in terms of fine sensorimotor skills to completely replace humans.[8]

The Kiva robotic sorting systems that Amazon has introduced in its warehouses (from 2013 onwards) can perform tasks that involve storing and moving whole cases of goods, but cannot retrieve individual products from the cases (the remaining manual tasks that are still open to humans). These robots are intended to save workers the time of walking through the warehouse to gather individual products to be shipped together in a composite order. Rather than having workers going through the aisles selecting items, a Kiva robot simply zips under an entire pallet or shelving unit, lifts it, and brings it directly to the worker packing an order. The robots navigate independently using a grid laid out by barcodes attached to the floor. Thus, while the humans work in one location, the Kiva robots maneuver bins of individual items at just the right time of the shipping process, and the humans pick and assemble the products to be shipped.[9]

As Amazon increasingly seeks to further automate its warehouses, it recently introduced Digit, a humanoid robot. The robot, which is 5 ft. 9 in. tall and weighs 143 lbs., can grasp and lift items (and carry up to 35 lbs.), walk forwards, backwards and sideways, and can also crouch. Amazon began testing this two-legged robot by having them shift empty tote boxes at some of its facilities in October 2023. Predictably, this ambitious drive to automate sparked fears about the effect on its workforce of almost 1.5 million humans. In a briefing at an Amazon facility in Seattle, Tye Brady, the head of Amazon Robotics, admitted that Digit would make some jobs redundant, but boasted that new jobs would also be created. Insisting that people are "irreplaceable" in the company's operation, he rejected the suggestion that Amazon could one day have a fully automated warehouse, as humans "are so central to the fulfillment process: the ability to think at a higher level, the ability to diagnose problems." It was, instead, a strategic opportunity for Amazon to scale a "mobile manipulator solution," such as Digit, to work collaboratively with regular workers.[10] It remains to be seen, of course, whether or not future advances in artificial intelligence cum algorithmic management will ultimately eliminate most of the remaining higher-level tasks for humans as well, leaving only non-automatable vestiges of manual work to a culled workforce.

Micromanaging Taken to the Extreme

However distasteful or harsh Amazon's smart practices are, the company is now leading the way in how to use smart tech in workplaces and on workers.[11] Similar kinds of smart tech—used for the same reasons of managing and exploiting workers—have been deployed to a great variety of jobs. One telling example is the trucking industry. Truckers who transport and deliver the goods to warehouses are subjected to intensive surveillance and managerial control through electronic logging devices (ELDs) installed in their vehicles. (It should be acknowledged that not all are directly employed workers but an increasing number are "independent

6. The Smart Self and Algorithmic Management Expanded and Intensified 99

contractors" experiencing the same kind of surveillance and control.) Not unlike the starter interrupt device, the ELD closely monitors each trucker's daily activities. The ELD dictates when, where and how the trucker can drive and stop.[12] Those dealing with these devices describe the constant scrutiny as a dreadful way to work/live. Despite the truckers' resistance, a new law in the United States (effective December 2017) requires that commercial vehicles must be equipped with an ELD.[13] Trucking companies and legislators contend that this intrusive tech makes driving safer, while truckers point to the loss in privacy, trust and independence they experienced in the past. For those truckers who live and sleep in their trucks on long hauls, the ELD is a form of total control. The next step is envisioned as full automation in the form of self-driving trucks.[14] The ELDs are also seen as a gateway to even more intrusive monitoring technologies, like SmartCap's EEG-monitoring hats, or Seeing Machines' computer vision-equipped inward facing cameras.[15]

Tracking employees in road transportation and delivery is becoming ever more intense and extensive. A worst case scenario from the perspective of worker control is the constant upgrading of the systems employed by UPS. Its package-delivery trucks are now equipped with around 200 sensors that monitor everything: location and speed; when the driver opens the bulkhead door; when he backs up; when his foot is on the brake; when he is idling; when he buckles his safety belt etc.; and, of course, how much time the driver is taking to deliver each package. A high-resolution stream of data, including all that information and the truck's GPS coordinates, flow back to the UPS office. This system, called "telematics," adds even more pressure on top of the UPS protocols the driver already has to follow. These entail strict guidelines derived from time-and-motion studies that will tell the driver the most efficient way to do everything: how to handle the ignition key; how to choose a "walk path" from the truck; and how to spend time while riding in an elevator. New drivers are called to account for a variety of small infractions. They have to justify bathroom breaks and any other deviations—"stealing time" in corporate jargon—that could impact their SPORH (Stops Per On-Road Hour) count.[16] At the time of writing, it was unclear how this telematics system would function in relation to ELDs, as UPS had received a renewal on its exemption from various provisions of the ELD mandate from the Federal Motor Carrier Safety Administration lasting until October 2027.[17]

This kind of intensified employee tracking can be found at FedEx as well. At some FedEx warehouses, workers have been forced to wear a computerized package scanner, strapped to their right forearm. This enables management to meticulously gather data about the individual worker's manual operations and track their speed.[18] In the fall of 2021, FedEx Express began installing driver-facing and front-facing cameras in its pickup and delivery vehicles. The company contended that it would help prevent fraudulent accident and unsafe driving claims, enhance safety, and serve as a training tool for new and seasoned professional drivers. But the move did not sit well with workers who saw it as a flagrant invasion of privacy and intrusive monitoring of their activities inside the vehicles they drive. Importantly, such cameras are gaining common usage industry-wide. Examples include ABF Freight System that since May 2019 has been installing outward- and driver-facing cameras in its vehicles, while Amazon in 2021 started rolling out vehicle cameras that face both the road and its drivers.[19]

Such violations of workers' dignity are not just limited to blue-collar jobs. In the service sector, computerized just-in-time scheduling has come to dominate the workplace. The software in question, used by virtually every retail and restaurant chain, analyzes data about sales patterns, weather forecasts and other sources to calculate the optimal schedule of shifts.[20] For the sake of maximizing efficiency and profit, this scheduling system denies employees a stable routine, making their work life highly precarious, with hours that may fluctuate wildly from week to week. Workers are often alerted to each week's work schedule with little notice, and changes made daily. They are put "on call" so the store management can summon them to come into work at any time. They also may be sent home in the middle of a shift to cut labor costs. Some employers even track the location of employees outside working hours, under the pretext of knowing who to call first about quickly coming into work to take up a shift.[21]

This is all in addition to the industrial rationalization of labor processes in the service sector that had already led to Taylorized work patterns over time. And in recent decades these have been taken to the next level of managerial control through the deployment of digital control systems offering supervisors opportunities for real-time employee monitoring that is unprecedented in scope.[22]

With regard to office workers—including those who work from home as freelancers, or as part of a regular employment agreement—there is a range of tools that enable managers to scrutinize their employees' every move. These include "productivity tools" such as WorkSmart, which can be installed on workers' computers, that not only captures regular screenshots but also takes photos of them every ten minutes through their webcam, using facial recognition software. From this and other data—including app use and keystrokes—managers derive a "focus score" and an "intensity score" that can be used to assess the value of freelancers to the company. There is also the real possibility that bosses, through stealth remote control, can read employees' email, check their web activity, scan their social media, and even track their keystrokes or mouse movements—if not personally, then algorithmically. This type of "productivity monitoring," which points to a lack of trust between employer and employee, is becoming more or less the norm; smarter tech harnessed for greater control and bigger profits. Resistance to this heavy workplace surveillance among workers has appeared in the form of automatic mouse movers to thwart prying bosses. A mouse mover is a tool that keeps the cursor jiggling if one steps away from the computer and shifts one's full attention elsewhere, thus preventing one's online status to go idle. (One must remember, however, that there might be legitimate reasons for being offline, such as caring for a child, going to the restroom, or simply taking short breaks to help recover from screen fatigue. And there is a lot of remote work—reading reports, listening to meetings etc.—that does not require moving the mouse.) These devices have existed for years, but their popularity soared as people began working from home during the coronavirus pandemic.[23]

Such productivity monitoring is likely to be a part, or an extension, of a bigger managerial control system. It is highly relevant here that, since the 1990s, office work at big companies has become subjected to so-called Business Process Reengineering, which has been roundly criticized for its alleged "totalitarian" nature of management control. Later versions of reengineering, Enterprise Resource Planning (ERP) and Corporate Business Systems (CBS) are even more troubling with their all-pervasive digital control structures across the firm extending to activities at all levels of management except the highest tier. Only the CEO and his/her senior colleagues dodge real-time surveillance.[24]

Overall, managerial control over workers in corporate life has come to rely—to a much greater degree than before—on the technological and organizational constraints built into the command-and-control systems that were introduced in many firms during the past decades. This is exemplified in recent years by the increased pace and further regimentation of office work by Corporate Business Systems, with their intensive targeting and monitoring through "performance evaluation" systems. However, one must not slip into an overdeterministic view of such technology and forget to account for the influence of human agency, particularly the possibility of management incompetence and worker resistance. Uneven and incompetent management system implementation may indeed stymie the effectiveness of these control systems. Worker resilience may also come into play, which is, of course, dependent on the countervailing power that labor has at its disposal, generally and more specifically in the workplaces in question.[25]

Worker Resistance to Digital Managerial Control

Individuals are more likely to engage in resistance against digital managerial control when they feel their occupational identity is threatened. Truck drivers, for instance, often view independence, or being "out on the open road," as a central component of their occupational identity. They have contested and resisted electronic logs as well as the newer electronic devices being installed in their vehicles that put them under increased surveillance. Another prominent example is the police, who are committed to their professional identity and authority, and are willing to close ranks if they feel threatened. Resistance to their own surveillance makes sense, in an ironic way—in that "they end up resisting the very tools they regularly use to surveil other people during the course of their work."[26] Police often view new developments with suspicion, believing technology is a means of deskilling (eroding craftsmanship), entrenching managerial control, devaluing experiential knowledge and threatening their professional autonomy.[27]

On the whole, unions have been less inclined to address issues of workplace control, including the ramifications of digital surveillance systems. Moreover, increases of temporary, part-time, and other forms of contingent work, have contributed to a loss of confidence in organized labor as the source of resistance to management-controlled surveillance systems,

There have been some attempts to unionize precarious workers carrying out platform labor. In the U.S., the New York Taxi Workers Alliance (NYTWA) has organized with Uber and Lyft drivers in New York.[28] Other examples are the Drivers

Union (affiliated with Teamsters Local 117) in the State of Washington; Massachusetts Drivers United; the movement for rideshare drivers' rights, Ride Share Drivers United, in California; and advocacy campaigns such as Gig Workers Rising funded by trade unions. Except for the latter, these endeavors have focused on private hire drivers rather than food delivery.[29] In the UK there is a somewhat longer history of organizing across both sectors. At Uber, this began in 2014, starting with worker networks involving the mainstream GMB union. Activities moved on to the alternative union, IWGB (Independent Workers Union of Great Britain), whose members are predominantly low-paid immigrant workers in London, including cleaners and transport platform workers (food delivery and ride hail) for companies like Deliveroo and Uber. Workers in Britain have also formed courier networks and some have organized with the IWW (Industrial Workers of the World).[30] The union IG Metall in Germany, which organizes workers by precarious contract rather than occupational status, has successfully done so. In the Netherlands, the national Federation of Dutch Trade Unions, FNV, has recently been trying to unionize bicycle couriers and delivery workers for companies such as Thuisbezorgd, Flink and Uber Eats through face-to-face appeals by union workers and the Instagram page of FNV Young & United.[31]

New Oppositional Practices

The prevailing control mechanisms of the platform economy are in part a return to industrial systems, particularly Taylorism and associated Gilbreth time-and-motion studies, morphed into a new way. As algorithmic management deepens the information asymmetries, traditional loci of labor struggle such as time, effort and wages merge with new disputes over access to information and algorithmic fairness. There is the potential here for new collective practices of shared aggregation and curation of individual data that enrich workers' understanding of work and hold the possibility of supporting coordinated responses to contemporary societal and industrial challenges surrounding well-being and productivity in the workplace and beyond. Maria Cecchinato and others contend that this creates opportunities not only for a greater understanding of the physical and affective impacts of contemporary work, but also for a bargaining framework concerning the terms under which that work is performed and remunerated in a world where clear measures and values of one's work and worth are increasingly abstract and out of reach.[32] This approach centers on a form of what has been called "sousveillance" (understood to be an inversion of surveillance)—"watchful vigilance from underneath," that is, the monitoring of management practices by and for workers rather than the other way around.[33] The focus here is on the potential of quantification technologies in the workplace to be used by employees for collective resistance against workplace exploitation..

Obviously, sousveillance is by definition a shared endeavor. Large monolithic entities like corporations and governments have the capacity to aggregate data at scale. Individuals can only achieve similar scale by pooling their data. Users and designers alone face limits to the extent of the changes they can effectuate in how the self-tracking technologies are deployed. In order to develop alternatives, it is therefore necessary for self-tracking groups of workers to form coalitions with peer production-oriented groups, open-source developers, crowdfunding communities and scientific research institutions.[34] Up until now, this has not occurred in any significant way, and the prospect for this to happen in the near future does not look favorable either.

There are other cases where workers target the heart of algorithmic management more directly. For example, extensive field work among food delivery workers in Berlin, Germany has revealed practices through which they attempt to "break" the algorithmic rules in use.[35] It was found that algorithmic management introduced three additional pressures in the labor process of these workers: the withholding of information; a lack of feedback mechanisms for workers; and methods of performance control that rely on data. This type of management can and does hinder organized resistance, but it can at the same time foster shared experience and needs that can only be addressed by collective practices of learning, solidarity and resistance.

In response, these workers developed a repertoire of resistance practices. First, they engaged in guessing the algorithm, attempting to understand the choices being made. These then became collective practices that workers engaged in together. Second, they found ways to game the system, seeking out ways to bypass the rules of the algorithm. From spoofing their GPS location in order to skip work without punishment, to exchanging profitable shifts between each other in the spirit of solidarity, workers used their technical knowledge to distance themselves from the company's disciplinary logic. They could also hold a log-off strike, an obvious response to the new model of work termed "logged labor."[36] The use of Slack, a cloud-based cross-platform instant messaging service, designed for other workplaces, could be easily repurposed even in an environment without physical bosses. Each of these practices helped workers create some space and autonomy in order to exercise a degree of control, not only as individuals, but also as a group. Third, they reframed the work, developing collective grievances at their meeting points away from management (or the algorithm in question).

These hidden practices of algorithmic resistance can be counted as labor resistance when they have a political intent. Guessing and gaming are often merely "coping strategies" without any intent to change the underlying power structure, but they do constitute resistance when they put the algorithmic regime under scrutiny.[37]

Similar resistance practices have been found among ride-hailing workers in the U.S. and elsewhere. Like workers in other parts of the gig economy, drivers at Uber and Lyft have responded to gamified algorithmic management by developing what Michael Burawoy has called "work games." The framework of his influential field study among factory workers, *Manufacturing Consent* (1979), helps to analyze how gig workers today respond to platform-initiated games. Burawoy argued that games provided an appropriate concept for understanding how and why the factory workers in question consent to their exploitation in the labor process. The unpredictability of receiving bonus pay—brought about by the piece-rate payment system—motivated workers to play the game of "making out." (In the context of work or a job, this phrase typically means managing, coping, or getting along. In Burawoy's study, "making out" refers more specifically to embracing piece-rate pay systems and workplace games that obscure their surplus labor being extracted as profits.) Recognizing that they could earn bonus pay by exceeding quotas of their piece-rate, machine operators improvised tools and sped up machinery to significantly increase their output. In addition to these tactics, making out was dependent on cultivating mutually advantageous relationships with auxiliary workers, such as the "scheduling man" who had the power to slow down the pace of work. This not only provided workers the possibility of optimizing their daily earnings, playing the work game also offered them social and psychological rewards such as prestige, sense of accomplishment and pride.[38]

In the case of Uber Drivers, two fundamentally different player modes have been recognized: "grinding" and "oppositional play." Grinding is a term used to describe how players of video games spend considerable time doing repetitive tasks as a condition for getting an eventual award. When drivers grind they consent to a mode of labor in which the ride-hailing platform conditions them to maximize their supply of labor by rewarding them for exhibiting compulsive behaviors. Oppositional play on the other hand involves a mode of labor in which drivers resist Uber's gamification features, using them in unintended ways to create and play work games. Drivers employ the interactive features of Uber's app in unexpected ways to gain a competitive advantage over the outcomes of their labor. In addition, they implement strategies of control, speed and skill to games that help them maximize their earnings, protect their labor supply, and gain some insight into Uber's opaque algorithmic management system. Ultimately, however, drivers consent to Uber's hegemonic regime in which the ride-hailing platform shapes the conditions of labor and uses its constant surveillance of drivers to squash dissent. Drivers who play work games, especially those that lead to customer complaints, run the very real risk that they will be deactivated from Uber's platform. And quitting may be the only option for drivers to opt out of the rewards program and, more broadly, Uber's gamified work setting.[39]

Online Microwork—the Ultimate Form of Digital Taylorism

Online microwork increased significantly with the further advancement of automation of existing jobs that still left specific tasks to be carried out by human workers. This led to the creation of a large global underclass of microwork platform laborers performing monotonous and repetitive tasks for low pay and no benefits—often in precarious conditions. Importantly, this includes the immense back office maintained by U.S. Big Tech companies, which use freelance and subcontracted workers in low-wage countries to perform the repetitive tasks that keep the machinery of online malls running, ranging from labeling data sets so that machine learning algorithms can learn to detect patterns from them, to scrubbing social media sites of obscene or offensive content.[40] The workers who reside behind the search engines, apps and smart devices, for lack of better alternatives, are compelled to clean data and oversee algorithms for very little money.[41] The feeds of Facebook and Twitter (since July 2023 rebranded X) may seem to wipe away violent content with automated precision, but decisions about what constitutes pornography or hate speech are not made by algorithms. A facial recognition camera seems, of its own volition, to spot a face in a crowd; an autonomous truck to drive without human involvement. But in reality, the magic of machine learning is the grind of sifting and labeling data. It is these badly paid, psychically damaging tasks carried out by human workers, not algorithms, that primarily make people's digital lives legible.[42] This shadow workforce, aptly called "ghost workers," is just as indispensable to the smooth operation of online malls as the relatively small number of direct employees who write the code and design the user interfaces. And most importantly, the internet makes it possible to manage a global archipelago of contractors at a distance, not only across physical distances, but also across the figurative distances of the divided workplace. This is because the same networks used to distribute work can also be deployed to surveil and supervise the workers doing it.[43]

The term "microwork" is used to describe tasks for which no efficient algorithm has been devised; it is work that still requires human intelligence to complete reliably. Microwork is considered to be the smallest unit of work in a virtual assembly line. Microwork sites allow contractors to split larger projects into radically short-time pieces of work. Contractors post these "human intelligence tasks" (HITs) to a microwork site such as Amazon Mechanical Turk, which appear on the screens of thousands of workers—or "Turkers," as they are known in this case—who jostle to compete the tasks on a piece by piece basis. Moreover, the platform takes a 20 percent cut from such transactions. Since the work is carried out remotely, workers never encounter each other except as digital avatars on online forums.[44]

Amazon Mechanical Turk is the first and still the best-known example of microwork sites, which have rightly been called "digital sweatshops" by analogy with sweatshops in the manufacturing industry. It started as a service available only to programmers employed by Amazon. Back in 2001, during the dot-com bubble, Amazon created the site to solve a simple in-house problem: its algorithms were failing to recognize many duplicate product listings. Realizing the tasks could be completed more efficiently by workers, Amazon decided to patent "a hybrid machine/human computing arrangement," namely, Mechanical Turk. Via an application programming interface (API), Mechanical Turk gave the company's programmers the ability to write software that automatically outsourced tasks too complex for computers. Recognizing a growing demand for cheap labor across a still nascent platform economy, Amazon publicly launched Mechanical Turk in 2005. The site's now familiar role—hosting contractors who farm out "human intelligence tasks" (HITs) to precarious workers—offered a prototype for the sites that followed.[45]

Though Mechanical Turk may appear as a labor broker—an intermediary that takes a cut for hosting exchanges between workers and employers—its real purpose is to provide data for Amazon Web Services (AWS). Each task completed on this platform automatically sends Amazon a precise data set about how it was completed. Just as Mechanical Turk allows Amazon to broaden the scale and scope of its data capacities, many smaller microwork sites have data-trading agreements that benefit larger platforms.[46]

Because workers on Mechanical Turk contribute data about the work process itself—the way workers behave, how they complete tasks, when and how often they log in, and the time it takes for completing tasks—the data can be fed back to the platform or even into algorithms used, for example, in Amazon warehouses, which require a range of behavioral data to effectively monitor and control worker performance. This side of today's digital Taylorism differs only in extent from the comparable management styles of the twentieth-century economy. The real difference lies outside of management regarding the use of data to enhance machine learning services. Hovering over every exchange between requester and worker, Mechanical Turk can, for example, funnel data from a short translation task into Amazon Translate, an automated neural network machine provided by Amazon Web Services (AWS). Here the primary function of Mechanical Turk becomes apparent; a hardly profitable (potentially unprofitable) labor platform cross-subsidizes Amazon's wider business operations as a logistics and software company.[47]

Mechanical Turk has been copied by competitors such as Appen, Scale and Clickworker, that offer the same potent combination of clean data and cheap labor to contractors ranging from academics to big online platforms like Facebook and Google. As brokers of labor arbitrage, these sites locate what Mike Davis has termed "surplus humanity"—sections of the global populace found outside of the economy proper—to sporadically fulfill the needs of Big Tech.[48] A quintessential feature of this online work is that microworkers are not treated as human beings but as computational infrastructure. Application programming interfaces (APIs) that connect "requesters" (the entirely anonymous "employers" hidden behind opaque interfaces) to workers are normally used by programmers to interact with computers. On microwork sites, however, requesters interact with humans posing as computers. These sites provide the illusion, of automation, or as Astra Taylor has described it, "fauxtomation," to make it appear as though the work has been automated.[49] Workers are hidden from view, overshadowed by the machinery, so that requesters, particularly larger platform clients, can uphold their marketing strategies without problems. The claim of Facebook, Google, Amazon and the numerous startups aiming to secure venture capital, is that their business model is radically lean, hardly dependent on the risky realm of labor, running almost entirely on the work of complex algorithms. For this purpose, platforms outsource their labor to keep it off their books and hidden from users, investors and customers, to appear more sophisticated than they actually are.

As Phil Jones adroitly puts it: "This data fetish—figuring automatic drones in place of data labelers, media feeds in place of moderators—conceals the hidden abode of automation: a growing army of workers cut loose from proper employment and spasmodically tasked with training machine learning."[50] Indeed, here are the darkest recesses of platform capitalism involving its hidden stratum of microworkers.

As sites like Mechanical Turk suggest, the automation of some service work is perhaps never likely to result in full mechanization, but rather in human-machine hybridity. In work which has historically proven difficult to automate, machine-learning squeezes out small productivity gains in terms of the partial automation of specific tasks and managerial functions, the hyperdivision of labor and just-in-time outsourcing. When certain tasks are automated, other tasks once geographically bound are set free to wander all over the world in search of opportunities for global labor arbitrage in the form of foreign outsourcing. Subsequently, work once properly waged is by default informalized, parceled into badly paid, erratic piecework, and detached from the regulatory frameworks that legislate pay and rights.[51]

Microwork is carried out by the rising numbers of workers in the informal sectors of countries in the Global South, in the Indian subcontinent, East Asia, Latin America and parts of Africa. In 2022, half of the global workforce competing for these microjobs was found in just three developing nations: India, Pakistan and Bangladesh.[52] Workers are often those found and trained in prisons, refugee camps and slums, the totally jobless or underemployed. Many of these workers are educated but have been cut off from the formal labor market.[53] Among the overeducated and underemployed workers of the Global North, the numbers of microworkers have risen significantly, too, further augmented by the coronavirus pandemic, when many young people started working this way remotely. For these workers, microwork is mostly a part-time pursuit to top up hours and stagnant wages.[54] For many across the world, however, microwork is a dreary and underpaid full-time job, as surveys by the International Labour Organization (ILO) have shown.[55]

Disappearance of Occupations

Like others who somehow manage to subsist on informal service niches, microworkers have no obvious occupation. Nevertheless, there is a tendency to use "microworker" as if the term describes a proper profession with routine and specific tasks, like "lawyer" or "doctor." This is partly due to the fact that the term "microwork" originates with Samasource, a platform to which refugees are no more than just useful worker bees for the AI industry hive. The term tacitly serves the interests of such actors, who along with institutions like the World Bank, wish to dignify an essentially degrading pursuit. But microwork is, by its very nature, highly contingent, irregular and essentially formless. Jeff Bezos has described its essence (in marketing Mechanical Turk) as "humans-as-service."[56] By evoking software as service—in an effort to disguise labor as computation—Bezos also captures the emptiness of a role that ranges over a multitude of tasks, often split off from other jobs.[57]

It is clear that microwork hardly represents a new source of "job creation," as suggested by enthusiastic reports and articles from the World Bank—"from millions of tasks to thousands of jobs" promises one such piece.[58] Such publications actually invert the logic of a system that aims to convert thousands of jobs into millions of tasks, which do not readily return jobs.[59] Microwork turns on the same illusion as the informal sector, which, Mike Davis contends, "generate jobs not by elaborating new divisions of labor, but by fragmenting existing work, and thus subdividing incomes."[60] The worker patches together a subsistence out of various bits of low-skilled labor, prised from the remains of other jobs. Something like an occupation does not by any means emerge from these economic split-offs.[61]

Moreover, whether as the labor of the refugee camp or prison or as workfare disguised as welfare, microwork offers a convenient way of putting a surplus of cheap labor to work, not only for profit but also for discipline. In the years following the 2007–2008 financial crisis, state governments across the U.S. contracted Samasource to train jobseekers for online labor, primarily in Rust Belt regions faced with ongoing state budget rollbacks and deindustrialization. The point of the program was to prepare the long-term unemployed for a brave new economy, where, instead of full-time factory employment and demanding managers, workers should expect contingent tasks and tyrannical algorithms. Unsurprisingly, these "training" programs often coincide with participants accessing online platforms, which makes it difficult to tell precisely where education ends and workfare begins. Though this may be an extreme example of how microwork operates in the U.S. and Europe, the Samasource program is telling in that it signals who in the Global North uses these platforms. Many of these workers have been drawn into capital's orbit but cast outside of formal labor markets. They comprise a redundant mass ranging from the partially employed—left without enough working hours to live on—to those left without a wage indefinitely.[62]

Maximal Exploitation of Microworkers

In his book *Work without the Worker* (2021), Phil Jones details the exploitative and often mean practices of microwork sites, which are in his view reminiscent of "Victorian capitalism but now reimagined for professionals and precariat alike, introducing piece-work across a range of once waged or salaried professions to push workers to merciless levels of intensity."[63] He considers this to be especially the case with Mechanical Turk, where rigorous standards of quality are often less important than brute speed. Although AI can already do many tasks listed on microwork sites, workers maintain the upper hand when it comes to pace.[64] With tasks paid by the piece, workers experience long inactive periods as they search for new jobs, which often means longer hours as a whole to make ends meet. More time is spent hunting for jobs than actually completing them. Moreover, the chronic "super abundance" of labor in the digital realm of most of these sites is strategically planned rather than the result of spontaneous development as in other sectors of the informal economy. Microwork sites are organized to attract greater numbers of workers than there are tasks available in order to ramp up productivity and drive down wages. This oversupply coupled with the lack of employment opportunities elsewhere, forces workers to spend an inordinate amount of time hunting for tasks that pay little more than a few cents.[65]

To make sure pace is maintained, most sites allow requesters to place time limits on tasks, which, if broken, result in reduced pay. Workers remain susceptible to the quirks of server volatility, poor connectivity and hostile requesters. On

Mechanical Turk, time restrictions are only indicators of how long a task could take, but because the restrictions are defined by requesters who are eager to cut costs, a task might be marketed as one dollar for fifteen minutes, but may actually take closer to thirty minutes to complete. A worker might remain unaware of this until ten minutes into the task, but once under way, backing out means giving up payment. Even tasks completed in the afforded time frame frequently go unpaid. Those deemed "bad quality" by requesters are more often than not rejected outright.[66]

On microwork sites, there are multiple "employers" over the course of a single day and they remain entirely anonymous—behind opaque interfaces—leaving the worker with no idea who they are working for. "Bad" requesters would be unable to refuse payment if platforms were not designed to allow infringements of the wage contract. Platforms operate under the guise of "neutrality" to protect their status as intermediaries, and refuse to enter disputes between workers and requesters. Furthermore, curated crowdwork sites like Appen and Lionsbridge attract long-term clients, but these clients are under no obligation to stay on the platform. Requesters can easily vanish without paying, while workers are forced to wait until they can cash in their wages, sometimes as long as thirty days after joining the site or until their payment balance reaches a certain sum.[67] This often means wages disappear before the worker has a chance to withdraw them. One of the more severe measures taken by microwork sites is to shut down the accounts of workers who protest or act in unorthodox ways deemed to undermine a site's rules—often without clear notice from the platform—which can mean all wages scored during the holding period are lost.[68]

Then there are score systems, which offer a veneer of objectivity, allowing requesters to numerically measure worker performance. But they are, in their way, as biased as account closures. Each worker receiving an aggregate score of ratings previously given by requesters is made public, so that other requesters can identify good or potentially bad performers. If a worker's ratings suffer from the entries of a particularly strict or hostile requester, his/her reputation plummets and opportunities to find more work diminish. Metrics tend to take on a tyrannical quality here, often representing the difference between finding work again and being cut off from work altogether.[69]

Alongside these stealth tactics are cruder forms of wage debasement. Many platforms pay exclusively in nonmonetary "rewards" in the form of Amazon gift cards, cryptocurrency, Walmart coupons or Starbuck vouchers, or points for online reward programs and videogame credits. Mechanical Turk is most interesting in this respect, given its size and geographical scope. Workers from across the globe use the platform, but only a limited number have access to payments via bank transfer. The majority must accept gift cards for rewards. While Amazon offers workers in most European countries the option of bank transfers, workers in countries from the Global South—such as Botswana, Qatar and South Africa—only receive gift cards and points. In these countries the platform basically comes to look like a digital company town, where tasks are completed for tokens that can only be spent on services and goods provided by Amazon.[70]

The Essence of Microwork and Its Obstacles to Worker Organizing Efforts

The way online microwork is set up amounts to what has been called "black box labor." The online microworker may know that training data is fed into the algorithm and that a decision comes out at the other side, but what goes on in between remains entirely opaque—a black box that is impenetrable to those outside its workings, for reasons of power and secrecy.[71] How the algorithm makes decisions—on what grounds, for whom and with what aim—remains hidden. As appendages to these algorithms—refining, enhancing and supervising their capacities—workers spend their days in this shadowy world. They are neither able to see the process on which they labor nor are they seen by those outside its parameters. This is how larger platforms want their labor: obscure to those doing it and invisible to the wider world.

The aim, however, is to conceal not only a wider labor process but also workers from each other. Platform interfaces provide no messaging services or profiles that workers can access. This is partly to counteract labor militancy but more fundamentally, to prevent the emergence of a real workforce, in any conventional sense. Numerous workers in contact with each other would increase the risk of a secret project—such as those of the U.S. Department of Defense, the National Security Agency (NSA), or the U.S. Immigration and Customs Enforcement (ICE)—being made public. And it would also threaten to dispel the algorithmic illusion—disguising workers as machines in bids to venture capital—and thus undermine the financial interests these sites uphold.[72]

Microworkers pose a particular challenge to labor organizers, whether looser worker associations or the more typical institutions of organized labor. Everything from the international geography of microwork to the pools of surplus labor on which platforms draw, makes their collective action an uphill struggle. Monthly or annual models of union membership run up against microwork's temporal dynamics, with workers joining sites daily and some staying for only brief periods. With "contracts" between microworkers and requesters lasting mere minutes, sometimes only seconds, wages are so volatile that membership fees are likely unaffordable. Even if unionization were financially feasible, unions so often relate to their members through identities of a professional or occupational nature that microwork offers no viable basis to organize.

Crucially, microwork often takes place in slums, refugee camps, prisons and occupied territories, places where unions fail to reach and organizing is not only dangerous, but may also be seen as a criminal activity by authorities. Even outside of these more extreme spaces, workers tucked away in bedrooms and internet cafes remain invisible to one another and to the institutions that might otherwise organize them. Workers are geographically dispersed, rarely if ever brought together in physical space.[73]

Organization thrives on a public dimension that microwork sites prevent, not only by geographical distance but also by software frontiers that limit worker contact, as discussed earlier. Such barriers restrict organizing to the terrain of online forums. Users of TurkerNation and MTurkGrind, as well as Turker-themed Reddit

6. The Smart Self and Algorithmic Management Expanded and Intensified 111

threads, engage in small-scale, non-antagonistic action such as raising funds for fellow workers.[74] Such action has been most effective when aimed at the architectures of specific websites. Turkers have developed Turkopticon, a website and browser plug-in that overlays the worker's screen and allows them to write reviews about requesters and publish them in real time. As a common infrastructure, Turkopticon also enables workers to connect with one another for mutual aid.[75] The simple fact of the tool's existence—letting requesters know they might be rated—itself acts as a deterrent against wage theft and other misdemeanors. But while the plug-in helps to discipline requester behavior, it is not built to transform the platforms themselves.

Worker resistance might also take the form of a digital blockade that obstructs the circulation of data. But the question is, what precisely would a blockade look like in the ether of digital capital? Even if large numbers of data labelers laid down their keyboards and mouse pads, it is likely that a large pool of workers would still remain, ready to take over the work of those on strike—similar to "scabs" in traditional strikes. These labor pools are not geographically limited—like on other online labor sites such as Uber and Deliveroo—but extend to every country where a given site operates, often numbering millions of workers and continually recruiting new ones. Anything less than action supported by a great majority of this pool would leave loads of tasks to be taken up by an intractable mass of remaining workers.[76]

As with action taken by other segments of the informal proletariat, microworkers will need to find ways of disrupting circulation that go beyond the mere abrupt refusal to continue working on the sites in question. Perhaps targeted sabotaging of tasks is a possible tactic. Its forms range from mass idling of tasks to sustained and widespread "machine breaking," a timeworn tactic that tends to tilt towards insurrectionary organization. The Luddites, British textile militants of the nineteenth century, were one such group, which "continually trembled on the edge of ulterior revolutionary objectives."[77]

Probably the best way to look at Luddism in the current context is as Phil Jones suggests:

> In the digital realm, such "machine breaking" may in fact be little more than a metaphor, so different would data vandalism look to Victorian loom-smashing. Mass derailing of data tasks would be more like a blockade, temporarily stopping data traffic, as opposed to actually destroying it—data, unlike a loom, is nonrivalrous, ubiquitous and endlessly reproducible. Nonetheless, there remains an anarchic allure to such action in a moment of algorithmic order. But like other kinds of online action, it would rely on a sufficiently large number of participants so that no one worker could be penalized. Unlike the historical Luddites, who were masked, disguised and smashed machines under cover of darkness, "smashing" algorithms would enjoy no such expeditious conditions. Monitored closely, workers fall under the watchful eye of platforms that can shut down action on the spot.[78]

So it appears that, ultimately, this avenue does not offer a promising way to force microwork sites to change their basic setup to the benefit of laborers.

7

Smart Homes and Smart Cities

*Hype About Benefits
Versus Drawbacks for Users*

Under digital capitalism, devices and appliances are not just commodities but have also increasingly become a means of producing data. The drive to create and circulate data as capital, influences the design of many new and "improved" household products.[1] When the imperatives of digital capitalism are translated into product design, it implies that every gadget becomes a new way to stealthily record personal information and send it to corporate servers; those companies, in turn, may sell the data to third parties. It also means that those companies maintain ownership and remote control over the embedded software—and therefore the device.[2] Ultimately, "the proliferation of smart, connected products will turn the home into a prime data collection node," writes designer Justin McGuirk, "in short, the home is becoming a data factory."[3] Yet it is highly unlikely that the people living with the smart home devices will themselves own (or even have access to) most of the data manufactured in their home as factory. So, even in their own homes, these consumers do not necessarily own the means and outputs of data production.[4] Unsurprisingly, there is a lot of hype about the benefits for users in the stories about the smart home being told by tech entrepreneurs, marketers, advertisers and experts, all stakeholders in the businesses involved. Of course, potential drawbacks are downplayed, denied or completely ignored. Therefore one must tread carefully to tease out the ramifications of smart tech at home, particularly regarding spying, hacking, and the collection and packaging of data by the firms concerned.

A home automation system monitors and/or controls things such as lighting, climate, entertainment systems, and appliances. It may also include home security such as access control and alarm systems. This means that door locks, thermostats, home monitors, televisions and other entertainment products such as speakers (often voice-controllable), cameras, and lights are all linked to this system. Even appliances such as refrigerators and coffeemakers can be controlled through one automation system. When connected with the internet, as they often are, such home devices are an important part of the Internet of Things, a network of physical objects that can gather and share electronic information. A home automation system typically connects controlled devices to a central smart home hub. The user interface for control of the system deploys either wall-mounted terminals, tablet or desktop computers, a smartphone application, or a web interface that may also be accessible off-site through the internet.

By now, a very large section of smart homes operates through digital assistants like Amazon's Alexa, Apple's Siri and Microsoft's Cortana. Users often interact with those apps-enabled devices through voice control, and the "assistants" can take commands, field questions, organize the user's calendar, schedule conference calls, or provide alerts. Though not specifically related to one's home, the digital assistants provide a broad range of controlling smart assets, their schedules and their statuses.[5]

There may arguably be some benefits for home owners utilizing home automation. Advocates of smart home technology most often mention that it provides homeowners with convenience and long-term cost savings, while "personal and family security" and "excitement about energy savings" are the primary drivers for the increase in use of this technology in recent years.[6] However, there may likewise be significant drawbacks of home automation that should worry people. Because many of the Internet of Things devices have microphones and cameras that are always on line, they are an easy target for hackers who can use them to spy on the inhabitants of the house. It has been proven that adept hackers can gain access to a smart home's internet-enabled appliances, including devices such as DVRs, cameras and routers, and can even bring down a host of major websites through a denial of service attack.[7] There is also the increased risk of data breaches; for example, data from home devices can show whether one is at home or not, aiding burglars. Perhaps more likely, ransomware could attack a particular device—for example, turning off the heat until the homeowner pays up. Another issue is that patches to bugs found in the core operating system of a smart home often do not reach users of older and lower-priced devices, while failures of vendors to support older devices with patches and updates leaves those devices, if active, vulnerable as well.[8]

Much attention has been drawn to invasions of privacy through internet-connected home security surveillance systems such as Vivint, ADT, SimpliSafe, Ring, and Wyze, that rely on video cameras to record what is happening. Amazon-owned Ring has perhaps become the most notorious for its video doorbells; vulnerabilities of this doorbell system quickly surfaced when a shocking event drew national media attention. In December 2019, a hacker accessed a Ring camera that a mother of four had installed in her children's bedroom for "peace of mind." Her 8-year-old daughter (who was alone in her room) was harassed by the hacker, who played a song from the horror film *Insidious*, used racial slurs and tried to persuade her to misbehave. It came to light that several other Ring users across the country had reported that their security systems were likewise compromised by hackers via the camera's two-way talk function. Amazon Ring's knee-jerk reaction was to blame the customer. A Ring representative told the little girl's parents that the incident was their fault for failing to set up two-factor authentication as an added security measure. The representative stated that what had happened was "in no way related to a breach or compromise of Ring's security." The "bad actors" behind the attacks "often re-use credentials stolen or leaked from one service on other services."[9] Ring repeatedly gave customers who had experienced similar hacking problems, the same blanket statement, blaming the victims for using weak passwords. At that time, the company appeared unconcerned about safeguarding its customers.

Amid growing public pressure, Ring announced in 2020 that it had strengthened security control measures, making two-factor authentication mandatory for all users and giving them the option to block most of their data being shared.[10] However, security experts accused the company of not doing enough to protect users' information and called for multiple changes to Ring's policies and device functionalities. The company continued to allow hackers to take over the smart cameras used on doorbells and in homes, as dozens of customers claimed in a class action lawsuit against Ring in December 2020. Victims allegedly were subjected to various kinds of harassment (including obscene language, racial slurs, blackmail and murder and sexual assault threats) after their in-home smart cameras were hacked, and sued the company over "horrific" invasions of privacy. The lawsuit built on previous cases, joining together complaints filed by more than 30 people in 15 families who said their devices were hacked. In response to these attacks, Ring "blamed the victims, and offered inadequate responses and spurious explanations," the lawsuit alleged. The plaintiffs claimed, too, that the company had also failed to adequately update its security measures in the aftermath of such attacks. Moreover, the lawsuit cited research from the Electronic Frontier Foundation and others that Ring violated user privacy by using a number of third-party trackers on its app.[11] Afterwards, nothing of significance happened until the Federal Trade Commission got involved in the case. This led to a settlement in May 2023, as part of a broader array of regulations of contemporary capitalism cum data colonialism, about which more in Chapter 9.

Amazon also faced questions in the U.S. Congress and elsewhere about the sharing of Ring data (including video footage) with police departments. By May 2021—within a time span of three years since Amazon purchased Ring—the company had brokered more than 1,800 partnerships with local law enforcement departments (about one in 10) across the United States. This meant that the police could request recorded video content from Ring without a warrant. Investigations showed that Ring also shared these data with the likes of Facebook and Google. Ring's pervasive network of cameras expands the dragnet of everyday pre-emptive surveillance—a dragnet that surveils anyone who gets into its field of vision, whether a suspect in a crime or not. Although the dragnet indiscriminately captures everyone, there are racial, gender and class-based inequities when it comes to who is targeted and labeled as "not belonging" in the residential space involved.

Current facial recognition technology, which has been denounced by AI researchers and civil rights groups (especially for its racial biases), could be applied to Ring-recorded content and live feed, and lead to excessive policing of race in residential space. It was also reported in 2021 that Amazon Ring was in fact building the largest corporate-owned, civilian-installed surveillance network ever seen in the United States. This always-on surveillance network extends even further when one accounts for the millions of users on Ring's affiliated crime reporting app, Neighbors, which allows people to upload content from Ring and non–Ring devices.[12]

Initially, Ring's cloud-based infrastructure (supported by Amazon Web Services) made it easy for law enforcement agencies to place mass requests for access to video recordings without a warrant.[13] Because Ring cameras are owned by civilians, law enforcement agencies were consequently given a backdoor entry into private video recordings in residential and public spaces that would otherwise be protected under the Fourth Amendment.[14] However, in January 2024, in response to fierce criticisms coming from civil liberties groups, Amazon Ring announced that U.S. police would now need a warrant to access user footage. At the same time, the company had disabled features allowing law enforcement to request footage directly from users. Up until then, public safety agencies including the police, were able to ask users to voluntarily share video footage from their Ring cameras rather than seeking warrants to obtain that user data from Amazon.[15]

Overall, increasing connectivity results in a growing loss of privacy, as these smart devices collect, package and share data with the manufacturer and others. It is a rich source of data about how the devices are being used, and increasingly, who is using them.[16] It is incumbent upon consumers to recognize how many different kinds of sensors are being brought into the house—whether it is microphones, video cameras or just devices—that capture all sorts of data about themselves and their guests. But the issue is not just about a single video doorbell or one internet-connected TV, and so forth; it is thinking about all of those devices working in concert, as Maurice Turner, deputy director of the Center for Democracy and Technology, emphasizes: "What information is being collected and who has access to either keep it, or more importantly, share it?"[17]

There is an appalling lack of data privacy protections and transparency, or even basic information about the data that is being collected. Smart devices need to gather certain types of data to work properly and improve their performance, which seems reasonable on its face. But the problem is that, in many cases, too much information is being collected and shared with third-party companies—with opt-in (rather than op-out) as the default setting on the device, often next to other default settings aimed toward collection and sharing.[18] For the sake of transparency and increased consumer awareness, digital security experts have suggested that the companies involved should improve disclosures about notice, choice and consent, and let consumers opt in—thus explicitly agreeing to that kind of data collection.

Health insurance companies were among the first to see the lucrative potential of smart tech in the home. With data shared by manufacturers, they are now closely tracking the information flowing from home medical devices such as heart monitors, blood glucose meters, and even Apple Watches.[19] A notable example is the CPAP machine—a doctor-prescribed mask worn while sleeping by millions of people with breathing problems; it also turned out to function as a spy for insurance companies, allowing them to track when and how long the mask is used.[20] Some patients found out that their insurer stopped covering the costs of the very expensive machine and supplies because they had failed to strictly comply with the prescribed use.

Insurers have also been partnering with tech firms to offer special deals like discounts on premiums to customers for installing smart home systems and allowing them access to the data produced. For instance, State Farm offers a discount on their home insurance policy for installing a Canary home security monitor. Liberty Mutual will send customers a free Nest Protect smoke detector and cut the cost of fire coverage, if they are allowed to monitor the device. Some insurers want to go further, thinking that coaxing home owners to wire their homes with internet-connected devices will generate a mass of lucrative new data that can make handling claims more efficient while creating a new relationship with the customer. With a feed of data from the customer's home, an insurer could nudge them to prioritize maintenance tasks and fix problems before they cause major damage. In 2016, Jon-Michael Kowall, assistant vice-president of innovation at USAA (United Services Automobile Association), announced he was aiming to create a suite of smart tech that acts like a "check-engine light for the home" whose data could then be used by the insurer. The general idea is to fill each customer's home with sensors that oversee everything, from leaky pipes to daily routines and send status reports to the insurance company. The data could then be used to notify customers about potential issues and what actions should be taken. On the other hand, the same data could also be used to profile some customers as more likely to engage in risky behavior and their premiums could be raised or coverage canceled altogether. It is not clear to what extent this kind of extreme monitoring has actually materialized so far, also given the fact that insurers' wild dreams of rewiring homes have been hampered by questions from at least one State Department of Insurance (Illinois) about privacy and security, as well as by incompatibilities between the smart devices from different companies.[21]

Many analysts predict that insurance will be the major business model underpinning smart tech, similar to the way advertising now bankrolls many web platforms—the likes of Google, Facebook, Amazon and so on.[22] This means that any tech financed by insurance, rather than by advertising or some other industry, will be designed to reflect insurers' interests. And by exerting influence over how smart tech is designed and deployed, insurance companies can wield extensive power over people. There is no doubt that the insurance industry will use the power of risk scoring, personalized pricing, and other innovations to increase its own revenues. A case in point is the unfair (and even illegal) practice of "price optimization," where insurers analyze non-risk-related data, such as credit card purchases, to target people with personalized prices that reflect how much they will pay, not their risky behaviors. Now such practices—and the sensitive data set they rely on—can be implemented through opaque algorithms, giving human actuaries plausible deniability when bias and deception is uncovered. And along with detailed data monitoring comes the power of behavior modification; the great power that insurers have over how people behave—ranging from individuals to multinational corporations and police departments—has been well documented.[23] The industry's ability to record, analyze, discipline and punish people often surpasses the power of government agencies.[24]

The trade-offs of "dataveillance" for discounts and control for convenience may look harmless. But discounts quickly become penalties once expectations about data disclosure shift from novel to normal. The transition from voluntary to mandatory enrollment is already well underway. As surveillance by insurers becomes more accepted, and disclosures become low-cost and routine, those who hold out are suspect.[25] Rejecting an insurer's request to audit one's daily life and domestic habits raises a red flag, which can trigger all kinds of negative responses by the insurer: hiking rates or denying claims because one was not using the required data-streaming devices; or canceling coverage altogether, if one refuses to wear, install and use the smart tech provided.[26]

In assessing the ramifications of smart home devices, it is important to recognize that an incrementalist view can be misleading and lead to habitual complacency. The problem is that smart technology is likely to creep along in various guises. For example, the specified purpose of a smart TV with an always-on microphone may be to capture aural input for voice-activated commands. Communicating such data outside the home may be necessary for machine learning and other techniques to improve the voice recognition technology. But one should expect data collection and function creep, as the always-on microphone can pick up much more data than actual commands. Using this information solely to improve voice recognition seems highly unlikely in the mid- to long term, as the data will almost certainly be collected and shared with third parties.

If one deploys a host of different smart devices in the home, aggregate effects are likely to emerge. Standing alone, each device might not transform the home environment, but together they might. The aggregate effects might resemble a tragedy of the commons in the sense that incrementally rational decisions lead to a net tragedy.[27] Obviously, tragedy in this case is not inevitable. Its occurrence depends on various contextual details, such as whether the devices interconnect locally or beyond the home, and how defaults are set for security, privacy and user engagement (for example, in setting parameters or programming options), among other things. The point is that creep phenomena play a role here as well.[28]

Connectivity sells products, so manufacturers continue to flood the market with all kinds of household appliances and wearable devices that are web-enabled. The question becomes: is the risk to one's privacy or safety worth the benefits? In some cases, is one connecting things via the internet that do not necessarily need to be linked? For example, is the risk of compromise worth the reward of being able to see the inside of one's refrigerator from across town? The question remains whether consumers really need all these smart home devices, which may not be used solely for the intended purposes, but may open a Pandora's box of abuse unbeknownst to them. There is also the deskilling potential of smart devices (which may lead to the dummification of their users), as well as environmental impacts, including energy usage. Indeed, the logic driving the developmental pathway of such devices privileges machine learning.[29] Placing a higher value on machine learning than human learning regarding task performance is the hallmark of today's Internet of Things, convenience-oriented devices.[30] A weighing of the risks versus the benefits of each of these devices or of all of them acting in concert, might very well lead to the decision not to install (or uninstall if already in place) specific smart home devices or even more drastically, the whole Internet of Things at home.[31]

Smart Cities, Surveillance and Control

The smart city is one of the most prominent of the "smart" concepts that has captured the public imagination. It is also one of the most consequential and politically significant, informing and shaping the work of urban planners, architects, infrastructure operators, real-estate developers and transportation officials, as well as mayors and entire industries. And like most things smart, the "smart city" is not reducible to a single meaning—a factor that may very well help to explain the rapid adoption and proliferation of this buzzword among professional elites.[32] The usage of the term "smart city" has increased exponentially over the last decade, so much so that it has overshadowed the much longer known term "sustainable city" to a great extent. Originally the concept was being used with an explicit focus on environmental sustainability in the 1990s through the Kyoto Protocol, but later leaned more towards information and communication technologies for transformative global urban agendas.[33]

It is hard to give a precise definition of a smart city, due to the breadth of technologies that have been implemented under its label. A recent attempt at a definition begins by describing it as a technologically modern urban area that uses different types of electronic methods and sensors to collect specific data. Information from that data is used to manage assets, resources and services efficiently; in return, that data is used to improve operations across the city.[34] This involves the harvesting of data (from citizens, devices, buildings and assets) that is processed and analyzed to monitor and manage traffic and transportation systems, power plants, utilities, water supply networks, waste disposal, criminal investigations, information systems, schools, libraries, hospitals and other community services.[35] It is necessary to adopt a broader definition (leaving much room for inclusion) in order to avoid the artificial limits imposed by the "official" smart tech industry itself. This makes it possible to consider services offered to and within cities by firms ranging from Google to Uber and Airbnb, which otherwise would not be included alongside the numerous self-described "smart city" products and solutions offered by the likes of Cisco, IBM, Microsoft or Siemens. Corporate giants like Google have been busy entering cities, pitching various products for "free" wireless internet that is connected to sensor-based apps, promising to "solve" the parking problem, for instance, and thus relieving residents of both stress and environmental waste. Cities may thus be caught in a vicious cycle; the more services they subcontract and the more infrastructures they privatize, the more assistance they require from companies like Google (parent company Alphabet), Facebook (Meta), Apple and Amazon to run whatever remains of resources and assets under public control. As noted earlier, only these U.S.-based corporations have the capacity to aggregate, mine and analyze a vast array of data while running sophisticated machine learning programs and predictive models to exploit artificial intelligence in order to deliver personalized and value-added services.

Capitalist firms like Google specializing in data extraction can present themselves as benevolent agencies determined to rescue the public sector. This narrative appeals to local governments as these companies position themselves alongside (and in competition with) the far more predatory consulting firms (e.g., Ernst & Young, Deloitte, PwC, KPMG) and smart tech divisions of businesses like IBM (moving away from their traditional model of hardware and software sales to selling services, including consulting), which have plundered city budgets by demanding cash rather than data in exchange for their services. For cash-strapped cities facing a fiscal debacle due to austerity measures, this arrangement is a much more attractive alternative. In most cases, data is something city managers do not account for, let alone measure, and therefore can easily give away in exchange for nominally free Wi-Fi offered to residents, or advanced traffic analytics software provided to city planners. Importantly, a firm such as Google is not collecting data merely to help sell advertising—in many cases, data collection has nothing to do with advertising whatsoever. Rather the data are used solely to accelerate development of advanced artificial intelligence technologies, helping Google and kindred firms to automate processes that currently require human input. Ultimately, whoever controls the means of producing the most data obtains the best AI, making others dependent on it and allowing AI to be fashioned as a service accessed on a permission basis. Such AI-powered services can then be used to further optimize how the city is run and operated.[36]

From the perspective of cities, there are various possible motivations behind opting for smart city solutions, roughly classified into two types: normative and pragmatic. The former refers to efforts to deploy technology to achieve ambitious and broadly accepted goals like promoting political participation among ordinary citizens, helping to personalize public services, and de-bureaucratize national and local governments. It is in this vein that some scholars and activists, most notably Anthony Townsend, author of *Smart Cities, Civic Hackers, and the Quest for a New Utopia* (2013), have promoted the smart city as an inherently decentralizing force, one that would break down bureaucratic and centralized systems of control and response. They point to smartphones—and the "empowering individualism" enabled through digital technologies and platforms—as the material means of this democratic force.[37] Obvious questions arise: to what extent are these ideas simply wishful thinking? And what is actually happening in most smart city projects? The possible drawbacks and risks of smartphones and social media usage, particularly infringements of citizens' rights and mass surveillance by the state and local governments, are not addressed by these authors. Moreover, smart cities, especially in developing countries, have largely failed to acknowledge the challenge presented by the "digital divide," that is, the social and economic inequalities resulting from who has access to communication technology and how s/he uses it.[38]

The second type of motivation spans a broad and rather heterogeneous set of objectives. Some cities opt for smart technologies merely because they promise to resolve problems pragmatically, which may be specific to that particular city: congestion caused by a crumbling road infrastructure and lack of repairs; a scarcity of good paying jobs in inner cities, which might disappear or, at least diminish in numbers, as smart money follows smart citizens into the smart and creative urban districts. Another problem is possibly an ineffective garbage disposal system that clogs the streets, in which the solution is sought in real-time, with immediate feedback loops involving clever sensors inserted into "smart trash cans" that tell passing trucks they need to be emptied.

Some cities want smart technologies because they promise big savings on the provision of similar or even better types of services during a period of budget cuts and harsh austerity. Other cities desire them because they want more security and policing, particularly on the eve of mega-events like the Olympics or the World Cup. Smart CCTV cameras—along with the sensors present in much of the built environment, and new techniques of predictive policing—allow cities to exercise targeted, effective controls over areas previously hard to reach and govern or control.[39]

The Current State of Affairs and Its Critics in the U.S.

The "smart cities" trend in America began with IBM's Smarter Cities Challenge in 2010. The company promised to award technology worth millions of dollars to cities that wished to upgrade their infrastructure. Among other things, that initiative established a highly competitive approach to urban innovation that pitted cities against each other in a bid to win free products and services from the private sector. The 2010s brought a wave of these competitions, backed by corporations that selected the "winning" cities to host pilot projects. Many philanthropic organizations, including Bloomberg Philanthropies and the Rockefeller Foundation, launched similar initiatives. And in 2015, the U.S. Department of Transportation used this same approach for its Smart City challenge, selecting one winning city, Columbus, Ohio (from the 78 that applied) to serve as a testing place for transportation technology.[40]

IBM has been a leader in the deployment of technology focused on public safety and law enforcement. The company has promoted its "smarter planet" strategy to centralize analysis of the interconnected bits of information coming from cities and embed them in systems and infrastructures to better control operations, capture and optimize the use of resources. IBM has developed an Intelligent Operations Center (IOC) enabling the optimization of critical information stored in disparate systems across multiple departments purportedly for the benefit of the city's population, economy and greater ecosystem. The IOC was implemented in Rio de Janeiro in 2010 for flood prevention and transportation management, as well as in Miami to manage football stadium operations, facilitate data-driven decision-making and predict crowd problems to minimize disruptions. IBM's Rio de Janeiro's Operations Center drew international public attention in 2014 when the city hosted the FIFA World Cup. To critics it demonstrated the quintessentially technocratic character of the smart city by its Orwellian "control room" that, with its ubiquitous sensors and cameras throughout the metropolis, enabled a "rational" response to any digitally recorded events.[41]

IBM Solutions' focus on law enforcement, predictive policing and crime prevention has led to the establishment of "Intelligence Law Enforcement Centers" and "Real-Time Crime Centers." In Atlanta and Chicago, for instance, IBM has introduced facial recognition, advanced video monitoring and other pervasive surveillance technologies to provide police with the information that enables them to detect crime patterns based on Big Data analytics.[42]

Importantly, the wave of excitement about "smart" technologies has resulted in many products traditionally classified as tools of surveillance and predictive policing being rebranded as essential components of the "smart city" package. For instance, Microsoft's CityNext program, which offers "public safety and justice solutions," specifically targets municipal police departments with its products and services. CityNext also includes several products that extend well beyond city problems. Its "prison and offender management" initiative, for example, is advertised as the solution to "track and manage offenders throughout the entire corrections system." Many of these solutions are hardly new and have received widespread criticism from criminologists and other social scientists, as well as political activists because predictive policing often reinforces existing social inequalities, as it feeds on biased data.[43]

Many of the early smart city initiatives in the United States were partnerships between tech firms and individual cities aimed at upgrading large urban systems for transportation, energy, waste and communications. Hardware, software, business services and connectivity companies formed alliances to offer citywide solutions. An alliance that AT&T launched in 2016 exemplified this approach. The company partnered with Cisco, Deloitte, Ericsson, GE, Intel, and Qualcomm in Atlanta, Chicago and Dallas. The stated goal was to develop smart city systems made up of a whole package of integrated products and services. This industry-led consortium model left little room for small firms and startups.[44]

Today, the picture looks quite different. A greater diversity of firms are exploring a wider range of revenue models and marketing strategies, including Civiq Smartscape (which sells communications network infrastructure), Nordense (embedded sensor networks for waste management), Soofa (information and wayfinding kiosks), and Urban Footprint (a mapping analytics platform and service). These newcomers are generally less focused on building citywide systems or upgrading physical infrastructures than on developing new digital services for a particular sector, or apps targeted at residents themselves. Many of these projects have been driven by tech companies accustomed to creating their own markets for emerging products. According to Jennifer Clark, writing in 2021 in the *MIT Technology Review*, almost all of those projects have failed to adapt technology "solutions" to the needs of individual cities and regions.[45]

The smart city concept—assiduously promoted by an entire industry of consulting firms, city fairs and smart city expositions—has already drawn considerable criticism. Critics have attacked the utopian visions behind the smart city for their unrealistic abstractions, lack of connection to the real world problems of real people, technocratic quest for domination over people's everyday urban experience, obsession with surveillance and control, and their inability to think in terms of putting citizens rather than corporations or urban planners at the center of the development process.[46] These critiques target in particular the mechanistic and functionalist character of the new smart cities. A smart city, as a scientifically planned city, would disregard the fact that real development in cities is often haphazard and participatory. Accordingly, the smart city concept is unattractive for citizens as it "can deaden and stupefy the people who live in its all-efficient embrace."[47] The richness and complexity of existing historic sites contrast sharply with the inadequacy of the modernist architecture and urbanism that is replicated within the technological determinist plans of the smart cities. Those cities are remarkably conventional in their twentieth-century modernist layout and planning, with rigidly defined and segregated zones of activity, and highly centralized forms of management and control.[48] It has rightly been noticed that the smart city movement suffers from a lack of critical thinking, dominated as it is by corporate interests, as well as a lack of clarity about how to define the smart city as such.[49] It is also a telling feature of the smart city movement that the concepts originated within technology businesses rather than from urban planning theory, as one would expect.[50]

Many of the earlier critiques of the smart city hit the mark and helped to connect those to previous campaigns against the excesses of technocratic urbanism led by the likes of Jane Jacobs with her seminal book *The Death and Life of Great American Cities* (1961).[51] New York's massive push for roads, bridges, and urban renewal during the mid–twentieth century, for example, instigated a backlash against "big plans" that persists to this day. The legacy of the Cross Bronx Expressway looms large in the collective memory of urban planners and is reignited as each new generation learns about Robert Moses, the powerful public administrator behind much of New York's transformation at that time. His name has become a metonym for bulldozing vibrant neighborhoods to make room for highways.[52]

However, most critiques fail to recognize that cities are also engines of capitalist accumulation, that they are economic as well as social actors, and that most processes occurring in cities are driven by economic and social forces that have been percolating for a long time.

The truth is that the most important formative context for most cities in the current era—at least in North America and much of Western Europe—has been that of neoliberalism or, more precisely, the transformation of the postwar Fordist-Keynesian compromise of embedded liberalism into the highly entrepreneurial and financialized urbanism that emerged and then expanded from the late 1970s onwards.[53] Over time, the technological infrastructures were brought more in line with the tenets of neoliberalism. For those who share this critique, "smart cities, as peddled by the likes of Cisco, IBM, Siemens, and Hitachi, are actually Trojan horses, disguised as urban 'saviors' but actually there to privatize public services and spaces."[54]

Cities, like other units of societies, experienced severe pressure to roll back some welfare state institutions, as well as roll out some policy innovations of their own. Two such processes are of particular importance here. The first concerns the delegation and subcontracting of responsibilities previously delegated to public institutions, to private players. The second entails the enlistment of private financial capital—mostly from pension funds, insurance firms, various alternative asset management funds—for the management, maintenance and construction of infrastructure, most of which operate at the local level. Both processes display significant links to the smart city, as both require an extensive infrastructure for gathering, analyzing and acting upon data to succeed and proliferate. This involves the intensive deployment of data analytics and measurement to assess whether specific targets or outcomes are being delivered, coupled with timely interventions to steer the course of actions towards those outcomes. None of this would be possible without an extensive infrastructure for tracking and controlling both physical and human resources, with quantification of performance paving the way for all sorts of other, even more advanced experiments to be constructed on this foundation.[55]

It is crucial here to recognize the growing importance of rankings, competitive tables and comparative scores. The origin of this trend can be found in the city indebtedness rankings by credit agencies like Moody's or Standard & Poor's, with cities competing for a favorable rating which determines their costs of borrowing. Nowadays, this function is further fulfilled by various rankings—measuring innovation, creativity, or even "smartness" itself—compiled by the urban-philanthropic (mostly pro-capitalist) complex of think tanks (e.g., Global Smart Cities & Connected Communities Think Tank), foundations (e.g., Bloomberg Philanthropies, Rockefeller Foundation), and NGOs. Together this complex determines the broader constraints and parameters within which cities compete. How cities perform on those secondary indicators in turn feeds into how investors view their competitiveness.

Related to this, many cities have been facing pressure to quantify the performance of the various constituent parts to make them more accountable, competitive and manageable, a phenomenon commonly associated with the rise of neoliberalism and its conception of an "audit society." This drive towards quantification, and the underlying mentality to rank everything, is only possible in a city capable of collecting, analyzing and processing vast amounts of data. Thus, the smart city agenda, along with the infrastructure of sensors and connectivity it promotes, also opens the door to the rankings-obsessed quantification celebrated by neoliberalism.[56]

Smart city deployments take place in a diverse set of local contexts that shape how they are conceived and produced. In India, the ambitious 100 smart cities program is part of a political, nationalist development agenda.[57] In China, smart cities are a key aspect of its government's fast urbanization, economic development and population management agenda.[58] In the UK, smart cities are about the marketization of services and the further embedding of a neoliberal urban order, together with the creation of exportable business opportunities, particularly consultancy expertise and technologies.[59] In Germany, smart cities are about the efficiency of urban governance and sustainable growth.[60] Therefore one would expect a large degree of variation and ambiguity on how smart cities are conceptualized. This also means that the fuzziness of the term "smart city" allows for serving different agendas.[61] Still, regarding the "varieties of smart urbanism,"[62] some broader generalizations can be made and a dominant political-economic agenda distilled. As Evgeny Morozov and Francesca Bria note, in Western Europe, North America and parts of South America, the smart city revolves primarily around infrastructure improvements to existing cities, while in Asia, India, and, to a lesser extent, China, one can find numerous examples of "smart cities" being built from the ground up. Early examples are: Masdar City in the United Arab Emirates, which sacrificed its zero-carbon features after the global financial crisis of 2008; Songdo in South Korea, which has so far remained a ghost town; and Dholera, India, where farmers have been dispossessed of their land in order to build the city.

While the dominant smart city discourse in the Global North is often synonymous with that of privatization of (existing) municipal services, in the Global

South, the discussion tends to be driven by imperatives of state-led urbanization and the formalization of previously informal industries and services. The latter tends to overlap with discourses of financial inclusion and entrepreneurship (as in India) or ecology and sustainability (as in China). In both these cases, Morozov and Bria contend, the term "smart" seems to have emerged as the least problematic term for a set of rather conventional neoliberal policies and prescriptions being reactivated against much less political resistance.[63] India is a major exception. There the 100 smart cities program has triggered a backlash, with activists and academics signaling that it fits perfectly with Prime Minister Narendra Modi's plans to make India more attractive to foreign capital, even if this leads to greater inequality, deregulation (particularly in the case of some cities designated as special economic zones), discrimination and misappropriation of public funds to cater to the needs and interests of the wealthy elites most likely to populate India's smart cities. These activities happen in tandem with the efforts of billionaires and corporations to build their own completely privatized cities (e.g., Lavasa or Gurgaon) in India.[64]

Democratic and "Common Goods" Alternatives to Smart Cities

It is clear that tech companies have increasingly been taking on administrative and infrastructure responsibilities that governments have long fulfilled. If smart cities are to avoid exacerbating urban inequalities, it is necessary for the citizenry to understand where the smart city projects are creating new opportunities and problems, and who may lose out as a result.[65] This means in particular taking a closer look at the technologies being deployed. It is noteworthy here that there is recently a move away from reifying the role of technology in tackling urban problems[66] and a recognition that the issues facing cities are not going to be fixed through technological solutionism, but necessitate a multifaceted approach in which technology is at best just one component.[67]

As some critics suggest, after over a decade of pilot projects and flashy demonstrations, it is still not clear whether smart city technologies can actually solve or even mitigate the challenges cities face. They claim that a lot of progress on the most pressing urban problems could come from better policies and more funding, issues that do not necessarily require new technology.[68] As Rob Kitchin underlines, addressing homelessness requires a complex set of interventions, of which technology might be one part, along with health care and welfare reform, tackling substance and domestic abuse, and a shift in the underlying logics of the political economy. In short, it will not be fixed with an app. An intelligent transport system that seeks to optimize traffic flow is not going to resolve congestion; it requires shifting people from cars to public transit, cycling and walking. In other words, a more holistic approach to urban issues needs to be taken, one that recognizes that "smartness" may or may not be a means of addressing an issue. And when technology does have potential to provide a solution, it still needs to be implemented together with other kinds of interventions, such as social, economic and environmental policies, collaborative planning, behavioral change, community development, investment packages, multi-stakeholder engagement, and so on.[69] The focus of attention and research then needs to be on urban issues and processes per se, not on smart urban technologies in isolation.[70]

There are some indications that the prevailing trend of smart cities has begun to wane. After reaching a peak in 2015, only one of the top five global suppliers deployed any smart city technology in 2020. The diminishing enthusiasm for "smart city" projects can be attributed to a growing backlash against Big Tech companies, in combination with the effects of the coronavirus pandemic.[71] Citizens are also increasingly concerned about surveillance and manipulation from tech companies that use the idea of a smart city as a way to integrate their products into nearly every dimension of urban life. The Alphabet spinoff, Sidewalk Labs, canceled plans to transform the Toronto waterfront into a tech-driven urban environment after protests from residents and concerns from regulators about data collection and digital management.[72] However, there are examples that show how cities can turn over the traditional smart city model and make technology work for citizens. In Europe, there are instances of citizen-led movements focused on reclaiming urban resources as common goods, advocating the collective management of public resources such as water, air, energy, housing, and health care under the broad label of "Right to the City." The latter originates in the alterglobalization movement, specifically the World Social Forums beginning in 2001 in Porto Alegre, Brazil. These movements have been mainly active at the city level, fighting against evictions, energy poverty, labor insecurity, and advocating for remunicipalization of public infrastructures. The intended public policies aim to contest a privatized smart city constructed from the top down and favoring foreign corporations; oppose monopolized ownership of intellectual property; and reverse the private appropriation of collectively produced value by rent-seeking digital platforms.[73]

One well-publicized example stands out here. Since being named European Capital of Innovation in 2014, Barcelona has developed over one hundred projects on Urban Platform, which hosts projects such as ubiquitous public Wi-Fi, upgraded traffic lights, telecare services and shared electric cars.[74] Since the election of a new mayor, Ada Colau (a former housing and anti-eviction activist) in 2015, the citizen platform, Barcelona en Comú (Barcelona in Common), has implemented new bottom-up forms of democratic participation that put people first in how technology is used. The Barcelona example has shown that egalitarian outcomes are not automatically produced by existing technology in itself but must be consciously designed into projects.

This shift in municipal governance was initiated by a new collaborative platform, Decidim, to create the government's agenda based on citizen proposals. This new form of citizen participation into government strategic planning enabled the city leadership to focus on issues that people care about, such as affordable housing, health care, sustainability, mobility, fighting climate change and creating more green spaces.[75] By starting with the social and environmental challenges and then seeing how technology could be employed to tackle these—ideally avoiding technological solutionism—this model shows how data infrastructure can be used to serve the interests of the people living there rather than those of tech companies.

Barcelona also participated in DECODE (Decentralized Citizen-Owned Data Ecosystems), a series of pilot studies there and in Amsterdam between 2017 and 2019, that tried out new technology that put people in control of their personal data. It involved a consortium of 15 organizations from across the European Union (DECODE project). In one of the pilot projects, Barcelona gave residents sensors to place in their neighborhoods, which were integrated into the city's sensor network gathering data on air quality, energy usage and noise pollution. The project involved anonymized data sharing to create public value from a data commons. The digital architecture of the platform established a new system of decentralized data governance and identity management that protected citizens' privacy and allowed them to decide which data they wanted to keep private and what they were willing to share.

These pioneering projects are setting the tone for new ways of thinking about collaborative uses of digital infrastructures and data. Public platforms operated by municipal authorities are a key carrier of this new trend. Experiments like DECODE have shown that a different institutional setup of digital infrastructure, public policy and citizen participation is possible and affordable, James Muldoon points out. He concludes optimistically: "The democratic control over digital infrastructure works and is ready to be scaled through greater investment in a digital future that is sustainable and democratic."[76]

8

Smart Technologies, City Surveillance and Predictive Policing

A high degree of system integration of information and communications technologies, particularly when combined with live closed circuit television (CCTV) feeds and advanced facial recognition software, has raised widespread concerns about privacy infringements and excessive surveillance in U.S. cities and elsewhere (most notably China). Key to the American scene is the military-industrial-communications complex, specifically those capabilities of the FBI and the Department of Homeland Security to stealthily surveil populations in tackling domestic terrorism and policing dissent. A closely related trend has come under increased scrutiny as well; smart technologies, combined with ever-improving drones and a new generation of policing robots, foster an environment of heavily militarized urbanism.[1]

The militarization of police in the U.S. has accelerated after the September 11, 2001, terrorist attacks. Generally, militarization means police departments are armed with military equipment like armored vehicles and assault rifles, submachine guns, flash-bang grenades, sniper rifles, and the deployment of SWAT (special weapons and tactics) teams. This trend is also associated with the use of intelligence agency-style information gathering, a more aggressive style of law enforcement through the use of military tactics that frame the city as an "urban battle space," and the employment of officers and analysts with military training.

Much of the upgraded arsenal adopted by police was initially designed and developed for military use before being applied to urban policing.[2] The catalyst for this redeployment of military equipment into civilian settings was the expansion of counterterrorism programs. But once smart systems of data collection and social control are available, they are likely to be widely applied for other purposes, which happened in the case of policing as well.[3] When it comes to police militarization, mission creep is the norm, which includes the use of military hardware and organizational structure, ranging from surplus military gear to SWAT teams.[4] It is the creep of military *software* into police operations that is vital to contemporary law enforcement.[5]

This involves a shift from militarized policing, where police are analogous to an army that occupies and patrols the city, to smart policing where police operate more like an intelligence agency that probes and analyzes the urban battle space. This is not an abrupt break from older models of policing, however—these two models coexist and cooperate—but rather a process of phasing in new tech and tactics that amplifies as well as transforms the power that police exert.[6] The occupying army model of policing relies on tactics like the stop-and-frisk program in New York City. Here police supposedly halt people randomly on the street, question them, and search them for contraband like drugs and weapons. The stop-and-frisk practices in the past were overwhelmingly directed at black and Latino men, who might be stopped multiple times on the same day.[7] The intelligence agency model of policing, on the other hand, turns to surveil-and-analyze tactics by using devices like the Stingray, the portable device that mimics a cell tower. Since people's smartphones are constantly searching for a signal, the Stingray tricks all nearby phones into connecting with it, extracts data from the phones such as text messages, and then connects them to a real cell tower. Police can conduct these "digital pat downs" on a much larger scale than physical searches and without the target even knowing their information has been harvested.[8]

A Closer Look at Smart Policing

In 2009, faced with federal, state and local budget cuts in response to the Great Recession, police departments in the U.S. began looking for ways to do more with less. Technology companies rushed to fill the gaps, offering new forms of data-driven policing as the preeminent means to attain efficiency and cost reductions. This coincided with the smart cities craze that erupted about that time. Then, in 2014, the killing of Michael Brown by police in Ferguson, Missouri, upended already fraying relationships between the police and the communities they served in many U.S. cities. Successive police killings of other unarmed African Americans in later years all sparked protests nationwide and calls for racial justice and police reform, most prominently by the Black Lives Matter movement. In particular, the widely reported police killing of George Floyd (captured on video) in Minneapolis, Minnesota, in May 2020 triggered broad-based condemnation along with demands for reform.

Policing was driven into crisis mode as societal outrage and calls for "defunding the police" threatened to delegitimize the existing police power structure. In response to widespread criticism and fueled by the continuing pressure to economize, police departments increasingly embraced tech companies' promise of big data efficiencies and the hope that their "data-driven" approach would be able to move beyond the all-too-human problems of policing. Predictive analytics and body-cam video capabilities were sold as objective solutions to racial bias. The public relations strategy worked to a great extent, which led law enforcement in several cities to adopt predictive policing and increase digital surveillance.[9]

New Orleans as an Early Case of Smart Policing

Among the U.S. companies making, selling and operating high-tech tools for smart policing, Palantir (originally Silicon Valley–based) stood apart from much of the competition in those early years. Co-founded by Peter Thiel (also co-founder of PayPal) in 2004, Palantir is a data-mining company that has been described as a "pioneer in predictive policing," which "knows everything about you."[10] Palantir received startup funding from In-Q-Tel, the CIA's venture capital firm, and landed contracts with federal agencies such as the Central Intelligence Agency (CIA), Federal Bureau of Investigation (FBI), Immigration and Customs Enforcement (ICE), and the Department of Homeland Security (DHS). It then branched out, selling its services to local police departments such as the LAPD and the NYPD, as well as commercial customers such as JPMorgan Chase.[11]

Palantir's "social network analysis" technique was originally created for counterinsurgency efforts in Iraq and Afghanistan (finding terrorists and predicting attacks), but later was redirected to track regular citizens by compiling personal profiles and plotting social connections. "The software combs through disparate data sources—financial documents, airline reservations, cellphone records, social media postings—and searches for connections that human analysts miss. It then presents the linkages in colorful, easy-to-interpret graphics that look like spider webs," according to Bloomberg.[12]

Jathan Sadowski describes the impact of Palantir's technology on residents of his then-home town New Orleans, where, since 2012, the company had been working with the city's police department. The city partnered with Palantir Technologies to analyze crime as part of Mayor Mitch Landrieu's murder reduction program, NOLA For Life. Importantly, the program lacked transparency, oversight and community support as top New Orleans political and community leaders were kept in the dark about this program and excluded from any discussions about its adoption and implementation. Because the program was presented as a philanthropic gift to the city by Palantir—combined with the mayor's strong unilateral powers—the agreement between Palantir and the city never had to pass through a public procurement process, which would have required the sign-off of the city council and provided an opportunity for public debate.[13]

Sadowski, referring to the 2018 investigative report by the journalist Ali Winston at *The Verge*, offers this grim assessment of the situation in 2020:

> Palantir's system targets far more than just known criminals. In a city that uses Palantir, if you ever had an interaction with a police officer or government department, or if somebody you know had interacted with them, then you have been captured by Palantir's surveillance and analytics systems. In a place like New Orleans, this covers essentially anybody—including me. It doesn't matter if you've been suspected, let alone convicted, of illegal activity.
> "It's almost as if New Orleans were contracting its own version of the NSA [National Security Agency] to conduct 24/7 surveillance of the lives of its people," a civil rights lawyer told Winston.[14]

Others strongly disagree with this assessment, however. Professor of Law Andrew Ferguson draws a positive overall picture of Palantir's system in New Orleans in his book *The Rise of Big Data Politics* (2017).[15] He relies on official reports of Palantir and the New Orleans city government, particularly the Mayor's Office of Justice Coordination. In a follow-up interview in March 2018, Ferguson stipulated that the deployments of Palantir beyond law enforcement—its role in helping the city identify streetlights needing repairs, at-risk individuals who could benefit from support services, or schools with "high-risk populations" needing additional resources from the health department—made the partnership between New Orleans and Palantir different than the company's work in other cities. Palantir initially rolled out its partnership with New Orleans in a far more holistic way than it did in Chicago or Los Angeles, Ferguson said. "The theory of 'focused deterrence' is that risk factors exist, and if you mediate or reduce some risks, maybe crime goes down. But that costs money. You need people to intervene, offer jobs or mentorships. New Orleans in the early years actually did that."[16] Ferguson questioned whether the city had kept up this effort to use Palantir's technology to do more than identify and arrest suspected criminals.

In December 2020, the New Orleans City Council passed a new law that regulates certain parts of the surveillance system and places an outright ban on specific pieces of surveillance technology, including facial recognition software and predictive policing. The ordinance passed a month after a local non-profit public newsroom, *The Lens*, reported that the New Orleans Police Department had been using facial recognition software (through partnerships with the FBI and the Louisiana State police) despite years of denial. The original ordinance, which had been delayed repeatedly in the course of that year, was drastically watered down from what was initially drafted. It would have created comprehensive approval processes, oversight protocols and regular reporting requirements for every part of the city's sprawling surveillance system. These stipulations were removed from the final version, however. The ordinance as passed does put outright bans on four pieces of technology—facial recognition, optical character recognition and tracking software, predictive policing, and cell-site simulators (Stingrays or IMSI catchers).

Palantir's agreement with New Orleans had been extended three times since 2012 and was not renewed when it expired in 2018. But other city-owned software and programs, including Motorola's Command Center Predictive Suite, and Mayor LaToya Cantrell's high-risk resident ID program, have had similar capabilities. At the time the new law was passed, it was unclear whether both of these fell under the ordinance definition, or if there were any other pieces of software that could potentially violate the law.[17]

The Road to Data-Driven Police Surveillance

Based on her field work among the Los Angeles police, Sarah Brayne outlined five key ways data-driven techniques have changed police surveillance. This provides a valuable framework for analyzing how smart policing has been rolled out in urban environments, and offers a good insight into its mechanics. While Brayne's field study focused on the LAPD, the shifts in smart policing have also occurred in other U.S. cities and are global in scale.[18] The framework is also helpful more generally in analyzing the continuities and changes within a variety of institutions adopting new data sources and technologies.[19]

The *first shift* concerns how risks are assessed and who (or what) does the assessment—from discretionary choices to quantified scores. When police respond to a call, they take into account a number of factors to determine how dangerous or risky the situation might be. Officers are trained to assess situations using formal protocols, but they also bring in their own cultural values and personal biases. Each officer determines their own response based on things ranging from what was said on the emergency call, where the situation is happening, the race or ethnicity of the people involved, the officer's mood at that moment, and countless other contextual details. This is why the risk assessment is discretionary; ultimately, it is the individual officer's own judgment which guides their action.

The goal of quantifying risk assessment is to make these judgments more objective—or at least more standardized—by calculating risk scores.[20] These scores might be calculated in a relatively straightforward way, such as by assigning point values to certain factors (e.g., contact with an officer is one point, a history of violent crime is five points, etc.), adding up the points assigned to a person or neighborhood, and then producing a risk score. With these scores, police can rank "chronic offenders," and keep a close watch on certain people and places. This was initially part of Project LASER (Los Angeles Strategic Extraction and Restoration), the goal of which was "to target with laser-like precision the violent repeat offenders and gang members who commit crimes in the specific target areas."[21] It was the culmination of the move to collect pre-crime data through field interview cards that officers on patrol would fill out that included personal information such as name, address, gender, height, weight, "descent" (presumably ethnicity), date of birth, aliases, probation/parole status, phone numbers and Social Security numbers. Additional information about "personal oddities," union affiliation, or gang or club membership could also be added on the cards.[22]

Other methods of risk quantification rely on algorithms that take in reams of data and distill them into a single number meant to inform an officer about how to behave when responding to a situation. For example, police departments have implemented a software program called Beware to generate "threat scores" about an individual, address or area. The program searches through proprietary consumer databanks to provide a rough predictive judgment about the 911 caller, the address or the neighborhood. The predictions initially involved color-coded threat levels (red, yellow or green), which provided some measure of risk assessment for officers responding to a scene. Beware's data comes from commercially available records such as criminal records, warrants and property records—basically, most of the data kept in consumer big data systems—and early versions of the program also incorporated social media data which might signal threatening statements or other violent comments as well as gang associations.[23] Risk assessments like these are used at all levels of the criminal justice system, by officers and judges alike. Regardless of the many problems and concerns raised by risk scores, their use continues to become normalized and widespread. It is hard to resist their appeal because they offer the veneer of objective authority and the utility of reducing complex events to a single number.[24] In addition to assessing people in the here and now, the algorithms in question claim to reveal the future.

The *second shift* concerns how the police address crimes, moving from a reactive response to predictive analytics. Patrol officers used to spend time walking a beat or cruising in their car, but rarely did they actually stumble upon a crime in progress. Instead they would wait for the dispatcher's call to direct them to the scene of a crime, usually after it had already happened. In the 1980s—supported by the "tough on crime" and "law and order" campaigns of politicians—police began deploying more proactive strategies to *prevent* crime. This included the use of "hot spot" policing, whereby the police focused their attention on specific places—predominantly those populated by the poor and people of color. That era also witnessed the rise of the "broken windows" theory,[25] which dictated that police should enact a zero-tolerance policy for "lifestyle crimes" like loitering, vandalism and jumping train turnstiles.[26] Now an officer walking a beat in a "hot spot" could actually stumble across a crime, however minor it was, and issue hefty fines or haul offenders to jail.

In the 1990s, New York City became the testing ground for an early version of smart policing called COMPSTAT, which combined an aggressive use of the broken windows theory with a statistical method of policing. Established in 1994 and overseen by Police Commissioner William Bratton, the COMPSTAT program used newly available computer systems to quantify and manage police work.[27] It was a multifaceted management model linking crime and enforcement statistics. It served as a crime control strategy, a personnel performance and accountability metric and a resource management tool. Crime data was collected in real time, then mapped and analyzed in preparation for weekly crime control strategy meetings between police executives and precinct commanders.[28] By measuring the performance of every precinct, COMPSTAT allowed the top police management to evaluate commanders and officers against benchmarks. Those who performed well were praised and celebrated; those who missed the mark were shamed and humiliated. COMPSTAT showed clear similarities with how, in the corporate world, key performance indicators are used to put heavy pressure on both workers and managers. These "statistical rituals" of evaluation, as sociologist Emmanuel Didier calls them, deeply influenced the way police did their jobs.[29] What mattered most was putting up good numbers and meeting benchmarks. This harsh culture of quantification created perverse incentives to overpolice, "juking the stats" (as a crime reduction strategy) and gaming the system, as portrayed in the HBO series *The Wire*.[30]

The next phase of smart policing seeks to move the temporal focus forward even further, from reactive and proactive to predictive policing. A direct extension of previous strategies, predictive policing applies the power of data-driven analytics and statistical models to tell officers where crime is likely to happen and who is likely to commit it. It involves directed surveillance, or the surveillance of people and places deemed suspicious.[31] The embrace of big data by police is part of a broader trend toward quantification and algorithmic risk assessment in the criminal justice system. What is new and important about it is the role of private actors in public policing.[32]

The LAPD was a pioneer in predictive policing. In 2002, William Bratton of COMPSTAT fame (and the former commissioner of the NYPD) moved to Los Angeles to become its chief of police. He oversaw the process of merging previously disparate information systems to create a more data-based picture of crime in the city. In 2008 he began working with federal agencies to assess the viability of a more predictive, place-based approach, and in 2011 the LAPD started using PredPol.[33] It became available as a commercial product provided by a company of the same name founded in 2012. PredPol uses a patented and once-proprietary algorithm predicated on the near-repeat model, which suggests that once a crime occurs in a location, the immediate surrounding area is at increased risk for subsequent similar crimes. PredPol takes a minimalist approach to data variables. The algorithm is fed three kinds of inputs about past crime as training data to predict future crime: type of crime, the place it occurred, and at what time. From that data PredPol predicts what parts of the city have a higher risk of crime so that police can focus their attention in those areas. It limits its predictions to quite small areas—500 by 500 square feet.[34]

PredPol's algorithm is rather simplistic compared to the surveillance systems and predictive models that some police departments are now deploying. By continuously collecting and crunching data, a host of companies like PredPol and HunchLab, and researchers at Rutgers University who developed RTM (Risk Terrain Modeling), aim to equip police with analytics that can tell them how as well as where to deploy resources in the most efficient, effective ways. The quest for more accurate predictions has led to more surveillance that feeds more data into more complex algorithms. It is a vicious cycle that brings about the continuous amplification of police powers.[35]

The *third shift* concerns the way police find and track information: from query-based systems to alert-based systems, which makes it possible to systematically surveil an unprecedented large number of people. Query-based systems concern databases that operate in response to a user query. With alert-based systems, in contrast, users receive real-time notifications when certain variables or configurations of variables show up in the data.[36]

The police have access to many databases that contain information not generally (or easily) available to the public. Obtaining that information used to require actively searching for it by submitting a request and sifting through the results. This could take considerable time and it often required having some idea of what information one wanted. Once the information was retrieved, its significance might not be readily apparent, so the detectives and technicians had to analyze the data, find patterns and connect the dots. Now, however, rather than actively searching and parsing through information, the police can actively feed data into systems, and then wait for the software to tell them what it means or when something relevant occurs. These automated systems make continuous data collection a core part of the police's job.

While queries are still very much central to police investigations, the trajectory here, too, is headed toward more automation. These systems make it much easier for the police to conduct surveillance. Suppose an officer wants to keep a close watch on a specific person's activities. Now the officer simply has to set up an alert in the database, which will then notify the officer anytime that person's data is put into the system. If an automatic license plate reader (ALPR), which captures the location and time of all vehicle licenses it sees, records the person's car leaving a specific geofenced area (that is, an area with electronic sensors that alert when crossed), the officer gets a notification in real time on his/her phone. Or if the person is randomly stopped on the street by another officer who then files a standard report about the interaction, the querying officer is likewise notified. This type of targeted tracking and real-time alerts of (possibly) suspicious activity can also be set up for specific addresses and areas.[37]

The *fourth shift* is the lowering of thresholds for being included in police databases so that individuals with no direct police contact are now also collected by law enforcement systems.

The general public expects there to be a record of people who have had direct encounters with the police and courts, whether that entails being arrested and "booked," being fined for a minor infraction, or calling to report an emergency. There is also the expectation that the police increasingly collect more information, when somebody has been taken into custody for a crime. Most people would be inclined to see this as a reasonable practice of information gathering and file keeping. But at the same time, most people would likely be outraged if the police started taking DNA samples, fingerprints and mug shots of every random person they come across. This is because there is a general consensus that certain criteria should be met for the police to justify the collection of certain types of data, from certain people, for certain purposes. These criteria, or thresholds for inclusion, set important limits on the police to ensure that they act in ways that truly protect and serve the public.[38]

However, the rise of smart policing has led to much lower thresholds for inclusion into police databases. As a result of these new capacities to analyze data and new imperatives to extract it, "the police increasingly utilize data on individuals who have not had any police contact at all," Sarah Brayne explains. "Quotidian activities are being codified by law enforcement organizations."[39] This includes, for example, all the cars driving past an ALPR camera being stored in a police database, and all the phones that connect to a StingRay device being logged by a police database. Basically, all the people, vehicles, addresses and phone numbers that are in some way connected to a "person of interest" are in the police's digital catchment area and subject to sweeping dragnet surveillance. In this case, the adoption of big data analytics is associated with fundamental transformations of police activity, whereas in the case of directed surveillance mentioned earlier, it is associated with mere amplifications of previous surveillance practices.[40]

The police use of big data analytics has implications for inequality; on the one hand, it may be a means by which to diminish persistent inequalities in policing. Data can be used to "police the police," replacing suspicions about racial minorities and exaggeration of patterns with less biased predictions of risk. However, this only happens to a limited degree, if at all, in most police departments. Far more impactful, on the other hand, are data-intensive police surveillance practices implicated in the reproduction of inequality in at least four ways: by deepening the surveillance of individuals already under suspicion; codifying a secondary surveillance network of individuals with no direct police contact; widening the criminal justice dragnet unequally; and causing people under intense police surveillance to avoid more generally institutions that collect data and are fundamental to social integration (to be further explained later). "Crucially, as currently implemented," Brayne concludes, "'data-driven' decision-making techwashes, both obscuring and amplifying social inequalities under a patina of objectivity."[41]

Critics argue that businesses as well as governments should adhere to the principle of data minimization; that is, only collecting, storing and analyzing the data needed for a specific purpose. Devices like ALPRs and data analytics like Palantir do exactly the opposite, however, by capturing everything and everybody. The organizations involved are driven by a principle of data maximization; recording and storing all data, from all sources, by any means possible, even if its use is not yet clear. Restricting data collection is "largely antithetical to the rationale of big data and the functioning of data markets which seek to generate and hoard large volumes of data to extract additional values," explains Rob Kitchin, an expert on smart cities.[42] The detrimental impacts of this kind of mass surveillance are well known.[43] Civil liberties are sacrificed for the promise of security. This surveillance criminalizes the everyday life of citizens by casting suspicion on even mundane activities. It undermines the very legitimacy and accountability required for a functioning democracy. When, in 2013, Edward Snowden raised alarms about the NSA's warrantless dragnet surveillance, there was no widespread public outcry that police departments in various cities functioned de facto as mini-NSAs. Unlike the NSA, however, the police are also in regular, intimate contact with the communities they surveil—with the power to harass, arrest, deport and even kill their targets.[44] This obviously makes their impacts on local residents more immediate and often at least as troubling as those of national security agencies.

Most important is the *fifth shift*: the integration of data from different sources into massive, centralized, searchable databases, with police now collecting and using information gathered from institutions not typically associated with crime control. In addition to collecting new data due to lower inclusion thresholds, the police are assessing and combining data held by various institutions—both public and private—that was previously kept separated, and might have been off-limits before to the police. This is part of a widespread practice called "data fusion"—the sharing and merging of data from multiple sources to reveal new information, patterns and correlations, which can then improve the accuracy of profiles and predictions.[45] This is a powerful technique because it can uncover private information and circumvent data security. In short, data fusion is a way of breaking down the firewalls between different databases, possibly revealing sensitive information that should not, or at least not easily, be accessed by governments and/or corporations.

The 80 federally funded "fusion centers" that were established across the country in the wake of 9/11 play a crucial role in this.[46] These centers are multiagency, multidisciplinary surveillance organizations that receive federal funding from the Department of Homeland Security and the Department of Justice. They were designed to enrich Suspicious Activity Reports (SARs) and send them to appropriate county, state, or federal agencies. SARs are key conduits into fusion center databases. They are tips and leads from law enforcement and civilians, regardless of suspicion of criminal activity. Mundane activities like using binoculars, drawing diagrams, or taking pictures or "video footage with no apparent aesthetic value" can be recorded on SARs—and made into fusion center data. Along with the SARs, fusion center data comes from criminal justice, public health, financial, motor vehicle, credit, immigration, fax, insurance, property, car rental, post and shipping, gaming and utility records sources. It also includes dossiers from third-party brokers.[47] Law enforcement is increasingly securing routine access to a wide range of data from *nonpolice* databases, which is one of the most transformative features of digitization of policing.[48]

As ever more information is digitized, and personal records linked across previously separate institutional boundaries, the scope of data that brokers can make accessible to law enforcement increases exponentially. In some cases, it is simply easier for law enforcement to purchase privately collected data than to rely on in-house data, because there are fewer constitutional protections, reporting requirements and appellate checks on the private sector. This can then be a way for law enforcement agencies to circumvent privacy laws. In addition, private data can be more up-to-date than police-generated data.[49]

Fusion centers are treated like a "one-stop shop," where police and federal agencies can access and analyze a staggering amount of data from many different sources.[50] Fusion centers disseminate and analyze data on suspicious individuals or activities, assist investigations and identify potential threats. They respond to Requests for Information (RFIs) from agencies across the country, including the DHS, FBI, CIA, and ICE. Although their original mission was counterterrorism, there was insufficient terrorist activity to keep fusion center analysts busy, so their mandate quickly expanded to include gathering and sharing information related to "all hazards, all crimes, all threats." This mission creep is not unexpected; smart systems of surveillance and control are set up for specific purposes, but once they are put into action they tend to be widely used for other reasons—both legal and illegal. It can be hard to keep track of this creep, however, as the watchers resist being watched themselves.[51]

The proliferation of digitized records makes it possible to merge data from previously separate institutional sources into an integrated structural system in which disparate data points are displayed and searchable in relation to one another, and individuals can be cross-referenced across databases. This integration facilitates one of the most transformative features of the big data world: the creep of criminal justice surveillance into other, non-criminal justice institutions. Function creep—the phenomenon of data originally collected for one purpose being used for another—contributes to a substantial increase in the data used by police. Law enforcement follows an imperative to collect as much data as possible, in part by securing routine access to a wide range of data on everyday activities from nonpolice databases.[52] With policing based on the principle of data maximization, all data are now potentially police data.

Many institutions now act like eyes and ears for the police, which diminishes the rights and deters the behavior of people who rely on these institutions but distrust the police. For example, Sarah Brayne analyzed U.S. national representative data from the National Longitudinal Study of Adolescent Health and the National

Longitudinal Survey of Youth. She found that individuals with criminal justice contact tend to stay away from important institutions where they could leave a digital trace. More specifically, those who have been stopped by police, arrested, convicted or incarcerated are more likely to avoid surveilling institutions such as medical, financial, educational and labor market institutions that keep formal records, i.e., put them "in the system."[53] So just by enacting constant and expansive surveillance, the police inadvertently deter people from accessing the services those institutions provide and are entitled to use.[54]

This system avoidance and unequal participation in institutions may have serious consequences regarding inequality. Given that involvement with the criminal justice system is highly stratified by race and class, the ramifications of system avoidance will likewise be disproportionately distributed, thus exacerbating preexisting inequalities for an expanding group of already disadvantaged individuals. Lack of attachment to (or involvement with) medical care, banks, schools and employment is associated with poorer outcomes for health, financial security, upward mobility and what is called the "desistance from crime" in criminal justice—the reduction in criminal behavior that occurs after a person reaches adulthood. As the criminologists Kevin Haggerty and Richard Ericson suggest, "efforts to evade the gaze of different systems involves an attendant trade-off."[55] That trade-off consists of full participation in society.[56]

Technocratic City Surveillance and Control

Together, as forces of amplification and transformation, the five major shifts in smart policing outlined above have set in motion radical changes in urban governance. In addition to targeting people and populations, the police are now analyzing profiles and patterns in which the subjects of surveillance are no longer merely human beings but rather the data streams that stand in for them. Suspects have become data points as much as they are people. Thus, smart policing has catalyzed the creation of totalizing systems that aim to capture the entire city involved by drawing everything into their databases and subjecting everybody to their power—to create profiles, calculate patterns and control people. Rather than imagining utopian visions of a smart "city on the hill" when thinking of the real smart city, one should think of systems like the Domain Awareness System (DAS), a joint venture between the NYPD and Microsoft launched in 2012, which is one of the world's largest networks of urban surveillance.

DAS links approximately 9,000 closed-circuit surveillance cameras for real-time monitoring of lower Manhattan. The video feeds go directly to a digital alert system that automatically tracks suspicious behavior (such as leaving a bag on the street). In addition, automated license plate readers record every car that enters the area. These 500 license-plate systems connect with DMV records, police watch lists, open warrants, terrorist databases, and all of the personal information associated with these databases.[57] A video recorded by DAS can be replayed to track the direction, location, and movements of a suspect, and the technology can even search for descriptions, such as "all people wearing red shirts near the New York Stock Exchange." Still photos of matching people can be pulled up with one search, tagged to location, time and date.[58] Officers equipped with smartphones and tablets have mobile, real-time access to DAS. This enables them to do things like pull up feeds from CCTV cameras, search a range of databases that are integrated into DAS, and set up automatic alerts when DAS detects "suspicious activity."[59]

This surveillance system is another product of the global "war on terror," in which the enemy can be anywhere and the battlespace is everywhere.[60] DAS integrates the shifts outlined above into the ultimate, unified smart platform, which contributes to what Jathan Sadowski adroitly calls the "captured city." He explains: "It is captured in two ways, by corporate and military surveillance systems that aspire to always watch everything, and by police who occupy and control urban space. The captured city is the result of public-private partnerships, which compound the worst excesses of both forms of power."[61] Other cities in the U.S. and around the world have looked at DAS—and similar types of control centers and analytics platforms built by IBM and Cisco—as models for how to govern urban society.

As the tech advances, increasingly more powerful hardware and software can be built on and integrated into these platforms. One such important upgrade already on the market is supplied by the company Persistent Surveillance Systems (PSS). The technology was originally created for use by the U.S. military in Iraq as a system to catch people planting the improvised explosive devices that killed American troops before being sold as a service to police departments.[62] It consists of a small airplane or drone equipped with an array of (initially low-resolution) aerial cameras that can capture an area of roughly 30 square miles and continuously transmit real-time images to analysts on the ground. It can fly for six or eight hours before needing to refuel (or recharge batteries), allowing it to monitor the city from above without interruption for long periods.

Aerial cameras provide the ultimate mass-surveillance tool. Flying high overhead, able to record whole neighborhoods for hours at a time, aerial cameras like PSS can watch crime in real time and record the patterns below of all vehicles, people and events. In combination with sophisticated audio capture, these aerial camera systems or drones provide the potential for cities to be surveilled in real time and for days at a time. Similar to the Domain Awareness System, police can roll back the tape and watch how the crime developed, where the suspects fled, and patterns of behavior. PSS planes have flown missions over Baltimore, Los Angeles, Indianapolis, Charlotte and other U.S. cities.[63]

There are other devices that can capture the city from the street level. A scandalous example of corporate-police partnerships is the rollout of Ring, a smart doorbell with a camera that records and stores video, as described earlier in relation to the smart home. Through secretive agreements with Amazon's Ring, numerous police departments across the U.S. have partnered with the company to supply discounted or free devices to citizens, sometimes actually using taxpayer funds to purchase Amazon's products. Amazon profits from data storage usage fees while the police have access to a distributed network of cameras that they can access through a "law enforcement dashboard" provided by Ring and connected to the Neighbors app. "While residential neighborhoods aren't usually lined with security cameras," reporter Alfred Ng points out, "the smart doorbell's popularity has essentially created private surveillance networks powered by Amazon and promoted by police departments."[64] In January 2024, however, Amazon introduced a significant policy change, as noted earlier. Since then the police needs a warrant to access user footage from Ring doorbells, and features are disabled that previously allowed law enforcement to request footage directly from users.

Most recently this development is being overtaken by a quiet but rapid expansion of law enforcement surveillance in which U.S. cities are buying and promoting products from Georgia-based company Fusus in order to access on-demand, live video from public and private camera networks. The company sells police a cloud-based platform for creating "real-time crime centers" and a streamlined way for officers to engage with their various surveillance streams, including predictive policing, gunshot detection, license plate readers and drones. For the public, Fusus also sells hardware that can be added to private cameras and convert privately-owned video into instantly accessible parts of the police surveillance network. In cities like Atlanta, Memphis, Orlando and dozens of other locations, police departments have been asking the public to buy into a Fusus-fueled surveillance system, trying to convince people and businesses to trade away privacy for a false sense of security, the Electronic Frontier Foundation (EFF) has stated.[65]

The Foundation cautions that deciding whether to expand police surveillance to every facet of people's lives should never occur without "strong community conversation, transparency, and real respect for procurement rules and the public's liberty."[66] The EFF also suggests that Community Control Over Police Surveillance (CCOPS) ordinances (about which more later) could be helpful here in empowering a community's residents, through their elected officials, to decide whether law enforcement may implement surveillance programs in their neighborhoods. These kinds of policies could help to protect residents from unwarranted surveillance, facilitate community conversation and set out processes by which police might seek to use these technologies, and if so, under what safeguards and reporting requirements.

In the domain of software, applications powered by facial recognition and artificial intelligence have become widely available, allowing police to integrate them into already-existing CCTV cameras and officer body cameras. Now every police officer on patrol can be equipped with a mobile, real-time facial recognition scanner (as is already happening in China), if deemed necessary and the money to purchase the equipment is there. In addition, along with services like Amazon's Rekognition, other tech companies have been developing artificial-intelligence-powered systems marketed as "Google for CCTV," which enable users to search through video footage with keywords.[67] Rather than going through countless hours of video to find a man wearing particular clothing or driving a certain model of car, police can now simply enter their query and find all the relevant clips. When taken together, this hardware and software aims to record, analyze and search every aspect or moment in the life of a smart city.

Police Usage of More Smart Technologies and Persistent Privacy Invasions

Importantly, the promise of objective, unbiased technology has not panned out. Scandals involving facial recognition, social network analysis technology and large-scale sensor surveillance serve as a warning that such technology cannot address the deeper issues of race, power and privacy that lie at the heart of modern-day policing. Andrew Ferguson remains hopeful regarding smart technology's promises. The lesson to be learned from the first era of big data policing is that issues of race, transparency and constitutional rights should be at the forefront of design, regulation and use:

> Every mistake can be traced to a failure to see how the surveillance technology fits within the context of modern police power—a context that includes longstanding issues of racism and social control. Every solution points to addressing that power imbalance at the front end, through local oversight, community engagement and federal law, not after the technology has been adopted.[68]

But the problem is that the technology is outpacing the regulations that are in place, and may continue to create similar problems for local communities, as police departments respond to new siren calls of big data surveillance. Regarding the technologies which police officers integrate into their day-to-day job, the following deserve further scrutiny, according to investigative journalist Jon Fasman, in an interview he gave to NPR in January 2021.[69]

- *Citizen Virtual Patrol*, a network of publicly owned cameras, used, for instance, in Newark, New Jersey. It allows at-home viewers to stream video from cameras placed around the city. The idea behind it was to allow residents to observe and perhaps testify to crimes from behind a veil of anonymity. Technically, the cameras do not show anything that an observer on the street would not be able to see. The system shows public streets; it is not aimed at anyone's home or apartment, nor peeps into anyone's windows. On the other hand, it does show people's yards, which have a slightly higher expectation of privacy, and it could provide some information that people would not want to become public knowledge. For example, watching a camera aimed at somebody's house allows anyone to see people coming and going (possibly carrying bags, suitcases or other things) and whether there is any activity or the resident is away from home. Viewers may even be able to identify the residents who live or work there. Indeed, anyone who logs in has the ability to observe an enormous swath of the city. Any single instance may be unobjectionable but when it comes to scale, this is something very different.
- Police use of *automatic license plate readers*, or ALPRs. These readers are little flat cameras attached to police cars that citizens would not normally notice unless looking for them. Mounted on the front or roof of the police car, ALPRs capture the image of each license plate as the police drive by and log geospatial data: where the observed car is and when it is parked there.

The information then goes into the database. Again the issue here is one of scale. There is nothing illegal about the police noting the license plates of cars parked in public, their location and time of parking. But if people looked look down their residential street and saw a police officer writing down every license plate all day, every day, they might wonder the reason why. The police department, too, might question the practice, if they had to allot manpower to the task.

The license plate images often wind up in databases shared by multiple police forces and supplemented by privately held ALPR companies. So questions arise: who should have the right to access that database and when? What sort of security safeguards does the database need? What happens in the event of a data breach? More importantly, do law-abiding citizens have the right to have their data deleted or must they simply accept that the police now amass records of everywhere they go, records they may then keep forever in many states?[70] The crucial issue is what law enforcement does with all the data of cars that are not associated with any crime or actively searched for. How long does it take for those data to be deleted? This varies from state to state: in New Hampshire it has to be deleted within three minutes; some states set 24-hour limits. Many other states set no limits at all and just compile these pictures in a huge database. Often these databases are poorly secured. So it is the combination of the collection at scale, the lack of regulations over how long that information is kept—and often the lack of security over how it is kept—that make this kind of technology very troublesome from a privacy and security perspective.

- *"ShotSpotter" technology*, designed to detect gunshots and dispatch police. (Besides the firm ShotSpotter, there are other gunshot detection firms such as EAGL Technology.) ShotSpotter is an acoustic sensor designed to detect the sound of gunshots. The system works through a combination of algorithmic and human operations. The devices look like little white diamonds or rectangles up on traffic light poles that have been trained to recognize what the company calls "loud, impulsive sounds" between 120 and 160 decibels. When the device registers such a sound, it sends an alert to the ShotSpotter headquarters where a human listens to it and figures out if that was indeed a gunshot or, alternatively, a metal door slamming or a car backfiring, for example. Once the device detects the gunshot (as confirmed by the human intermediary), an algorithm notifies the local police department, and conveys the numbers of shots, location and time, and that officers should be dispatched to the scene.[71]

As Fasman found out during his investigation, many people within the communities in question suspected that this technology was being used to eavesdrop on private conversations, which, to the best of his knowledge, was not the case. He concluded that this was an instructive lesson in how deploying technology can be used to improve or worsen relations between law enforcement and the communities they serve. In too many instances, police departments approve the purchase of ShotSpotter devices and deploy them without consulting the communities they police, failing to keep an open line of communication with the people involved, who often have a history of mistrust of law enforcement that has built up over decades. In other words, police need to be very careful about how they roll out the technology and need to go the extra mile to gain the public's trust.[72]

ShotSpotter has met intense criticism in recent years for its methodology and the impact of its technology on communities of color. A study by the MacArthur Justice Center, whose findings were released in May 2021, found that Chicago's use of ShotSpotter was inaccurate, expensive and dangerous. (Chicago was then one of ShotSpotter's two largest customers, accounting for 18 percent of its annual revenue in 2020.) The vast majority of alerts generated by the system turned up no evidence of gunfire or any gun-related crime; there were over 40,000 unfounded, high-intensity police deployments in Chicago in 21 months, which were focused almost exclusively on predominantly Black and Latino communities.

ShotSpotter claimed to be 97 percent accurate, but had not released any scientifically valid study to substantiate that figure. There were also no studies testing whether ShotSpotter could reliably distinguish between the sound of gunshots and other noises like firecrackers, backfiring cars, construction noises, helicopters and other loud impulsive sounds. ShotSpotter data also fed into the city's predictive policing technology. The inflated gunfire statistics generated by ShotSpotter in the abovementioned neighborhoods in fact skewed how the police would allocate its resources in the future, because the city was not taking careful steps to eliminate the effects of unfounded ShotSpotter alerts. These statistics could also create a false "techwash" justification for racialized patterns of policing.

This study was the basis of an amicus brief by the MacArthur Justice Center filed in support of a motion by the Cook County Public Defender that challenged the scientific validity of ShotSpotter's system's gunfire reports that prosecutors have attempted to use as evidence in criminal prosecutions. The amicus brief was submitted on behalf of a coalition of community-based organizations concerned about the impacts of ShotSpotter on overpoliced and underresourced communities on the city's South and West sides. Nevertheless, the City of Chicago's three-year contract with ShotSpotter that expired August 19, 2021, was then extended.[73] Furthermore, an investigation by *The Guardian* in collaboration with research and civil liberties organizations, including the Lucy Parsons Labs and the Oregon Justice Research Center, was telling. In May 2023, it disclosed how in the case of Portland, Oregon, ShotSpotter (recently rebranded to SoundThinking) had worked with local top police officials to try to secure a contract, effectively circumventing parts of the public procurement process.[74] Of course, there are probably more instances of such improper practices elsewhere, unbeknownst to the general public, which will come to light sooner or later. It also is important to note that, in October 2018, ShotSpotter had acquisitioned HunchLab to make the move towards AI-driven analysis and predictive policing, expanding the company's platform with ShotSpotter's Mission "to deliver data-driven patrols and help deter crime."

- *The "Stingray," or IMSI Catcher* (international mobile subscriber information) that collects phone data, similar to a cell tower. A Stingray mimics a cell tower and connects to people's phones within its reach and geolocates them. Then all of the metadata on those phones, that is the non-voice call data, can be read. This includes texts the phone users might send, websites they might browse, who they called and for how long they talked, without even knowing the actual content of the conversation. Increasingly, the system is deployed by court order, but that has not always been the case. Even when deployed by court order against a specific subject, the problem is that data from every other phone in that area is harvested as well. This happens on a regular stakeout, too. If the police are staking out a suspect, they see all kinds of people walking past. But the crucial difference is that they do not retain the data from all those pedestrians, while in the case of data collected by Stingray, this data is often kept for a longer time than is necessary. This is a case in which there is no question that Stingrays can help police catch serious criminals. But so would repealing the Fourth Amendment, which prohibits unreasonable searches and seizures.[75] In this case, too, there need to be clear regulations over when such data can be used and what happens to the data collected incidentally during those digital stakeouts.[76] Furthermore, there is the important issue that Stingrays are being used to surveil and control participants in legally sanctioned protest demonstrations.

In his book *We See It All: Liberty and Justice in an Age of Perpetual Surveillance* (2021), Jon Fasman explores the privacy issues related to the police tactics in question—especially as regulations vary from state to state. The rules are all over the place with a lot of this technology, because it is new and also changes rapidly. And even when there are regulations, there often are no penalties for violating them nor strong enough penalties (assuming that the rules are enforced). Fasman, too, remains critical of predictive policing programs, which ingest an enormous amount of historical crime data and then, based on that data and past practice, indicate the locations likely to be at elevated risk for crime on a given day—and where patrol officers need to be deployed. He shares the concern that historical crime data is not an objective record of all crimes committed in a city; it is only a record of crimes that the police know about. And given the historic pattern of overpolicing minority and poor communities, these programs run the risk of essentially solidifying past racial biases into current practices.[77] For that reason, some critics have even argued that, because of their racist implications, predictive policing algorithms should be dismantled altogether.[78]

Resistance and Countermovements Against City Surveillance and Smart Policing

The issue of surveillance and its differential outcomes remains a crucial component of the critique of smart cities. Smart city projects are necessarily political, from the decisions on what data will be collected to the geophysical placement of sensors to collect that data.[79] To resist datafication and surveillance risks, communities need strong political organizing at the local level, as well as the ability to conduct monitoring and evaluation themselves. Transparency, oversight and community support for smart policing programs can only be accomplished this way.

Privacy advocates have been organizing against data collection in certain areas. As Andrew Ferguson reports, community objections to drone flights in Seattle shut down police drone use. Community objection to predictive policing in Oakland prevented its initial adoption, and similar concern has arisen in Baltimore and St. Louis about the use of Persistent Surveillance Systems. This pattern recurs, but not all communities have the political will to organize and protest. Those places that do not have the organization or capacity to protest will be less successful in blocking surveillance, and the result may be a patchwork of surveillance that tends to capture more from disempowered communities than from well-organized ones. Data holes may exist in these more protected areas, further altering the fairness of data collection and use. After all, if police respond only to where the data exist, then police will target those same disempowered communities more. Indeed, Persistent Surveillance Systems first tested its aerial cameras on Compton, California, and West Baltimore, Maryland, two of the poorest and most racially segregated areas in the country.[80]

Xerxes Minocher and Caelyn Randall single out the importance of local community groups' activism in achieving bans on facial recognition technology used in U.S. cities.[81] Amnesty International has been calling for a total ban on the use, development, production, sales and export of facial recognition systems for mass surveillance purposes by both states/governments and the private sector. As part of this campaign (and as an example of the ability to perform independent monitoring and evaluation), in 2021 the organization completed a crowdsourced project identifying 15,000 surveillance cameras in New York City, and signaled their inequitable distribution across the city. In February 2022, Amnesty International and partners revealed their research findings, which indicated that New Yorkers living in areas at greater risk of stop-and-frisk by police are also more exposed to invasive facial recognition technology. The analysis showed how the New York Police Department's vast surveillance operation particularly affects people already targeted for stop-and-frisk across all five boroughs. Moreover, in the Bronx, Brooklyn and Queens, a higher proportion of non-white residents corresponded with a higher concentration of facial recognition compatible CCTV cameras.

The unwillingness of the NYPD to address this problem was evident, perhaps best illustrated by the following example. In September 2020, Amnesty International USA filed a public records request under New York's Freedom of Information Law to obtain NYPD records on its surveillance of the Black Lives Matter protests in 2020. The request was rejected by the NYPD along with a subsequent appeal. In July 2021, Amnesty International and S.T.O.P. (Surveillance Technology Oversight Project), a privacy and civil rights group, filed their joint Article 78 lawsuit against the NYPD for refusing to disclose its records. Soon afterward, the New York Supreme Court ruled that the NYPD broke the law in withholding this information, which was a damning indictment of their lack of transparency and accountability to the public. The Court ordered the NYPD to disclose thousands of records on how the police force procured and used facial recognition technology against Black Lives Matter protesters. This information would enable New Yorkers to better understand how political dissent was policed in the city and help S.T.O.P. prevent more abuses in the future.[82]

New York citizens' surveillance fears revived when, in April 2023, Mayor Eric Adams began rolling out robot dogs (quadrupedal military robots), which were advertised as helping New York police investigate high-risk areas. Such technology has been criticized as inaccurate and disproportionately used against minorities. Adams already faced criticism for escalating police powers, including reinstating a notorious plainclothes unit, and pledging to increase the use of facial recognition technology—actions that flew in the face of recent calls for a facial recognition ban by Amnesty International and its allies. The mayor insisted that the success of the controversial robot in response to a recent garage collapse should convince critics that such devices could improve safety in the city. But critics remained skeptical; Donna Lieberman, executive director of the New York Civil Liberties Union, told *The New York Times*: "While deploying robots is, of course, appropriate in situations like this, that doesn't eliminate the need for transparency about this and other technologies that may have the capacity to engage in massive surveillance and routinely collect large amounts of private data on millions of New Yorkers."[83] The robot dogs were eventually withdrawn after a backlash over their use in public housing and in response to a home invasion in the Bronx.[84]

With all this attention focused on New York City, it should be noted that Atlanta has become the most surveilled city in the United States, in large part thanks to a program called Operation Shield. Motorola Solutions, which provides high-tech surveillance equipment for U.S. prisons, the U.S.-Mexico border and the Israeli-occupied West Bank, designed and implemented the surveillance system. Operation Shield currently involves more than 12,800 private and public interconnected cameras monitored by the police—the highest number per capita in the country. The Atlanta Police Foundation (APF) played an important role in helping to finance the program. Police foundations, which sprang up during the big homeland security push after 9/11, are nonprofits which raise private money from individual and corporate donors that is funneled to police departments with little oversight or accountability. Today, the APF is known for being funded by corporations like Home Depot and Wells Fargo.[85] Research shared with *The Guardian* uncovered also a web of connections between the Atlanta Police Foundation and private equity firms regarding the "Cop City" project. This private equity involvement adds an additional layer of opaqueness to the already severe accountability and transparency issues at play when it comes to policing in the U.S.[86]

It is also in Atlanta where, from December 2022 through March 2023, there was a heavy police crackdown on community protests against a $90 million project to build a police and fire department training facility in the South River forest (the indigenous Muscogee Creek people call it Weelaunee forest). The area is one of Atlanta's largest remaining green spaces, previously earmarked for a public park. Called "Cop City" by its opponents, it would be the largest facility of its kind in the U.S. It was first planned in 2017, but only gained real traction following the 2020

Black Lives Matter protests. Opposition to the project was centered on a range of concerns such as unchecked police militarization and the clearing of forests in an era of climate crisis.

The crackdown led to the police killing of an environmental activist during a police raid triggered by the occupations of the forest, disruption of construction, and "domestic terrorism" charges against 61 protestors. They were charged under Georgia's Racketeer Influenced and Corrupt Organizations (Rico) act, an extension of a federal law created under the Nixon administration to crush the Italian-American mafia.[87] Constitutional law experts have determined that, while some activists did engage in acts of sabotage to protect the forest, it is absurd to consider their activities as constituting a criminal organization, unless one considers all protest movements illegal. The indictments basically do just that—lump together acts like passing out flyers, receiving reimbursements for glue and food, providing legal support, raising money to bail others out of jail, and literally writing the letters "ACAB" (meaning "All Cops Are Bastards") into an amorphous nonsensical conspiracy. In reality it was a large network of organizers and activists, from faith-based and environmental groups to socialist parties and collectivist anarchists that got together to protest and protect the forest. They used a range of tactics, from occupying the land to knocking on the doors of neighbors to inform them about the construction. Because the prosecutors cannot name a clear command structure like one might do with a criminal mob, the indictment of the Stop Cop City activists is focused on the alleged anarchist ideology of some of the protestors and their desire to create a better world. The indictment lists things like "mutual aid," essentially inter-communal charity, as if they are acts of terrorism or equivalent to the extortion of store owners for "protection." Essentially, it should be seen instead as a blatant attempt to destroy a grassroots social movement.[88]

Work on the project continued apace, resulting in clear-cutting, and eventually, cement being poured on a 171-acre tract of land in the South River forest.[89] The crackdowns on its opponents continued as well. In February 2024, SWAT-style raids on Cop City activists' homes were conducted during pre-dawn operations by agents and officers of the Federal Bureau of Alcohol, Tobacco, Firearms and Explosives (ATF), Atlanta police, Georgia state patrol officers and FBI agents. It was the third operation of this kind in residential areas of Atlanta and nearby unincorporated DeKalb county since the movement began in 2021, and the first in which the ATF played a prominent role. The latest operation came after weeks of Atlanta officials promoting a campaign to catch activists linked to arson against construction and police equipment. All the while activists had been committing more acts of sabotage along with tactics of nonviolent civil disobedience.[90]

Subsequently, Atlanta police carried out around-the-clock surveillance for months, on people and houses linked to opposition against the police training center.[91] Tactics included following people in cars, blasting sirens outside bedroom windows and shining headlights into houses at night. The police established themselves in four neighborhoods, concentrating on about 12 houses—including those that were previously raided—with marked and unmarked cars parking near them, driving slowly by and leaving when approached by residents. It has been noted that this low-tech type of surveillance and related police behavior has precedents dating at least to the civil rights era.[92]

Public Attempts at Containment of Mass Surveillance

As a possible solution to the problems of mass surveillance more generally, the ACLU has been advocating—through their effort called Community Control Over Police Surveillance or CCOPS—to enact legislation at the state and local level that would prohibit police departments or other agencies from acquiring or using surveillance technologies without public input and the approval of elected representatives.[93] As of early 2020, more than thirty cities and counties had passed or were working on CCOPS legislation derived from a model bill created by the ACLU. The bill is designed to ensure that citizens, rather than police, have the final say over the type of technology the police use and how they use it. It calls for city councils to approve the purchase, acquisition and deployment of any military or surveillance technology by any city agency. Usually, that will be municipal police forces, but it can also be fire departments (sensors and cameras to detect fire drills at landfills, for instance) or municipal transit services, that often install cameras on buses and trains and use data to record and track commuting and travel patterns.[94]

The bill mandates that agencies create usage policies for their technology, regularly report on their compliance and make the reports public. That does not mean that agencies are left without the tools they need, or are required to seek approval every time they want to deploy any sort of technology. The city of Oakland, California for instance, has an "exigent circumstances" exception that has allowed law enforcement to quickly deploy drones to search for suspected shooters, or to surveil large public events, such as a victory parade for a local sports team. So police departments can actually use whatever technology they deem necessary to use, as long as they report what, why and how it was deployed in a timely manner. This exemption obviously creates an enormous loophole regarding usage policies. But without it, departments would be hamstrung—and probably more reluctant to agree to worthwhile oversight, according to Jon Fasman. He admits that such an exemption clause in formal usage policies is open to abuse, but regular audits should mitigate persistent or systemic abuse, which is probably an overly optimistic view.[95] In Oakland's case, an advisory commission attached to the city council holds hearings, keeps abreast of relevant state and federal legislation, and provides annual reports.

The same system may not work everywhere. Oakland has a mix of characteristics—a deep-seated suspicion of law enforcement, a history of activist politics, and a tech savvy populace—that exist in few places elsewhere. But at least seventeen states, cities and counties have chief privacy officers, including solidly Republican states, such as West Virginia, Tennessee, and Arkansas. Their precise duties vary, but broadly, they purportedly ensure that state agencies protect their citizens' private data (although it is not always clear exactly how). Most likely, some states created these functions not in response to public outcry, as Oakland did, but out of liability concerns (data breaches, as the private sector has shown, can lead to hefty lawsuits).

Regardless of the motivation, Fasman seems convinced that bodies like Oakland's Privacy Advisory Commission are needed at the local level alongside states' attempts to protect citizen data.[96]

The issue of privacy and civil liberties violations has become even more pressing with the recent arrival of the highly sophisticated facial recognition software developed by the tech firm Clearview AI (founded in 2017). According to Kashmir Hill in his revealing book about the rise of this secretive surveillance company, *Your Face Belongs to Us* (2023), Clearview AI has developed software via a database expected to contain some 100 billion "faceprints" from people's online photos (harvested from websites without permission) over the course of 2023. Almost anyone in the world whose photo and name has ever appeared on the internet allegedly can be identified.[97] If unchecked, it would enable covert and remote surveillance of Americans on a scale unlike anything ever seen before. In fact, Clearview did what companies like Google and Facebook could technically do, too (e.g., collect an enormous number of online photos and apply artificial intelligence to them), but did not dare to put into practice—the specter of China's dystopian world of facial recognition systems everywhere was too frightening. When Clearview AI (unintentionally) first went public in 2020, it was revealed that tens of police departments and criminal investigation services in the U.S. were already deploying this company's facial recognition software to identify suspects who appear in camera footage.[98]

The ACLU monitored this development very closely; in May 2020 they sued Clearview AI on behalf of groups representing survivors of domestic violence and sexual assault, undocumented immigrants, current and former sex workers, and other vulnerable communities allegedly harmed by facial recognition surveillance. The lawsuit argued that Clearview repeatedly violated the Illinois Biometric Information Privacy Act (BIPA), the groundbreaking Illinois privacy law adopted in 2008 to ensure that Illinois residents would not have their biometric identifiers—such as a fingerprint, faceprint, or iris scan—captured and used by private entities without their knowledge and permission. This lawsuit had wider implications at the national level as well.

In a settlement of this case two years later (May 2022), Clearview agreed to a set of restrictions that ensured the company followed BIPA provisions. The central tenet of the settlement restricted Clearview from selling its faceprint database not just in Illinois, but across the U.S. The company was permanently banned from granting paid or free access to its facial recognition database to private companies and private individuals nationwide, subject to narrow exceptions mentioned under the Act. The ban was further extended to include any state or local government entity in Illinois (including law enforcement) for a period of five years. This meant that within the state, Clearview could not take advantage of BIPA's exception for government contractors over that period. Among the provisions, one was especially telling. It explicitly stated that Clearview would end its practice of offering free trial accounts to individual police officers without the knowledge or approval of their employers.[99]

It is hard to overstate the power that integration of facial recognition software and aerial surveillance hardware affords to police departments and government agencies. Aside from the potential for abuse, the "normal use" of these systems is already a serious threat to democratic rights. All the while, tech corporations make a hefty profit by selling the advanced tools needed to invade privacy, target individuals and groups, and violate civil liberties; this alongside extracting even more data and profit for themselves.[100]

This discussion of rampant city surveillance and smart policing leads back to the overarching question explored throughout this book: how can the negative ramifications of digital capitalism cum data colonialism be curbed, if at all?

9

Moves to Rein in Capitalism Interlocked with Data Colonialism

There is an emerging industry dedicated to advising people how to better deal with platform capitalism. The proposals for digital media literacy in question focus directly on people's everyday handling of the technologies that constitute the infrastructure required for the extraction of profit from human life though data mining. They involve, among other things, the advice to take regular breaks (unplugging), change privacy settings, or deploy a search engine like DuckDuckGo that offers multiple types of privacy protection. A common recommendation is to keep oneself informed about the risky features of the various technologies as they develop further and innovations are introduced. This kind of media literacy may have some benefits in the short run, but, as Couldry and Mejias insist, media literacy alone is not sufficient to resist data colonialism.[1] Like all notions of literacy, media literacy relies on the virtuous disposition of the subject, which ignores how data colonialism works to erode the autonomy of the subject. In any case, media literacy hits upon some practical limits; when no data traces are left, basic economic activity online becomes increasingly difficult and an attempt to make an online purchase on Amazon, for example, is blocked arbitrarily by the company's algorithm.

Regulation and legislation are important areas of agency. As outlined earlier, among the many interlocking components of data colonialism's construction, public policy plays a foundational role that constitutes the major, if not the only, lever of power really capable of tempering capitalist digital enclosure and the relentless drive to commodify information. It is, after all, the very set of instruments that private and public actors deployed to lay the legal foundations of data colonialism in the first place. By now it is crystal clear that industry self-regulation and the "notice and choice" privacy paradigm are utter failures. When pressed, online platforms will continue to roll out transparency tweaks, privacy dashboards and other superficial changes. Such interventions, likely to be applauded by their public relations teams as a job well done, are actually trivial and insignificant. These companies will not do anything voluntarily that might undermine their core business model of unaccountable surveillance—that is, unless democratic society gives them no choice. As

Matthew Crain has argued, the only solution to a problem of this magnitude is a political program that confronts the business model head on.[2]

The EU's General Data Protection Regulation

Government attention towards the dangers of data colonialism increased significantly after the exposure of the Facebook-Cambridge Analytica data scandal that erupted in early 2018. In response to the misuse of digital mass surveillance data, governments of several countries took preventive measures. The European Union swiftly imposed regulations on misuse of big data through the introduction of a new law, the General Data Protection Regulation (GDPR), which came into effect in May 2018. The GDPR offered the most significant legal challenge to the discourse and practice of data colonialism up until then.[3] The very first sentence of the law challenges the idea that markets and technologies have made privacy irrelevant in the age of big data. It states unequivocally that "the protection of natural persons in relation to the processing of personal data is a fundamental right."[4]

The EU approach differs fundamentally from that of the U.S. in that companies must justify their activities within the GDPR's regulatory framework. In this context, Frischmann and Selinger make the following interesting observation about the different fundamental premises regarding privacy issues in Europe and the United States:

> Privacy is important on both sides of the Atlantic. But it is conceptualized, valued, implemented, and prioritized differently in the US and the EU. In the US privacy is often conceptualized instrumentally and valued in terms of its contribution to welfare. There are exceptions, as in the context of health care. By contrast, in the EU, privacy is conceptualized as a fundamental right. Privacy is the identity and integrity of individuals as human beings and, thus, it is given higher priority than in the US.[5]

The GDPR requires data collectors to give an ordinary-language account to every user about what data they hold and for what purpose. This approach assumes that the lack of such information damages the fundamental right of each person to control the boundaries of his/her information, hence the need for an immediate alert when an activity potentially breaches those boundaries. The GDPR's practical requirements interrupt the supposed "naturalness" of data collection from individuals by requiring them to be informed of what is happening with their data.[6]

The GDPR regulations include: a requirement to notify people when personal data is breached; a high threshold for the definition of "consent" that puts limits on a company's reliance on this tactic to approve personal data use; a prohibition on making personal information public by default; a requirement to use privacy by design when building systems; a right to erasure of data (= "the right to be forgotten"); and expanded protections against decision making generated by automated systems that impose "consequential" effects on a person's life. The regulatory framework also imposes substantial fines for violations, which can be as high as 4 percent of a company's global revenue, and it allows for class-action lawsuits in which users can combine to assert their rights to privacy and data protection.[7] With the passage of the GDPR, the EU was clearly at the vanguard of citizen protections.

In 2018, it was unclear whether this new law would disrupt data colonialism in practice. While it changed the rules under which data colonialism operated, it left unchallenged the commercial purposes for which data is collected. There was also the question about the limits of the "informed consent" principle upon which the GDPR relies; this is dependent on how situations of consent are configured in everyday life. It was not clear how this principle would play out in situations where consent to data collection and processing is overruled by the bargaining power of employers or insurers. Moreover, powerful platforms such as Google were arguably well positioned to bear the costs of accommodating their business to GDPR rules.[8]

As long as ownership and control over the platform itself are not affected, companies can find ways to work around constraints like the GDPR. These regulations tend to hit smaller companies harder, further extending the market dominance of big players, because they are the most capable of complying with the regulations. A focus on data breaches such as the Facebook-Cambridge Analytica scandal also normalizes the exploitative practices of the companies by assuming that there is a safe and acceptable standard they could aim for within their existing business model.[9] Moreover, it seems unrealistic to expect the legal institutions of market societies to be the site at which general resistance to platform capitalism's expansion can be put effectively into practice.

It is predictable that geopolitical battles over competing regimes of privacy and regulation of data flow—as, for example, between the European Union, the United States and China—will likely be a major factor in how data colonialism develops on a global scale in coming years.[10]

U.S. Attempts to Protect Privacy and Personal Data

In the United States, too, a preoccupation with internet reform has begun to take root among policy makers. The recent attempts at such reform are by no means monolithic; the particular proposals and points of emphasis vary widely. Still, one can discern two tendencies, which are often combined in practice. The first involves formulating new rules about how companies are allowed to behave, or enforcing existing rules; the second involves antitrust measures, outlined further below. A prime example of the first tendency is the California Consumer Privacy Act (CCPA), a state law passed in 2018 that gives residents certain rights regarding the collection and processing of their personal data.

The CCPA, which took effect on January 1, 2020, is partly modeled on Europe's GDPR, and is serving as a blueprint for other states to follow. Residents are given the right to learn what data companies have collected about them; to ask companies not to sell that data; and to request its deletion. The data can come from any source, including the internet, databases and paper forms. The law defines information broadly so that everything from one's browsing history to personal characteristics such as race and marital status are covered. The definition also covers biometric and location information. Other states have elected less strict versions, however. And even California's law did not adopt all of GDPR standards, such as requiring companies to have a valid reason for processing data and minimizing the amount of data they collect.[11] At the time of writing, 12 other states (Colorado, Connecticut, Delaware, Iowa, Missouri, Montana, New Jersey, Oregon, Tennessee, Texas, Utah, Virginia) had passed similar legislation, while 16 states had such bills under consideration.[12] If this trend continues, companies operating in the U.S. may potentially have to comply with 50 different state privacy laws, which appears untenable. Understandably, there has been growing pressure to enact a single GDPR-style privacy law at the federal level, which would override state laws.

Anti-Monopoly Measures

The second trend of internet reform entails attempts to reduce the market power of the big corporations. This is the focus of the New Brandeisians, a group of anti-monopoly advocates who have become influential within the Democratic Party and includes some Republicans as well. They take their inspiration from Louis Brandeis, a leading liberal jurist of the Progressive Era (and member of the U.S. Supreme Court from 1916 to 1939), who believed that monopolies posed a threat to democracy. His solution was a Madisonian system of checks and balances designed to disperse corporate power and promote fair competition.[13]

The New Brandeisians push for a range of measures in order to crack down on tech monopolies. In some cases, they want to break up large firms into smaller ones; for example, one proposal calls for forcing Facebook to split off WhatsApp, Instagram and its ad network. In other cases, these reformers are willing to accept a certain degree of "bigness" and seek to constrain such "natural monopolies" through regulation. Above all, they want markets to be more competitive, believing that more competition will bring a number of benefits, from a wider distribution of wealth to reduced corporate influence over the political process.

However, antitrust enforcement is more often the result of business competitors' complaints about the dominant firm in a market rather than pressures from public opinion or empirical reality, as the Microsoft case of the 1990s illustrates. By 2012, the monopolistic features of the tech giants were again drawing the attention of the Federal Trade Commission (FTC), the Justice Department and the U.S. Congress. Several of those companies—e.g., Apple, Amazon and Facebook—were embroiled in various lawsuits and negotiations. Google in particular, was being pressed by the FTC over whether it "has abused its dominance by manipulating search results, making it less likely that competing companies or products will appear at the top of the results page."[14]

From a Progressive perspective, such monopolies should either be publicly owned or, at the very least, be heavily regulated to prevent abuses, especially as they often tend to monopolize certain public functions. If these companies could be effectively reformed into competitive businesses, that route should be considered, too, but the strong monopolistic pressures in this sector made that unrealistic.[15] According to Robert McChesney's evaluation of the state of affairs in 2013, corporate political power had basically eliminated the threat of public ownership as well as credible regulation in the public interest. The regulation that remained, antitrust or otherwise, was done as much to guarantee the existence of profitable firms and industries as it was to protect the public interest threatened by commercial monopolies. The existing range of political debate did not allow to question the propriety of these tech behemoths; only a little squabbling at the margins could take place.

None of the U.S. government's activities toward the new giants was even remotely life-threatening for them. Even at their most rigorous, U.S. or European antitrust regulators seemed to show little concern about the markets as long as there were two or three players with double digit market share. The regulators seemed to be giving in to the market reality of contemporary capitalism, except for the most blatant monopolies.

The single greatest antitrust threat to Google and the other tech giants came from the European Union. By 2012, the EU was in tense negotiations with Google over its monopolistic practices. Google then had 85 percent of the European search market, some 15 percent more than its U.S. position. The FTC was in regular contact with the EU officials. Google's concern was that a formal European antitrust case against Google might embolden American regulators and give them powerful ammunition.[16] In this Google proved to be correct, as these regulators would indeed take on the company in later years. The European Union did three investigations of Google, stretching over more than a decade. By September 2022, the antitrust fines that the EU's competition commission imposed on Google (based on these cases) totaled €8.25 billion.[17]

By then, U.S. legislators and regulators had begun showing much greater interest in taming the tech giants in one way or another. The Federal Trade Commission and the U.S. Justice Department started investigations into Google, Facebook, Apple and Amazon during the Trump administration. In October 2020, the House Judiciary Committee concluded a sixteen-month investigation into those companies with a hard-hitting report that recommended various New Brandeisian reforms. By the summer of 2021, a legislative package of six bills designed to strengthen antitrust enforcement was working its way through the House. Meanwhile the New Brandeisian law professor Lina Khan[18] had been selected to chair the FTC and the antitrust lawyer Jonathan Kanter was chosen to lead the Department of Justice's Antitrust Division. Moreover, President Biden had issued an executive order that directed more than a dozen federal agencies to pursue pro-competition initiatives.[19] By early 2022, all the bills in the legislative package had been voted out of Committees in both Chambers of Congress with bipartisan majorities. All had broad support from voters, knowledgeable experts, regulators and the White House. Ultimately, however, none was taken up by the full House and Senate. The prime suspect in killing the various bills was Big Tech itself. The four companies in question spent an estimated $250 million on lobbying, advertising, public relations and stepped-up campaign contributions to their supporters. Financially, this represented only about 1/10 of 1 percent of their combined annual profits, but on a political scale, it was an overwhelming show of force.[20]

Nevertheless, there have been some important developments on the U.S. antitrust front under the existing legislation. In October 2020, the U.S. Department of

Justice and the attorneys general of 35 states, along with Guam, Puerto Rico and the District of Columbia, first filed a complaint against Google. In this landmark antitrust lawsuit, Google was accused of being a "monopoly gatekeeper for the internet," using "pernicious" anticompetitive tactics to maintain and extend its monopolies. These tactics included a series of business agreements that effectively locked out competition. The Justice Department also challenged an arrangement in which Google's search application was preloaded—and could not be deleted—on mobile phones running its Android operating system. The company allegedly spent billions of dollars in partnerships with companies like Apple to make sure its general search engine was the default browser on their device.[21] The Department of Justice was basically arguing that Google—which had a share of between 90 and 95 percent of the search market—had maintained its monopoly by cutting off almost every other avenue by which consumers might find a different search engine, making sure they only saw Google wherever they looked.[22]

The last time the Justice Department had taken on an aggressive monopolist was in 2001, when it sued Microsoft for illegally tying its Internet Explorer browser to Windows as part of a campaign to destroy Netscape, maker of the first distinctive commercial web browser. Now the Department accused Google of similar tactics, such as illegally tying the company's search engine to its Android smartphone operating system and its Chrome browser. The government was also seeking to break up the company, just as it once sought to break up Microsoft.

There are more striking parallels between the two cases, as John Naughton, a leading expert on the social, political and cultural impact of internet technology, has pointed out. In 2001, Microsoft had 93 percent of the global market for operating systems, while in 2023, Google had 92 percent of the market for its search engine. In the 1990s, Microsoft had been slow in appreciating the significance of the web and arrived late to the market with a mediocre browser—Internet Explorer—inferior to the Netscape alternative. At the time, manufacturers of personal computers could not get a license to install Windows on them without bundling it with Explorer, thus making the Microsoft browser the default.

In the recent lawsuit against Google, the company was accused of a similar default-setting practice. Where it had the necessary power as with the Android system that it controls, or its dominant Chrome browser, Google's search engine was the default. Where Google lacked ownership, the company took recourse to money, for example, paying $10 billion a year for privileges such as making Google the default engine on Apple iOS.[23]

There was another big difference between the 2001 Microsoft case and the Google one, whose trial began in September 2023—the absence of media coverage. The first case was widely covered by mainstream media at the time, whereas the Google case has drawn relatively little media attention. This may be partly due to the lack of public interest in antitrust matters today. Naughton thought of another explanation: Google's demand to keep as much evidence as possible out of the public eye, to which the presiding judge deferred to a great extent. Naughton rightly stated that, for a self-proclaimed democracy, this was not the way to check unaccountable corporate power.[24]

Google also became the subject of another important antitrust case, this time focused on its digital advertising business. A federal antitrust complaint was filed against Google in January 2023, when the Department of Justice joined an additional lawsuit against Google by attorney generals from more than three dozen states and territories (a number that had expanded from initially eight states: California, Colorado, Connecticut, New Jersey, New York, Rhode Island, Tennessee, and Virginia). The lawsuit concerned allegations that the company abused its dominance of the digital advertising business by using anticompetitive, exclusionary and unlawful means to eliminate or severely diminish any threat of its rivals in this regard. The Justice Department accused Google of unlawfully monopolizing the way ads were served online by excluding competitors, which included its 2007 acquisition of DoubleClick and the subsequent rollout of technology that locked in the split-second bidding process for ads that get served on web pages. The Justice Department asked the court more specifically to compel Google to divest its Google Ad manager suite, including its ad exchange ADX.[25]

Meanwhile, Google had agreed to pay $700 million and make several other concessions to settle allegations that it had been stifling competition against its Android app store. Google was accused of overcharging consumers through unlawful restrictions on the distribution of apps on Android devices, and pocketing unnecessary fees for in-app transactions. All 50 states, the District of Columbia, Puerto Rico and the Virgin Islands joined the settlement. The disclosure of its terms came in December 2023 in documents filed in San Francisco federal court, shortly after a federal court jury rebuked Google for deploying anticompetitive tactics in its Play Store for Android apps.

The settlement with the states included $630 million to compensate U.S. consumers funneled into a payment processing system that state attorneys general alleged drove up the prices for digital transactions within apps downloaded from the Play Store. Like Apple did in its iPhone store, Google collected commissions ranging from 15 to 30 percent on in-app purchases—fees that state attorneys general contended drove prices higher than they would have been had there been an open market for payment processing. Another $70 million of the pre-trial settlement would cover the penalties and other costs that Google was being forced to pay to the states.

Google also agreed to make other changes designed to make it easier for consumers to download and install Android apps from other outlets besides its Play Store over the next five years. It would refrain from issuing as many security warnings, or "scare screens," when alternative choices were being used. Further, the makers of Android apps would gain more flexibility to offer alternative payment choices to consumers instead of having transactions automatically processed through the Play Store and its commission system. Providers of apps would also be able to offer lower prices to consumers who chose an alternative to the Play Store's payment processing system.[26]

At the closing of this high-stakes lawsuit in early May 2024, Justice Department lawyers argued that Google's preeminence as a search engine is an illegal monopoly bolstered by more than $20 billion spent each year by the tech giant to lock out competition. Much of the case revolves around the extent to which Google derives its strength from business dealings with device makers like Apple and Samsung and web browser companies like Mozilla (which runs Firefox) to make Google the default search engine preloaded on smartphones and computers.[27]

In early August 2024, a federal judge ruled that the tech giant had built an illegal monopoly over the online search and advertising industry. Judge Amit Metha's ruling specifically found that Google broke antitrust laws by striking exclusive agreements with device makers like Apple and Samsung, in which Google would pay billions of dollars to ensure that its product was the default engine on their phones and tablets. During the trial it was revealed that Google paid companies more than $26 billion ($18 billion of which went to Apple) in 2021 alone to remain the default option in Safari. What happened next would determine whether Google would be forced to make sweeping changes to how it does business or whether it could successfully defang the ruling on appeal. The ruling in U.S. vs Google did not contain any remedies for the company's illegal monopolization, and the Justice Department did not seek specific penalties when it argued the case. There would be a separate trial to determine what remedies the government should enforce against Google, which could range from tweaks to how it handles contracts to breaking up the company entirely.

One possible outcome is that Mehta rules the company can no longer make such deals. This would allow Google to remain the default research engine if device makers opt for it, but would block the multimillion-dollar payments Google has made to guarantee this status. Another result could be what has been playing out in the European Union since regulators began forcing businesses to comply with its Digital Markets Act. When users log on to a service from tech companies like Google, Apple or Microsoft, they face a "choice screen" where they are prompted to select which browser they would like to use.[28]

The second antitrust trial pitting Google against the U.S. Department of Justice began on September 9, 2024. This lawsuit centered around Google's acquisition and application of digital ad tech. Website publishers looking to make money from advertising rely on this technology to act as a kind of middleman. Google's services allow sites to sell ads on their pages and advertisers to buy ad space that reaches potential customers, while Google takes a sizeable cut of the ad dollars from both sides. The Justice Department singled out several of Google's acquisitions to argue that the company now dominated every facet of digital advertising. Google bought the ad tech company DoubleClick in 2007 for $3.1 billion, which provided the tech company with an online marketplace for publishers looking to sell ad space. The DOJ alleged that DoubleClick now controlled over half the ad market for open-web display transactions. Over the next few years Google acquired two other companies, Invite Media and AdMeld, which gave it access to advertisers looking to buy ad space and the ability to connect them with publishers. These deals resulted in Google controlling both the supply and demand sides of online advertising as well as the point of exchange where those sides meet. The Justice Department claimed that Google had built a monopoly through a series of ruthless anticompetitive maneuvers. These included eliminating rivals through acquisitions or exclusionary practices that amounted to wielding an illegal monopoly over the industry. The lawsuit alleged that Google used its dominance to deliberately overcharge advertisers, while keeping at least 30 cents of every dollar that flows to website publishers through its advertising technology. Google's ad-selling tools for publishers controlled 87 percent of the U.S. market. The Justice Department said that overall, Google's technology for selling ads across the web brought in about $31.7 billion in 2021. That portion of its business contributed only a small fraction of the company's profits. But the Justice Department argued that Google's dominance over placing ads online resulted in higher prices for advertisers and publishers. Google had also hurt specific industries, like news publishers, the government contended. This was because Google took a portion of the price each time those publishers sold their ad space, making it harder for them to stay in business. A government win could require the sale of DoubleClick and other ad tech acquisitions, forcing Google to change its behavior when selling ads online.[29]

Then there was the case against Facebook. In January 2022, a federal judge ruled that the antitrust lawsuit against Facebook (now Meta, its parent company's name since October 2021) could proceed after the FTC's first attempt at targeting the company's alleged monopoly power was dismissed for lack of evidence in June 2021. In its amended complaint, the FTC argued that Facebook's illegal monopoly power had hurt innovation, had been detrimental to user privacy and data protection and subjected users to excessive advertisements, with little choice or control over what ads they were served. This time the judge found that federal regulators had offered enough proof to argue that Facebook's acquisition strategy—particularly its takeover of Instagram and WhatsApp—was driven by a "buy or bury" ethos. In other words, Facebook was allegedly gobbling up competitors in order to maintain an illegal monopoly. The judge did narrow the scope of the lawsuit, though. The accusation that Facebook's policies around interoperability—the ability to smoothly move between competing social networks—were unduly restrictive, was not allowed to move forward. This was based on the fact that Facebook had abandoned a key platform policy around interoperability in 2018.[30] It was expected to take several years for this case to run its course.

U.S. Antitrust Legislation Reinvigorated

The FTC under Lina Kahn has returned the agency to its original remit, fully activating the tools and legislation that Congress had charged the FTC with administering. The arsenal includes the Sherman Act (1890), which cracks down on monopolies and anticompetitive interstate trade practices, and the Clayton Act (1914), which prohibits anticompetitive mergers and acquisitions. There is also the Federal Trade Commission Act (1914), which gives the FTC broad authority to prosecute companies that engage in "unfair methods of competition." While the agency is limited in the remedies it can seek—it can issue cease-and-desist orders, but no fines or criminal penalties—it has a broad range of powers to regulate corporations, including the ability to make rules that make enforcement stronger.

Since the 1970s, however, the FTC's focus had been on "efficiency" and "consumer welfare," neither of which is based on the laws governing the agency.[31] This implied that the government would only take action against a company over anti-competitive practices, if consumers were harmed by increased prices. Winning an antitrust case revolved around proving that a company's dominance hindered competition and raised prices for consumers, or limited consumers' choices. That approach, Lina Khan and others have argued, allowed tech companies to build de facto monopolies by giving away their products for free or at such low prices that no one else could compete. With companies such as Google and Facebook, this concerned mainly "free" services where users paid with their attention and personal data. That attention was valuable for advertisers and the large volume of data also gave Google an edge regarding the development of new applications such as artificial intelligence, for instance. In the case of Amazon, Khan argues that keeping prices low has allowed the company to amass a large share of the market and stifle competition.[32] So the FTC has worked to broaden the agency's mandate, focusing on how companies wield power to distort markets and hurt the overall economy.[33]

Amazon became a major FTC target, beginning with privacy violations and the like. In May 2023, the FTC settled with Amazon for more than $30 million over violations that prohibit unfair or deceptive business practices in relation to two lawsuits. The first case concerned Amazon's Ring video doorbell; the images could be easily viewed by Amazon employees and the account data of Ring customers went all over the internet.

In the complaint, the FTC argued that Ring deceived its consumers by failing to restrict employees' and contractors' access to its customers' videos, using these videos to train algorithms without consent and failing to implement security safeguards. According to the complaint, these failures amounted to flagrant violations of users' privacy. Ring allegedly also failed to implement standard security measures to protect consumers' information from two well-known online threats—in cryptography called "credential stuffing" and "brute-force" attacks—despite warnings from employees, outside security researchers and media reports. (Credential stuffing involves the use of credentials, such as usernames or email addresses and the corresponding passwords obtained from a user's breached account, to gain unauthorized access to user accounts on other systems through large-scale automated login requests directed against a web application. A brute-force attack consists of an attacker submitting many passwords or passphrases with the hope of eventually guessing correctly. The attacker systematically checks all possible passwords and passphrases until the correct one is found.)

In addition to the mandated privacy and security program, the proposed order required Ring to pay $5.8 million in fines, which would be used for consumer refunds. The company would also be required to delete any customer's videos and face embeddings—data collected from an individual's face—that were obtained prior to 2018, and delete any work products it derived from these videos. The proposed order also demanded that Ring alert the FTC about incidents of unauthorized access or exposure of its customers' videos, and notify consumers about the FTC's action.[34] The settlement between Ring and the FTC was announced in November

2023. As part of the settlement, the FTC also ordered Ring to delete all data and algorithms derived from the unlawfully accessed videos. Additionally, Ring was required to establish a new privacy and security program that limits employee access to customer videos, with exceptions primarily in law enforcement situations.[35]

The second case was about smart speakers linked to Amazon's digital assistant Alexa, which collected personal data and audio data without permission, using that information to train the algorithms for its speech recognition AI system. This suit alleged Amazon violated the FTC Act and Children's Online Privacy Protection Act by illegally retaining thousands of children's information through their profiles with the Alexa voice assistant. Next to a $25 million civil penalty, Amazon was prohibited from using children's voice information and geolocation data—that were subject to deletion requests—for creating or improving any data product. Amazon was also required to delete inactive child accounts on Alexa, notify users about the government action against the company and of its retention and deletion practices. Regarding the rather low fines for this Goliath. it should be noted that the FTC's ability to pursue monetary relief for consumers was limited by a 2021 U.S. Supreme Court ruling that narrowed the scope of financial compensation it could impose.[36]

September 2023 marked a sea change, when the Federal Trade Commission, along with a bipartisan group of attorneys general from 17 states, filed a sweeping antitrust lawsuit against Amazon, accusing the company of illegally staving off competition to become one of the most powerful companies in the U.S. This landmark case represented a significant—even existential—threat to Amazon's dominance in the online retail industry. Over the course of the complaint, the FTC repeatedly depicted the retail giant as a company that willfully stifled competition while restricting consumer choice in order to maintain its monopoly. More specifically, the FTC and the states involved alleged that Amazon imposed anti-discounting measures that prohibited merchants who sell products on the platform from offering lower prices elsewhere on the internet, and intimidated third-party sellers to use its expensive fulfillment services, requiring them to employ the company's delivery and fulfillment system in order to qualify for its popular Prime subscription service. Amazon also prioritized its in-house line of products (including its own copycat products) over others. The FTC further accused Amazon of enrolling millions of consumers into its paid subscription Amazon Prime service without their consent and making it difficult for them to cancel.[37]

At least for the time being, the FTC has not sought to break up the company, but instead asked for a permanent injunction from a federal court that would prohibit Amazon from engaging in its unlawful conduct. The FTC complaint also asked for the court to consider "any preliminary or permanent equitable relief, including but not limited to structural relief, necessary to restore fair competition." Structural relief in this context generally means that a company sells an asset, such as part of its business.[38] This lawsuit, too, is expected to take years to complete; the trial is to start in October 2026.

On March 21, 2024, Apple became the subject of a wide-ranging antitrust lawsuit in which the U.S. Department of Justice, joined by the attorneys general of 15 states and the District of Columbia, accused the tech titan of engineering an illegal monopoly in smartphones that wards off competition, stifles innovation and keeps prices artificially high. The suit argues that Apple rose to its current powerful position thanks in part to the 1998 antitrust case against Microsoft and that another milestone antitrust correction is needed to allow future innovation to continue.

The lawsuit takes aim at how Apple allegedly molds it technology and business relations to shake more money out of consumers, developers, content creators, artists, publishers, small businesses and merchants. That includes diminishing the functionality of non–Apple smartwatches, limiting access to contactless payment for third-party digital wallets and refusing to allow its iMessage app to exchange encrypted messaging with competing platforms. The company's app store also charges developers up to 30 percent of the app's price for consumers.

The suit specifically seeks to halt Apple from undermining technologies that compete with its own apps—in areas including streaming, messaging and digital payments—and to prevent the company from continuing to draw up contracts with developers, accessory makers and consumers that let it obtain, maintain, extend or entrench a monopoly.[39] The case aims to penetrate the digital fortress that Apple has assiduously built around the iPhone and other popular products such as the iPad, Mac and Apple Watch to create what has been referred to as a "walled garden" where its intersecting hardware, software and services can seamlessly offer user-friendly harmony, albeit exclusively for Apple devices' users.

Central to the case is whether or not Apple's strategy of blocking rival companies from accessing various proprietary features, such as its iMessage instant messaging service and Siri virtual assistant, constitutes anticompetitive practices. The case will also examine whether Apple's practice of making its devices easily integrate with each other but not with non–Apple products, creates unfair limitations that block competitors from the market.

At the announcement of the lawsuit, the DOJ indicated that any potential remedy was on the table for Apple, implying that even breaking up the company was a possibility. The case is expected to take several years to conclude—with a verdict not to be reached until 2026, meaning the case could easily drag on with appeals.

The EU Digital Markets and Digital Services Acts

Regarding antitrust measures and other regulations of the Big Tech firms' conduct, the EU advanced further with the introduction of the EU Digital Markets Act (DMA) and Digital Services Act (DSA). The DMA intends to ensure a higher degree of competition in the European digital markets by preventing larger companies from abusing their market power and by allowing new players to enter the playing field. The DSA aims to update the European Union's legal framework regarding illegal content on intermediaries, transparent advertising and disinformation. The DMA was signed into law by the European Parliament and the European Council on September 14, 2022, and became applicable (for the most part) on May 2, 2023.[40] The European Parliament approved the DSA on July 5, 2022, while the European Council gave its final approval on October 4, 2022.[41] It went into effect on August 25,

2023; affected service providers had until January 1, 2024, to comply with the DSA's provisions.

Once implemented, the DMA establishes a list of obligations for designated "gatekeepers," that is, the largest digital platforms operating in the EU, with more than 45 million users. In cases of non-compliance, there are enforceable sanctions mechanisms, including fines of up to 10 percent of the worldwide annual turnover, or up to 20 percent in case of frequent infringements. The designated gatekeepers have a maximum of six months after the Commission's decision to label them as violators, to ensure compliance with the obligations and prohibitions laid down in the DMA. The list of obligations include protections on combining data collected from two different services belonging to the same company (e.g., Meta's Facebook and WhatsApp); provisions for the protection of platforms' business users (including advertisers and publishers); and legal instruments against the self-preferencing methods used by platforms for promoting their own products (e.g., preferential results for Google's products when using Google Search). The list also includes articles concerning the pre-installation of some services (Android); regulation related to bundling practices; provisions for ensuring interoperability, portability and access to data for businesses and end-users of platforms. According to the European Commission (the EU's executive branch), the main objective of the DMA is to regulate the behavior of the Big Tech firms within the European Single Market and beyond. The Commission aims to guarantee a fair level of competition ("level playing field") on the highly concentrated digital European markets, which are often characterized by a "winner takes all" configuration.[42]

The DSA aims to harmonize different national laws within the EU that have emerged at the national level to address illegal content, especially in Germany, Austria and France. The law applies to any digital operation serving the EU, forcing those companies to be legally accountable for everything from fake news to the manipulation of shoppers, Russian propaganda, and criminal activity. It applies to large and small operators alike, but the rules are tiered, with the toughest obligations pertaining to seventeen companies including Facebook and Amazon that have been designated as "very large online platforms," and two "very large online search engines," Google and Bing. A firm that does not comply with the law can face a complete ban in Europe or fines running up to 6 percent of its global revenue.

Relevant details are as follows:

- Platforms are obliged to combat the sale of illegal products and services, which will affect Amazon and Facebook Marketplace, among others;
- New measures are designed to crack down on illegal content—including Russian propaganda, interference with elections, hate crimes and online harms including harassment and child abuse. These measures are also meant to ensure that fundamental rights, including freedom of expression and data protection, are recognized by law and safeguarded across Europe;
- Platforms are prohibited from targeting children with advertising based on their personal data and cookies. Big social media firms are required

to redesign their systems to ensure a high level of privacy, security and the safety of minors and to prove that they have done so to the European Commission. Platforms have to redesign their content recommender systems to reduce risks to children. They must also carry out a risk assessment of negative effects on children's mental health and present it to the Commission;
- Social media companies are not allowed to use sensitive personal data, including race, gender and religion, to target users with advertisements;
- A ban on "dark patterns." For shoppers, this is protection from everyday interfaces employed to manipulate users into buying things they do not need or want. Under the new rules, online shops must abandon their reliance on "manipulative practices to exploit consumers' vulnerabilities or trick them."[43]

Furthermore, users will be offered clear avenues to report illegal content, goods or services on online platforms. They will be informed about, and can contest removal of content by platforms. Users will also have access to dispute resolution mechanisms in their own country.

Tech companies are banned from ranking their own services more favorably than others and must stop making it difficult to uninstall pre-loaded software and apps.

The Commission's voluntary code of practice, which is envisioned as a "nursery slope" to prepare internal systems for the new regulatory regime, had already been endorsed by 44 tech firms at the time of the law's enactment, August 25, 2023.[44] From then on, the DSA regulation was compelling more than 40 online giants including Facebook, X (formerly Twitter), Google and TikTok to better police the content they deliver within the EU.

Recent Enforcements of EU Digital Markets and Services Regulations

At the time of writing, the various EU regulations were beginning to have significant impacts as various cases demonstrate. First there was Meta's business model that came under attack following a ruling that its legal justification for targeting users with personalized ads broke EU data laws, the GDPR in particular. The move could force the owners of Facebook and Instagram to ask users to "opt in" to having their data used for targeted ads. This should be seen against the backdrop of Meta's advertising-based business, which was already under pressure after Apple introduced a privacy policy change for iPhones, in April 2021, that required apps to seek users' permission to track their online activity in order to serve them personalized data.[45] But it is questionable how much of Apple's marketing message that it respected user privacy, was actually true. Apple was facing a number of court cases for failing to live up to its privacy claims, and researchers had found that the company collected data analytics even when privacy controls were set to prevent this.[46]

In January 2023, Ireland's Data Protection Commission (DPC) fined Meta a total of €390 million, after the EU's data authority rejected the company's argument that users agree to receive ads based on their personal data when they enter into a "contract" with its social media platforms via the terms and conditions they sign. Predictably, Meta argued that it would appeal against the decision and that it was "incorrect" that personalized ads could no longer be offered without users' consent.[47] Meta also responded by constructing its own systems inside the apps it builds, for example, creating the ability to buy products directly from Facebook (by allowing businesses to set up shop right into the app), thus reducing the need for third-party tracking.

In May 2023, Ireland's DPC fined Facebook's owner, Meta, again, this time for €1.2 billion, for mishandling user information, and ordered the company to suspend the transfer of data from the EU to the U.S., and move the data that was stored overseas to servers in Europe. At that time, the fine set a record for a breach of the EU's data protection regulation. (It superseded by far the previous record GDPR fine of €746 million imposed on Amazon in 2021.) The suspension of Facebook data transfer was not immediate and Meta was given five months to enact it. The DPC punishment in this case relates to a legal challenge brought by the European Center for Digital Rights (NOYB)—a non-profit organization led by the Austrian privacy activist Max Schrems—over concerns resulting from the Edward Snowden revelations that European users' data was not sufficiently protected from U.S. intelligence agencies when it is transferred across the Atlantic. Meta responded that it had been "singled out" by the DPC despite many other businesses using the same data transfer mechanism, and that a dangerous precedent was set for countless other companies transferring data between the EU and the U.S.[48]

As Meta continued to violate privacy rights of users on its various platforms, EU authorities became more determined to intervene. A July 2023 ruling by the European Court of Justice (EJC) stated that under the GDPR, Facebook could not justify using personal details to target people with personalized ads unless it received their consent first. As Meta grappled with this regulatory pressure, the company considered charging EU users a monthly fee of €13 to access an ad-free version of Instagram or Facebook on their phones, and a €17 monthly charge to use Instagram and Facebook without adverts on desktops. Meta's ad-free plan would give users the choice of continuing to access Facebook or Instagram for free with personalized ads, or paying for ad-free versions. The plan was to be introduced in November 2023, since Meta had until the end of that month to comply with the ECJ ruling. Regulators were allegedly looking at the size of the fees and whether these were too expensive for people who did not want to be targeted by adverts. Max Schrems, the European campaigner against Meta's data practices, said the proposals were tantamount to paying for fundamental rights, which could not be sold. It would mean that only the rich could enjoy these rights at a time when many people were struggling to make ends meet.[49]

Nevertheless, Meta went ahead with its plan, albeit with a lower subscription rate for desktop users than initially suggested. On October 30, 2023, Meta announced that, starting in November, it would offer ad-free versions of Facebook and Instagram in the EU. Users on desktop browsers would be charged €9.99 ($10.50) a month, while Apple iOS and Android users would pay €12.99. The company said that the higher charges reflected the commissions charged by the Apple and Google app stores on in-app payments. The fee would cover all linked Facebook and Instagram accounts only until March 2024, when Meta would begin charging €6 for each additional account on the web and €8 for smartphones. Unsurprisingly, the company once again defended its primary way of making money by stating "we believe in an ad-supported internet, which gives people access to personalized products and services regardless of their economic status."[50]

Meta would continue to offer versions of Facebook and Instagram with personalized advertising, but under EU data privacy rules, those social media platforms had to gain explicit consent before tracking a user for advertising purposes. In return, Meta pledged to comply with those rules. So in the EU's 27 member countries (plus Switzerland, Norway, Iceland and Liechtenstein) users aged 18 and older would still have the choice of continuing to use Facebook or Instagram with ads. Only users who paid for ad-free accounts would be protected against violations of their privacy rights. Meta suggested that it was thus complying with the ban on personalized advertising. This was far from the reality, since the data of millions of users without a paid account were still collected and used for personalized ads on Meta apps. Full compliance with the EU privacy rules would mean the complete switchover to contextual advertising.[51] Apparently, Meta was not willing to go that far, as this would seriously undermine its primary revenue model.

Relevant developments regarding the enforcement of the Digital Markets Act (DMA) and the Digital Services Act (DSA) should also be noted. On March 25, 2024, just two weeks after the DMA had come into force, the EU announced it had launched investigations of Apple, Alphabet and Meta for potential breaches of the new law. It was the first formal action since the three companies were designated as "gatekeepers" and had been given until early March to comply with a tighter set of regulations than other companies in the EU space. The European Commission's Vice President and the EU's competition chief, Margrethe Vestager, said they were looking at potential breaches related to the following issues:

- concerns about "steering": whether recent measures by Apple and Google imposed limitations that hindered app developers from informing users about offers outside their app stores free of charge;
- whether Alphabet, Google's owner, favored its services such as Google shopping, Google Flights and Google Hotels over rivals in search results on its search engine, and whether it discriminated against third-party services on Google search results;
- whether Meta's introduction of subscriptions to no-ads versions of Facebook and Instagram in Europe in November 2023 complied with DMA rules, and whether the company should offer free alternative options;
- whether Apple was allowing users to easily uninstall software applications on its iOS operating system, to change default settings on iOS or access choice screens allowing them to switch to a rival browser or search engine on iPhones as DMA rules prescribed.

The law requires the six tech gatekeepers—Alphabet, Amazon, Apple, Meta, Microsoft and ByteDance (owner of TikTok)—to comply with guidance to ensure a level playing field for their competitors and to give users more choices. The European Commission said that they had been in discussions with gatekeepers for months to help them adapt, and that they could already see changes happening on the market. But the Commission suspected that the measures taken fell short of effective compliance under the DMA. They were not convinced that the solutions enacted by Alphabet, Apple and Meta respected their obligations for a fairer and more open digital space for European citizens and businesses. The Commission was also taking steps to investigate Apple's new fee structure for alternative app stores and Amazon's ranking practices on its marketplace. They aimed to conclude the investigation within a year, the timeframe set out under the Act.

Non-compliance with the DMA could result in hefty fines of up to 10 percent of the companies' annual global turnover, rising to 20 percent for repeated infringements. Annual revenue at Apple in 2023 was $383 billion, while at Alphabet it was $307 billion and at Meta, $134 billion.

The European Commission might also use non-financial punishments allowed under the DMA, such as forcing the sale of parts of a business or putting bans on acquisitions, to force the gatekeepers to comply with the Act. But the Commission seemed to be reluctant to make use of these other powerful sanctions.[52]

Apple was the first big U.S. tech company making changes in response to the new digital market rules that forced it to modify its practices. Even though it initially appealed the DMA ruling, by the end of January 2024 the company announced historic changes to its iOS mobile software, App Store, and Safari browser in the EU. In an effort to placate the regulators in Brussels, the adjustments were incorporated into the new version of operating system iOS 17.4. Beginning March 7 (when the DMA came into effect), Apple would allow iPhone users to access rival app stores and download their apps. Apple also made it easier for consumers to switch to different default options besides its own Safari browser. The changes also included slashing the fees paid by companies using the App Store to sell digital goods and services from 30 percent to 17 percent. But Apple introduced new charges in Europe, including a "core technology fee" (CFT) of $0.50 on developers for each first annual install of their app over a one million threshold, regardless of whether it was through the App Store or an alternative. This would also apply to free apps, but not apps distributed by the government, education or non-profit organizations. Apple would charge an additional 3 percent fee to app developers that use its payment processor.[53]

In the EU, Apple had no choice but to allow alternative app stores, yet wanted to retain a large degree of control. The various steps that aspiring app store providers would have to meet, were onerous. Like in multiple other instances, the company sought to introduce stumbling blocks to make change difficult.[54] In this case, the company stated that it tried to protect the European user as best it could, arguing that installing its software outside of its own ("official") download store was risky because the company could not then screen for malicious code or privacy invasions. Therefore it had to make sure that each app would be checked for malicious software.

For this purpose, the company had created a system to monitor all iOS apps, review and authorize alternative app stores and track alternative payment systems.[55]

Apple was also under an obligation to allow "sideloading" (i.e., the ability to download apps directly from a website), but it interpreted this obligation in an extremely narrow way: *"sideloading refers to downloading iOS apps outside of an official app marketplace—and in the EU, users will have the option to access alternative marketplaces that offer apps for download."*[56] In other words, direct downloading of apps from the web would be ruled out (even though this is perfectly possible from PCs, including the iMac). In addition, Apple introduced a host of controls that would obfuscate its obligations under the DMA. For instance, apps would have to be officially notarized, with notarization being defined as "a baseline review that applies to all apps, regardless of their distribution channel, focused on platform policies for security and privacy and to maintain their device integrity."[57] If that was the case, it was not clear why notarized apps could not be downloaded from the web.

Apple further lost its exclusive right on contactless payment at stores; previously iPhone users could only use Apple Pay. The company had no choice but to allow app developers to use an alternative payment service provider (PSP) within their app or link users to a website to process payments. But it made this option subject to various conditions that would disincentivize app developers from using alternative PSPs. Apple had decided to make the use of such options difficult by forcing app developers willing to use alternative PSPs to go through multiple steps. It also introduced warnings designed to scare away users from using these apps. Moreover, app developers using alternative payment options still had to pay Apple's full commission. Finally, and most importantly, developers were not allowed to "offer both In-App Purchase and alternative PSPs and/or link out to purchase to users in their App Store app on the same storefront."[58] So much for user choice.

No wonder then, that by June 2024, the European Commission had come to the preliminary conclusion that the iPhone maker had failed to comply with obligations to allow app developers to "steer" users to offers outside its App Store without imposing fees on them. The Commission was set to charge Apple over allegedly stifling competition on its mobile app store, the first case brought against a tech company under the Digital Markets Act. Apple could still take actions to correct its practices, which could then lead EU regulators to reassess any final decision. Meta was expected to be charged as well, in relation to its ad-free subscription for Facebook and Instagram in the EU. Regulators were also still investigating whether Google parent Alphabet was favoring its own app store, and Facebook owner Meta's use of personal data for advertising.[59]

In early August 2024, Apple changed its policy in the EU to allow developers to communicate with their customers outside its App Store. Apple said that developers would now be able to communicate and promote offers that were available anywhere, not just on their own website, from within their app. However, Apple introduced two new fees—an initial 5 percent acquisition fee for new users and a 10 percent store services fee for any sales made by app users on any platform within the 12 months of the app installation. The new fees would replace the reduced commission for all digital goods and services sold through the App Store. As indicated above, the European Commission had earlier criticized the fees charged by Apple for facilitating via the App Store the initial acquisition of a new customer by developers, stating they went beyond what was strictly necessary for such remuneration. Predictably, a Commission official said that they would assess Apple's eventual changes to the compliance measures, also taking into account any feedback from the market, notably developers.[60]

In the previous year, Google had already faced charges from EU regulators that it violated Europe's antitrust laws. In June 2023, the European Commission brought a case against the company, focusing on Google's dominance over the online advertising market, and argued that parts of the company should be split up.[61] This was in addition to the billions of fines that the EU regulators had levied against Google over antitrust violations in recent years, decisions still under appeal at the time of writing. Moreover, in February 2024, Google was hit with a €2.1 billion ($2.3 billion) civil lawsuit by 32 media groups, alleging that they had suffered as a consequence of the company's practices in digital advertising. Those groups included publishers in Austria, Belgium, Bulgaria, the Czech Republic, Denmark, Finland, Hungary, Luxembourg, the Netherlands, Norway, Poland, Spain and Sweden. The media companies involved claimed to have suffered losses due to a less competitive market as a direct result of Google's alleged misconduct. Without Google's abuse of its dominant position, the media companies would have received significantly higher revenues from advertising and paid lower fees for ad-tech services, their lawyers argued, "Crucially, these funds could have been reinvested into strengthening the European media landscape." They cited the French competition authority's €220 million fine against Google on its ad-tech business in 2021, as well as the European Commission's charges in 2023 to buttress their clients' claims. "If there is a follow through to the regulatory scrutiny, Google may need to curtail its practices and provide more consistent, predictable pricing to its advertising customers," said Gil Luria, an analyst at the U.S. firm D. Davidson & Co., an employee-owned broker-dealer firm.[62]

On December 18, 2023, social media platform X became the first subject of a formal EU investigation over potential breaches of the Digital Services Act, which focused on the failure to block illegal content, inadequate measures against disinformation, and the lack of transparency. The European Commission had made the decision to launch formal proceedings against the company weeks after X was asked to provide evidence of compliance with the DSA provisions designed to eliminate hate speech, racism and fake news.[63]

TikTok, too, came under increased scrutiny; in February 2024, the European Commission opened formal proceedings against the company over potential DSA infringements. The investigation was looking at areas including protection of minors, maintaining records of its advertising content and whether TikTok's algorithms led users down damaging content "rabbit holes." The investigation into child safety included age verification—the minimum age for using TikTok is 13 years—and the default settings used for children's accounts.[64]

Apple was then facing an expected fine of €500 million over its abuse of power regarding practices in the music streaming app market that allegedly violated the DSA. The European Commission had been investigating whether Apple blocked music streaming services such as the Swedish company Spotify from informing users about cheaper ways to subscribe outside Apple's App Store. The Commission found that the tech company disadvantaged users contractually by restricting app developers from openly promoting cheaper services. Music streaming developers were also not allowed to inform the users inside their own apps of lower prices for the same subscription on the internet. Nor were these developers allowed to change links to customers in their apps to their own websites that charged lower prices. Spotify argued that the restrictions benefited Apple's rival music streaming service, Apple Music. Spotify and other app providers had been longstanding critics of Apple's App Store, which, they argued, stifled competition by charging a 30 percent fee on apps and in-app purchases. Those extra costs made competition with Apple's own streaming service virtually impossible and also drove up prices for consumers.[65]

The European Commission imposed a fine of €1.8 billion (nearly four times higher than expected) on Apple in early March 2024, in a move to show it would act decisively against tech companies who abuse their dominant position in the market for smartphones and online services. Max von Thun, the Europe director of the Open Markets Institute, which researches the impact of corporate monopolies, said the large fine "sets a positive precedent which the EU would do well to draw on in future enforcement actions against tech giants."[66] This may be true, but the €1.8 billion (nearly $2 billion) fine levied against Apple amounted to merely 1 percent of its gross profit over the year 2023, and less than one per thousand of its U.S. stock exchange value at the time—indisputable proof of the tech company's enormous financial-economic clout.[67]

However, it was only a few weeks later that the U.S. Department of Justice launched its sprawling antitrust case against Apple mentioned earlier, which was expected to seriously impact the company's core business, and might ultimately lead to its break up. Moreover, Apple along with Alphabet and Meta then came under enhanced antitrust scrutiny in the EU as well.

10

Taking Up the Challenges of Regulating Artificial Intelligence

With the arrival of more advanced AI systems, whose deployment could potentially have serious harmful—if not catastrophic—effects, the need for regulation became ever more pressing in the eyes of concerned policymakers around the world. Following the DMA and DSA Acts, the European Union was the first to develop a proposal for a comprehensive regulatory framework on artificial intelligence, which was tabled in April 2021.[1] The draft Artificial Intelligence Act (AIA) focused on the specific utilization of AI systems and their associated risks. It laid down a classification for AI systems with different requirements and obligations crafted from a risk-based assessment; some AI systems presenting "unacceptable" risks would be prohibited outright. A wide range of "high-risk" systems would still be authorized but subject to a set of requirements and obligations to gain access to the EU market. AI systems presenting only limited risk would be subject to very light transparency obligations. The European Council agreed the EU Member States' general position in December 2021, while the EU Parliament voted on its position in June 2023. EU lawmakers then started negotiations to finalize the new legislation, which was a very arduous path. From the outset, there was a heavy lobby by U.S. tech companies such as Microsoft, Google and fast-rising newcomer OpenAI defending their interests in the European market.[2]

The EU Artificial Intelligence Act proposal was considered an important first step in regulating AI in Europe and, indirectly beyond the continent, given its extraterritorial reach. It must be emphasized, though, that some of the most problematic AI systems were excluded from the regulation, notably those used for military purposes, such as drones and other automated weapons. It was also possible that other applications, such as the fusion of AI with existing mass surveillance capabilities, could be permitted where authorized by law. This would leave the door open for their use in law enforcement. Obviously, such loopholes for AI-driven state surveillance were troublesome from the perspective of human rights and privacy protection.[3] The draft regulation also did not apply to public authorities in a third country, nor to international organizations, or authorities using AI systems in the framework of international agreements for law enforcement and judicial cooperation.

Consequently, much leeway remained for deployment of such AI systems to surveil citizens as part of military intelligence and national security data gathering, which could be highly problematic in light of the Snowden revelations (among others) regarding the malicious operations of what Yasha Levine has called the "military-digital complex."[4]

At the very last moment, in early November 2023, negotiations on the draft text abruptly stalled after EU member state representatives learned that France and Germany, supported by Italy, were opposed to a compromise on foundation AI models designed to produce a broad range of outputs. Those three countries no longer wished to regulate these AI systems, which are able to create texts or images and can be used by companies in applications such as the ChatGPT or Bard/Gemini chatbots. Earlier, in June, dozens of CEOs of big European corporations, such as Siemens and Heineken, had already sent an open letter to European politicians that expressed their "serious concerns" about the AI Act, fearing "loss of competitiveness and technological sovereignty."[5] France and Germany in particular pushed back hard because they realized the possible implications of the EU proposals for their up-and-coming domestic AI companies Mistral AI and Aleph Alpha, respectively.[6]

The objecting countries suggested "mandatory self-regulation through codes of conduct" for foundation models of AI. According to the joint paper agreed between the three governments, developers would have to define model cards to be used to provide information about a machine-learning model. These cards would include relevant information to understand the functioning of the model and how it learns, its capabilities and limitations, and would be based on best practices within the developers' community. An AI governance body could help to develop guidelines and check the application of model cards. Initially no sanctions should be imposed, but if violations of the code of conduct were identified after a certain period of time, a system of sanctions could be set up.[7] Critics argued that such an approach might leave the EU with unenforceable rules for the most powerful and potentially harmful AI systems, raising concerns about the overall efficacy of regulation. The stark differences in perspectives on this issue created a deadlock that, if unresolved, could thwart the entire negotiation process for the Artificial Intelligence Act.

In attempting to prevent a derailment of the AI legislative process, the EU Commission responded quickly and circulated a compromise proposal concerning foundation models on November 19, 2023. It distinguished between "general-purpose AI *models*" and "general-purpose AI *systems*." The first type was defined as "an AI model, when trained with a large amount of data using self-supervision at scale, is capable to [competently] perform a wide range of distinctive tasks regardless of the way the model is released on the market." The second type was defined as "an AI system based on an AI model that has the capability to serve a variety of purposes, both for direct use as well as for integration in other AI systems."[8] Evidently the term "foundation models" was not used but replaced by "general-purpose AI models." The document also stipulated that all general-purpose AI models had to maintain current technical documentation, as outlined in model cards. Likewise, the compromise draft provided for codes of practice regarding transparency measures for these models. In a working document of the EU Parliament, circulated on November 24,

2023, EU parliamentarians stated that they could generally get behind the idea of codes of practice, but only if their purpose was to complement the horizontal transparency requirements set for all foundation models and if the codes of practice were drafted by small and medium-sized enterprises, civil society and academia.⁹

AI Regulation in the U.S.

There was momentum building towards AI regulation in the United States, too. On October 30, 2023, President Biden signed an executive order that contained directives issued by the White House regarding the safe use of AI. The new order moved the U.S. toward more comprehensive AI governance and went beyond previous voluntary agreements with AI companies. It built on prior Biden administration actions, such as the list of voluntary commitments that multiple large tech companies agreed to in July of that year and the Blueprint for an AI Bill of Rights released one year before. Additionally, the policy followed two other previous AI-focused executive orders—one on the federal government's own AI use and another aimed at boosting federal hiring in the AI sphere. Unlike those previous actions, however, the newly signed order went beyond general principles and guidelines; a few key provisions did require specific action on the part of tech companies and federal agencies. Under the new order, tech companies would be required to share test results for their AI systems with the U.S. government before they could be released on the market. The government would also set stringent testing guidelines, whereby the administration would work with allies and partners abroad on a strong international framework to govern the development and use of AI.¹⁰

The AI directives issued by the White House included the following:

- Companies developing AI models that pose a threat to national security, economic security or public health or safety must share their test results with the government. It was a first, though limited, step toward mandated transparency from tech companies. This rule was expected to apply to the next version of OpenAI's GPT (Generative Pre-Trained Transformer), the large language model (LLM) that powers its chatbot ChatGPT, launched in December 2022;
- The National Institute of Standards and Technology (NIST) government agency will develop guidelines for so-called red-team testing, that is, when benevolent hackers work with the model's creators to preemptively ferret out vulnerabilities. Such tests will be applied across the board, with the Departments of Energy and Homeland Security addressing risks involved with AI and critical infrastructure, for example;
- Official guidance on watermarking AI-made content will be issued to address risk of harm from fraud and "deepfakes";
- New standards for biological synthesis screening—to identify potentially harmful gene sequences and compounds—will be developed to mitigate the threat of AI systems helping to create bioweapons.

Beyond these mandates, the executive order primarily created task forces and advisory committees, prompted reporting initiatives and directed federal agencies to issue guidelines on AI within the next year. The executive order covered eight areas that were outlined in a fact sheet: national security, individual privacy, equity and civil rights, consumer protections, labor issues, AI innovation and U.S. competitiveness, international cooperation on AI policy, and AI skill and expertise within the federal government. Within these umbrella categories were sections on assessing and promoting the ethical use of AI in education, health care and criminal justice.

The White House said it would also accelerate the development of AI standards with international partners—without explicitly mentioning the EU, though. On the same day the executive order was released, the G7 group of nations published a code of conduct for organizations developing advanced AI systems. The Biden administration had also been pressuring lawmakers for AI legislation, but a heavily polarized U.S. Congress made little headway in passing effective bills.

Civil liberties and digital rights groups (including the Center for Democracy and Technology, and the Electronic Privacy Information Center) largely lauded the executive order as a positive necessary first step. There was the question, however, whether federal agencies would accomplish the ambitious list of tasks (with set deadlines for each of them) on time. If one did not have the human capital and particularly the forms of technical expertise needed for all of this, it would be very hard to get these kinds of requirements implemented consistently and expeditiously, according to Daniel Ho, a professor of law and political science at Stanford University who studied AI governance.[11]

Some critics saw the executive order as lacking real enforcement teeth, since much of it seemed to be centered around recommendations and guidelines.[12] While the executive order went some way toward codifying how AI should go about building safety and security into their systems, it was unclear to what extent it was enforceable without further legislative changes.

According to some experts there were notable loopholes; the order said nothing specifically about protecting the privacy of biometric data, including facial scans and "voice clones" (audio deepfakes). More enforcement requirements around evaluating and mitigating AI bias and discriminatory algorithms were needed. There were also gaps when it came to addressing the government's use of AI in defense and intelligence applications, as well as persisting concerns about the use of AI both in military contexts and for surveillance. Lastly, the executive order on its own was insufficient for tackling all the problems posed by AI. Executive orders are inherently limited in their power and can be easily reversed. The order itself called on Congress to pass data privacy laws; specific legislation regarding the private sector was needed for multiple facets of AI regulation.[13]

Some organizations who focus on surveillance voiced their serious concerns about the executive order's flaws, most notably Albert Fox Cahn of the Surveillance Tech Oversight Project, who claimed that the approach taken in the order would still allow further AI abuses. For one, the White House order relied on AI auditing techniques that could "be easily gamed by companies and agencies," he said. "The worst forms of AI, like facial recognition, don't need regulation, they need a complete ban," he added. "Many forms of AI simply should not be allowed on the market. And many of these proposals are simply regulatory theater, allowing abusive AI to stay on the market," he concluded.[14]

The EU Artificial Intelligence Act Finalized

On December 8, 2023, the EU Council and the European Parliament's negotiators reached an agreement on the proposed AI Act. This was largely a compromise between earlier, stricter consumer protection-focused versions favored by the Parliament and Commission and more lenient, industry-friendly terms favored by the Council. France, Germany and Italy had sought late-stage changes aimed at watering down parts of the bill. These EU states (with the support of Hungary, Finland and Poland) had lobbied to split off the most powerful AI models from the Act and only impose "self regulation," arguing that stringent rules would thwart innovation in Europe.[15] This effort was strongly opposed by representatives of the European Parliament. The result was a compromise on the most controversial aspects of the new law—one aimed at regulating foundational AI models, and another that sought broad exemptions for European security forces to deploy artificial intelligence. This AI Act, the first binding law of its kind in the world, would finally pass the EU Parliament on March 13, 2024, and then required the formal endorsement of ministers from EU members states.[16]

The AI Act includes a two-tier system of guardrails for general-purpose AI (GPAI) systems, such as the foundation models that underpin the boom in generative AI applications like ChatGPT. These models are large systems that can competently perform a wide range of distinctive tasks, such as generating video, text, images, conversing in lateral language, computing, or generating computer code. The deal reached on foundation models includes some transparency requirements for what is referred to as "low-tier" AI. These involve drawing up technical documentation, complying with EU copyright law and disseminating detailed summaries about the content used for training. In a nod to the EU's environmental sustainability efforts, this also includes how much energy was used to train the models.

For "high-impact" GPAIs with so-called "systemic risk," there are stricter obligations. These are foundation models trained with vast troves of data along with advanced complexity, capabilities and performance well above the average, which can disseminate systemic risks along the value chain. If these models meet certain criteria they will have to conduct model evaluations, assess and mitigate systemic risks, conduct adversarial testing, report to the Commission on serious incidents, ensure cyber security and report on their energy efficiency.[17] Obviously, this implies a rejection of the total exemption from obligations for GPAIs and foundation models that Mistral AI and its lobbyists had been pushing for. Ultimately, however, the legislation does give broad exemptions to open-source models, which are developed using code that is freely available for developers to alter for their own products and tools. Because these systems are already more transparent, they are subject to less obligations, unless they are labeled as "high risk." The move is likely to benefit open-source AI companies in Europe that lobbied against the initial law proposal, including France's Mistral AI and Germany's Aleph Alpha, as well as Meta, which released the open-source model LLaMA (Large Language Model Meta AI). Yet the European tech lobby gave the new legislation a lukewarm response. The most important lobby club, DigitalEurope, released a statement about how difficult it had become to comply with the rules; they actually had to spend more on lawyers instead of hiring new AI developers.[18]

Then there are several practices prohibited in the EU AI Act. These include: (1) "cognitive behavioral manipulation"—a broad term for technologies that interpret behaviors and preferences with the intent of influencing people's decisions; (2) the "untargeted scraping of facial images from the internet of CCTV footage"; (3) emotion recognition in the workplace and educational institutions, which could be used by companies to discipline, rank or micromanage employees; (4) public and private social scoring systems used to classify or evaluate people based on social behavior or known or predicted personal characteristics. The prohibited practices further include: (5) "biometric categorization," a practice whereby characteristics such as skin tone or facial structure are used to make inferences about gender, sexual orientation, race, or even the likelihood of committing a crime; (6) "some cases of predictive policing for individuals," which have already been proven to have racially discriminatory impacts; (7) AI used to exploit the vulnerabilities of people due to their age, disability, social or economic situation.[19]

In addition, the agreement entails a series of safeguards and narrow exceptions for the use of remote biometric identification (RBI) systems in publicly accessible spaces for the purpose of law enforcement. These will be subject to prior judicial authorization and for strictly defined categories of crime. Retrospective (non-real-time) RBI systems can be used only in the targeted search for a person convicted or suspected of having committed a serious crime. Real-time RBI will have to comply with strict conditions, and its use will be limited in time and location, for the purposes of targeted searches for victims in cases of abduction, trafficking or sexual exploitation; prevention of a specific and present terrorist attack; or the localization and identification of a person suspected or having committed one of the specific crimes mentioned in the regulation (terrorism, trafficking, sexual exploitation, murder, kidnapping, rape, armed robbery, participation in a criminal organization, and environmental crime).[20]

This means in effect that police and national security bodies in the EU will be prohibited from using real-time biometric data driven by artificial intelligence in most circumstances if they do not have judicial authorization. The ban on this kind of surveillance will apply in public and private places, ranging from parks to sports grounds, except in the event of specified serious crimes, a terrorist threat or urgent searches for victims or perpetrators. Even then, police will require approval first from a judge or independent administrative authority. Only in the most exceptional circumstances (such as a live terrorist threat) will police be able to switch on AI biometric tools without a judge's permission. However, under the new rules, law enforcement must still obtain authorization within 24 hours and provide the appropriate authority with "a prior fundamental rights impact assessment." According to

EU officials, these safeguards are meant to avert predictive policing, which EU parliamentarians feared could be used alongside racial profiling to discriminate against individuals.[21] It should be acknowledged, however, that governments of EU member states can still set up systems for general facial recognition, such as cameras that can monitor people everywhere in public spaces, which may then be used when deemed appropriate and allowed judicially.

The agreement also clarifies that the new regulation does not apply to areas outside the scope of EU law. Most importantly, the agreement confirms that the AI Act will not apply to systems that are used exclusively for military or defense purposes. The agreement likewise provides that the regulation will not apply to AI systems used for the sole purpose of research and innovation, or for people using AI for non-professional reasons.[22]

European digital privacy and human rights groups had been pushing representatives of the EU Parliament to hold firm against the pressure by EU member states to carve out broad exemptions for their police and intelligence agencies, which had already begun testing AI-fueled technologies. Because of this overwhelming pressure, the absolute ban of real-time facial recognition that the EU Parliament wanted, did not materialize. Civil society groups reacted skeptically—raising concerns that the agreed limitations on state agencies' use of biometric identification technologies would not go far enough to safeguard human rights. Amnesty International expressed its disappointment, emphasizing that an "absolute ban" was "really necessary, because no guarantee could prevent the harm done to human rights by facial recognition."[23] This issue was even more pressing because the AI Act does not include an export ban on AI technologies that could carry out banned activities. Digital rights group EDRi, which was among those pushing for a full ban on remote biometrics, said that while the deal contains some "limited gains for human rights," it looks like "a shell of the AI law Europe really needs."[24] Rights groups were also concerned about the lack of protection from AI systems used in migration and border control. "Whatever the victories may have been in these final negotiations, the fact remains that huge flaws will remain in the final text," said Daniel Leufer, a senior policy analyst at the digital rights group Access Now.[25]

According to the deal agreed, penalties for non-compliance can garner fines that range from €7.5 million or 1.5 percent of a company's worldwide turnover (whichever is higher) for giving incorrect information to regulators, to €15 million or 3 percent of worldwide turnover for breaching certain provisions of the act, such as transparency obligations. A fine of €35 million or 7 percent of turnover can be levied for deploying or developing banned tools. Fines for errant smaller companies and startups will be more proportionate.[26]

As things now stand, the outcome of the negotiations about the European AI Act is much more favorable for producers' innovation interests when weighed against the protection of citizens. After all, only the very powerful AI models will be covered by the new rules, because they constitute a high risk for society. Oddly enough, "high risk" is defined in terms of size (numbers of end users) and the computing power that is necessary for training these models,[27] and not in terms of potential risks for fundamental rights, health and security. Under the new rules, smaller

AI models remain unaffected. This leads to a curious situation in which, under the EU AI Act, a handful of U.S. tech companies have to ensure that their systems comply with the European rules and values, while European models like Mistral and Aleph Alpha remain largely unregulated. This is not only because they are not as big as their globally operating rivals, but also because more flexible rules apply for the open-source models they deploy.

The Fast-Changing AI Landscape

OpenAI has played a prominent role in what has come to be known as the "AI spring"—the recent period of rapid progress in the field of artificial intelligence—starting with its release of the text generator ChatGPT. When OpenAI launched ChatGPT in December 2022, the company's major investor, Microsoft, greatly increased its financial stake, enhancing its commitment to $13 billion, acquiring a 49 percent stake in the company and the right to 75 percent of OpenAI's profits. Microsoft also ensured that it would be OpenAI's sole cloud provider, locking in millions of dollars of value given the computational costs involved in running generative AI products. While advertised as a partnership, the deal smacked of a "killer acquisition" that gave Microsoft unparalleled access to a tech unicorn[28] that was on track to attain a multibillion dollar valuation.[29]

Microsoft is one of just a handful of gatekeeper firms—further including Alphabet, Apple, Amazon and Meta—that have the necessary computing power, access to data and the technical expertise needed to develop advanced AI systems. Their control of the AI development sector gives these companies the ability to dictate terms and fees, and fend off challengers. An example is Microsoft's limiting the availability of OpenAI's API (Application Programming Interface)—a way for two or more computer programs or components to communicate with each other[30]—to other search engines, and threatening to cut off access to its internet-search data if those rivals used it to develop their own AI chat products. Microsoft also charges other cloud providers higher fees for purchasing and running its software outside of its cloud system Azure, making it both expensive and technically difficult to switch, since data is often not interoperable across systems. And Microsoft has already been able to integrate OpenAI's technology into its consumer-facing products, productivity tools and business services, despite safety concerns expressed by its employees and warnings that it was not yet ready for integration into Microsoft's Bing search engine.

This should be cause for deep concern for policymakers focused on AI safety, governance and innovation. Yet amid the flurry of activities in the U.S. and Europe to ensure the development of responsible and safe AI, the harms and risks of massive concentration in the generative AI ecosystem were largely ignored and sidelined. Big Tech justified the rapid and reckless rollout of generative AI by seeking to convince policymakers and the general public that prioritizing speed over safety was an inevitable part of technological development and crucial to innovation, especially if the U.S. wanted to compete with China. This narrative allowed these corporations to divert attention away from the dangers posed by concentration at key points in the AI value chain, as well as their failure in addressing the multiple harms brought about by their online platforms.[31]

Developments in the field of AI continue to advance at seemingly breakneck speed. In December 2023, Google unveiled a new AI model that it claimed outperformed ChatGPT in most tests, and displayed "advanced reasoning" across multiple formats, including an ability to view and mark a student's physics homework. The model called Gemini, developed by the London-based Google unit DeepMind, came in three versions and was "multimodal," which meant it could comprehend text, audio, image, video and computer code simultaneously.[32] Gemini, to be folded into Google products including its search engine, was initially released in more than 170 countries including the U.S. in the form of an upgrade to Google's chatbot Bard. However, the Gemini Pro-powered version of Bard was not yet released in the UK and the European Economic Area (which includes the EU and Switzerland), as Google first had to seek clearance from regulators.[33]

In all of these cases, too, data used to train the AI model had been taken from a wide range of sources, including the open web. The same applies to "image generators" such as OpenAI's Dall-E and Midjourney—AI systems that use deep learning methodologies to generate digital images from descriptions in natural language, that enthralled people with their simulacrum photos and graphic art.[34] In February 2024, OpenAI launched Sora (the Japanese name for "sky"), a tool that went much further than these still image generators. This new ("text-to-video diffusion") model had the capacity to instantly create realistic video footage from text prompts involving the user's subject and style instructions.[35] Basically, it could turn a brief text description into a detailed high-definition film clip up to a minute long. This development led to feverish speculation about its possible impact on film and TV productions in particular.[36] However, the excited commentary flooding the internet did not notice the serious flaws of OpenAI's new video generation tool; a closer examination revealed that it did not understand physical reality, according to John Naughton. This time OpenAI was uncharacteristically candid about the tool's limitations; it might, for instance, "struggle with accurately simulating the physics of a complex scene." This was evident in the odd, inexplicable way an object might move in the space it occupied, which indicated that Sora did not understand the physics of objects in motion. Sora could also be rather confused about cause and effect: "a person might take a bite out of a cookie, but afterward, the cookie may not have a bite mark"; it might also "confuse spatial details of a prompt, for example mixing up left and right," and so on.[37]

Obviously, there is still a very long way to go before such a technology will reach (if ever) the level needed to create truly photorealistic depictions of human beings along with moving objects operating in the real world. Experts have questioned whether the large-language-model approach shared by all frontier AI systems might be hitting its limits here. For example, Meta's chief scientist, Yann LeCun, responding to a claim from Elon Musk that artificial general intelligence was going to arrive within the next year, said that it was just not happening. "We have AI systems that can pass the bar exam, but they can't clear your dinner table and fill up the dishwasher. We have systems that manipulate language, and fool us into thinking that they are smart, but cannot understand the world."[38] Instead, he suggested, researchers needed to work on what he called "objective-driven" AI with the ability to reason and plan about the world, rather than just work on words alone.

This was in April 2024, when the AI race heated up once again as OpenAI, Google, Mistral and Meta all released new versions of their frontier AI models. This coincided with an explosion in corporate demand for AI chips, in which these and numerous other tech companies were partnering with AI chipmaker Nvidia, which was seen as the leading provider of chips best suited to powering AI. Its fortunes were interpreted as a bellwether for the AI transformation under way; the company experienced gigantic growth as the AI boom showed no signs of slowdown.[39] The apparently unstoppable rise of Nvidia, seen by John Naughton as part of an AI investment bubble going through the five familiar stages of such bubbles—displacement, boom, euphoria (the stage the AI world's inhabitants were now in), profit-taking and panic—was ultimately destined to go only in one direction, from boom to bust.[40]

In May 2024, OpenAI announced the release of its new AI system, GPT4o (the o stands for "omni"), a virtual assistant that can carry out real-time voice conversations and has improved the quality and speed of ChatGPT's international language capabilities, along with an ability to upload images, audio and text documents for the model to analyze. Google then gave a preliminary view of a new AI assistant called Project Astra, as well as updates to its Gemini language model with far-reaching implications; it created the capability for AI-generated snapshots or summaries, called "AI Overviews," that aim to provide a concise answer to a user's search query. They appear at the top of Google search results when the system determines that generative AI can provide helpful information for complex queries that may require multiple searches.[41] A major limitation of AI Overviews is the potential for generating inaccurate, nonsensical or even harmful information. This is because the underlying large language models are prone to "hallucinations," in which they generate made-up information that is not based on facts, especially when there is limited high-quality data available for a given query, as critics have pointed out. Such "hallucinations" are due to the fact that generative AI answers questions by making statistically informed guesses. While these guesses are often correct, sometimes they are wrong. The result can be artificially generated information that bears little relationship to reality, such as explanations or images that seem superficially plausible, but do not actually provide the correct answer to the question.[42]

The big secret of generative AI is that it is only as good as its training data, and this data is frequently biased. Most AI models are trained on data that is scraped from the internet, which has large gaps and is unevenly distributed with regard to language, race and gender. Whatever efforts have been made to counteract this tendency, these are often still producing (at least partially) unreliable results. Bias and ahistorical ignorance is a problem at the best of times; when it is built into a technology that many people think of as impartial and reliable because it is computerized, it becomes an even bigger issue.[43]

AI systems like the one powering AI Overviews also struggle to fully comprehend the context and nuances of language, which can lead to misinterpretations and factual errors even when the source material is factually accurate. Moreover, AI Overviews can inadvertently incorporate information from satirical, humorous or user-generated content that lack credibility, leading to inaccurate or nonsensical responses.[44]

These issues became painfully evident when at the new feature's launch, it returned some weird answers ranging from bizarre to dangerous. Users were told, among other things, that glue is useful for ensuring that cheese sticks to pizza; that one could stare at the sun for up to 30 minutes without harm; and that geologists suggest eating one rock per day (presumably to combat iron deficiency). This led once again to Google facing public embarrassment after the release of a new AI product. Initially, Google said it was removing at least some of the problematic results manually (a laborious process for a site that fields billions of queries per day). Later it announced that it would reduce the scope of the type of searches that generate AI Overview results as well as limit the inclusion of satirical or humorous content.[45]

Google framed the problems with AI Overviews as mostly a series of marginal cases, but several AI experts have commented that these problems are indicative of wider issues surrounding AI's ability to assess factual accuracy. They argued that at least some of the problems with the new generative-AI-powered search answers may not be fixable anytime soon.[46] The fundamental limitations of large language AI models in understanding context, avoiding hallucinations, and reliably synthesizing information from credible sources, remain significant challenges.

Copyright Issues and Journalistic Implications of Generative AI Built into Search Engines

In all of these developments of advanced AI tools, masses of uncredited, unpaid-for human work was being harvested from the internet and repurposed by generative AIs. Publishing and creative industries continued to protest against AI companies' use of copyrighted content available online to build models. They were also beginning to mount a counter-offensive. News outlets in particular were concerned that the AI tools would spread misinformation attributed to them and use their content with no incentive to click through to check the original source. At the end of 2023, *The New York Times* was the first major publisher to sue OpenAI and Microsoft for copyright infringement over the use of its content to train generative AI and large language models, a move that could placate the newspaper publisher with billions of dollars in compensatory damages. The suit also called on the two tech companies to destroy any chatbot models and training data that used copyrighted material from the *Times*.

It was the latest in a string of similar cases, including a class-action lawsuit brought by more than a dozen fiction writers (organized by the Authors Guild) and a lawsuit by the Getty photo archive, Getty Images, against London-based Stability AI (a leading purveyor of AI-generated images) over alleged copyright breaches, both filed in September 2023.[47] In addition, a group of visual artists organized an open letter that called for restrictions on the "vampirical" practice of image generators. There were more open letters including one that called for a six-month pause on the development of any new AIs.[48]

In February 2024, OpenAI and Microsoft faced a new round of lawsuits from news publishers over allegations that their generative AI products violated copyright laws and trained illegally using journalistic work. Three progressive U.S. outlets—the Intercept, Raw Story and AlterNet—filed suits in Manhattan federal court demanding compensation from the tech companies. The outlets claimed that the companies in effect plagiarized copyright-protected articles to develop and operate ChatGPT, which had become OpenAI's most prominent generative AI tool. They alleged that ChatGPT was trained to ignore copyright and proper attribution, while failing to notify users when the service's answers are generated using journalists' protected work. At about the same time lawyers for OpenAI filed a motion to dismiss parts of the ongoing *New York Times* lawsuit, arguing that its services were not in meaningful competition with the newspaper.[49] Many news media are blocking outfits like OpenAI from harvesting data from their sites. Google and Microsoft are harder to stop; at this point, it is unclear what can be done. Copyright law may be inadequate in tackling the ongoing piracy, but publishers are keeping their options open concerning legal action.[50]

AI companies' defense for using copyrighted material tends to lean on the legal doctrine of "fair use," which allows use of content in certain circumstances without seeking the owner's permission—an argument used by OpenAI in its initial response to *The New York Times* lawsuit. Likewise, in a submission to the British House of Lords communications and digital select committee, OpenAI stated that it was impossible to train large language models such as its GPT-4 model—the technology behind ChatGPT—without access to copyrighted work, while insisting that "legally copyrighted law does not forbid training."[51]

Meanwhile, according to a recent stocktaking in *TechCrunch*, OpenAI has spent millions of dollars licensing content from news publishers, stock media libraries and more to train its generative AI models—a budget far beyond that of most academic research groups, nonprofits and startups. This includes contracts with large news corporations such as AxelSpringer, NewsCorp, the *Financial Times*, *Le Monde*, and Associated Press. Meta went so far as to consider acquiring the publisher Simon & Schuster for the rights to e-book excerpts, but ultimately this publisher sold to private equity firm KRR for $1.23 billion in 2023. Stock media Shutterstock has concluded deals with generative AI developers ranging from $25 million to $50 million, while Reddit allegedly made hundreds of millions of dollars from licensing data to companies such as Google and OpenAI. It appears that only a few platforms with abundant data accumulated organically over the years—such as Photobucket, Tumbler, and Q&A site Stack Overflow—have not signed agreements with generative AI developers. It is obvious that smaller players cannot afford these data licenses, and therefore will not be able to develop or study AI models. There are a few independent, not-for-profit efforts to create massive datasets anyone can use to train a generative AI model.[52] It begs the question, however, whether or not any of these open efforts can keep up with Big Tech. As long as data collection and curation remains a matter of resources, the answer is likely to be negative—at least not until some research breakthrough levels the playing field.

There are other ramifications for news outlets and their readers regarding the coming integration of generative AI into search engines that should be mentioned here. This could entail the most catastrophic threat to professional journalism so far, in an industry that has barely recovered from the flight of classified advertising to online platforms in the 1990s and the exodus of most of the rest of advertising revenue to Google and Facebook since 2004. Right now, if one goes to Google or Bing, for example, and searches for information on a specific news item or other subject, the standard procedure is that one receives a list of search results, some of which could be links to mainstream media articles. If one clicks through, one's visit will be counted and that traffic will attract advertisers and therefore generate revenue that pays for the journalists and their work.

Google and Microsoft's Bing offer the option of searching using AI, but it is not the default. If this changes, which may happen soon (Google is already rolling out its new system of AI Overviews with an expected 1 billion users by the end of 2024) one's search will yield not a list of sites to click through to, but rather an article about the search news item or some other subject written by a robot, using the content of news media companies and other sources. This will still involve prioritizing listings of search results to the benefit of the search engine owner's corporate interests. There may be links to the sources, but they will be much less prominent; they may not even appear at all. This will make it much harder, if not impossible, for individual readers to distill their own syntheses or summaries regarding the news (or other subjects they are interested in) from online sources of their own choosing as in the past.

Until now, publishers have been able to rely on significant volumes of traffic coming from the blue (URL) links that appear under many queries. But Google's AI Overviews often obscure these links, requiring users to click to see them, or simply abstracting them away in an automatically generated summary. The new system still relies on web-based information, but it does not reward the creators of that information with users' visits.[53] Of course, there is also a problem for Google here. It supplies advertising to many of the web pages that will lose all that traffic, as visits to those pages disappear. But because the company maintains a stranglehold over much of the digital advertising market, it seems to be betting that it can ride out the transition and overcome any hurdles by drawing on its many other sources of revenue. Google also plans to test the inclusion of ads labeled as "sponsored" within AI Overviews.[54]

Outlets that are free are particularly vulnerable. If a large part of the traffic from search engines is lost—and advertising revenue becomes minimal—making journalism sustainable becomes even tougher than it already is.[55] What is left for publishers is largely direct visits to their own home pages along with Google referrals. If AI Overviews take away a significant portion of the latter, it could mean less original news reporting, fewer creators publishing cooking blogs, for example, or other thematic blogs, fewer product reviews or how-to guides, and a less diverse range of informational sources.[56] It is hard to see why people would bother to contribute their expertise if their posts are not visited by seekers of information and instead just become fodder for AI to regurgitate. By making it even less inviting for humans to contribute to the web's collective pool of knowledge, Google's summaries could leave its own and everyone else's AI tools with less accurate, less timely, and less interesting information.

Antitrust Investigations of Big Tech's Economic Grip on AI Development

The huge economic influence of major tech companies on the development of AI attracted scrutiny from U.S. regulators and those outside the tech industry, too. In January 2024, the Federal Trade Commission (FTC) launched an inquiry into recent investments and corporate partnerships—and their competitive impacts—involving generative AI developers and major cloud service providers. The antitrust-driven inquiry focused on three multi-billion-dollar investments and associated agreements: Microsoft's and OpenAI ($13 billion); Amazon and Anthropic ($4 billion); and Google and Anthropic ($2 billion).[57]

In June 2024, the U.S. Justice Department and the Federal Trade Commission reached an agreement on antitrust investigations into what both agencies now saw as the main protagonists in the AI market: Microsoft, OpenAI and Nvidia. The Justice Department will focus on investigating whether or not Nvidia, the leading maker of chips that train and operate AI systems, had broken antitrust laws. Meanwhile, the FTC will scrutinize OpenAI and Microsoft, OpenAI's biggest investor and a major financial backer of other AI companies. The FTC is also investigating whether Microsoft structured a recent deal with startup Inflection AI to avoid an antitrust inquiry.[58] The head of the Justice Department's antitrust division, Jonathan Kanter, stated that the agency will look at the AI sector "with urgency" for potential antitrust violations and monopolistic practices. He emphasized the need for swift action to prevent powerful tech companies from dominating and controlling the rapidly growing AI market. Kanter expressed concerns over the significant influence wielded by tech giants like Microsoft, OpenAI, Google, Amazon, and Nvidia in the AI industry. He vowed that the DOJ would examine "monopoly choke points and the competitive landscape" in AI technology, including areas such as data usage, cloud computing, computer chips, and hiring practices.[59]

The latest interventions by the FTC in tandem with the U.S. Department of Justice appear to include looking at corporate strategies to steer and increase control over AI development and the AI ecosystem as a whole, which had remained mostly outside of regulators' scope up until now. This deserves further explanation.

Traditionally, antitrust authorities have concentrated on investigating potential mergers and acquisitions among companies in the same market. This is likely to be expanded in the AI era because apparently unrelated acquisitions can be motivated by hopes of accessing technology or talent or as a means to further expand and enter new business lines. A prime example is Microsoft's acquisition of Nuance Communications for $19.7 billion completed in 2022. Nuance offered a cloud-based solution for medical transcriptions, a business where Microsoft had no comparable product. Today, Nuance products are offered as part of Microsoft's Software as a

Service (SaaS) on its Azure cloud, including a very popular SaaS that offers a voice-to-recognition product specialized in radiology.[60]

As Cecilia Rikap has pointed out, another reason why big tech firms seek out smaller firms is to use corporate venture capital investments to steer AI development and get privileged access to what startups are working on. The relationship between Microsoft and OpenAI is another case in point, along with Microsoft's investment in the French startup Mistral. Such operations frequently escape regulations because, more often than not, startups receiving big tech funding are still in the early research and development stage. So they are not selling in any market yet or offering solutions that are complementary instead of substitutes to those offered by Big Tech.[61] For instance, Microsoft does not offer its own large language models (LLMs) on its cloud Azure but other companies such as Meta and OpenAI do. It also sells processing power and other computing services complementary to or based on OpenAI technology. LLMs are trained, prompts are accessed and new applications based on existing LLMs are offered as cloud services. Consequently, the cloud giants—Amazon, Microsoft and Google—are eager to invest in AI startups working on new AI models or applications of these models. This also ensures that these emerging companies build their architecture and run fully on the existing clouds controlled by the giants, which benefit financially from the arrangement. Every time an LLM runs on these platforms, the cloud provider earns not only a fee from those LLMs offered as services but also additional charges from processing power. And when the Big Tech cloud firm also owns a stake in the company providing the LLM, as in the case of Microsoft with OpenAI, it receives a share of its profits, too.[62]

Importantly, Microsoft also uses the fact that it is investing in competitors to OpenAI to avoid antitrust scrutiny of potentially collusive behavior. Its diversification also highlights its capacity to control startup companies broadly and create an ecosystem around its cloud. By fostering competition among complementors,[63] Microsoft pushes them to further focus on efficiencies related to specialization instead of attempting to diversify and become potential rivals. Such placement of organizations within the AI stack reinforces an innovation pattern characterized by a small, stable core of lead firms and a turbulent periphery of many smaller players. This strategy facilitates the systemic value capture from innovation produced in the peripheries by those at the core, thus cementing long-term intellectual monopolies that disproportionately capture rents from innovations that, at best, the giants have only partially developed.[64]

A related strategy is the control beyond ownership exercised by Amazon, Microsoft and Google through their clouds over any organization, Big Tech-funded or not. The public cloud is a global sourcing architecture where organizations pay for accessing computing services. Within the tech sector, those paying often use cloud services connected to their own software to create new services that will be offered on those same clouds. For instance, startups developing applications based on AI models rent AI services from Big Tech clouds that assist the development of those AI field-specific applications.

Moreover, rather like Facebook and other firms that achieved dominant positions by offering their services for free to users, cloud giants offer many services for free to attract more clients. Cloud credits represent extremely low additional costs for Big Tech. Most computing services rely on similar lines of proprietary code that can be resold many times over. In terms of processing power or storage capacity, having small projects consume only a tiny portion of Big Tech's gigantic infrastructure has very low opportunity costs.

Such credits can be seen as venture capital money that is targeted to specific uses, directing new firms to work on building AI applications on top of the models controlled by Big Tech and their satellite companies. These computer vouchers encourage migration to the cloud. For a company the size of Amazon, Microsoft, or Google, expanding their marketplace by providing credits to startups that not only use their cloud but will probably end up offering their services on their clouds, is both a source of direct revenues and long-term consolidation, since the larger the marketplace, the more likely that other companies will also choose that cloud.[65]

An additional aspect is particularly sensitive in the AI space. In pre-competitive markets, where what matters is dominating the technology and pushing the frontier forward ahead of others (even before defining a concrete business mode), capturing talent can mean virtually dismantling companies. The same three cloud giants are aggressively headhunting talent from startups. This practice not only avoids the scrutiny of regulators but is also a cheaper bet. These companies acquire only the assets deemed valuable for reinforcing their intellectual monopoly. This tactic came to light when, in March 2024, Microsoft recruited the three founders and many of the employees of the AI startup Inflection AI. This was a company that had previously received Microsoft's venture capital. (Two of the founders of Inflection AI had also been the co-founder of Google's unit DeepMind and one of its chief scientists. They had left Google, founded Inflection AI, and then went to Microsoft.) Tracking the paths of key talent in AI tells a much more complex story about the ways in which Big Tech companies control the field rather than merely focusing on company relations.

A final strategy that is becoming even more prevalent as it develops, are collusive agreements among leading AI corporations. It is true that these are often kept secret, but leading companies are more openly disclosing the signature of "strategic partnerships" that contribute to concentration and the perpetuation of those at the top. In September 2023, for example, it was announced that Oracle will operate and manage its Oracle Cloud Infrastructure services directly within Microsoft's data centers. Likewise, the Germany-based global software company SAP has its own public cloud, but also offers its Enterprise Resource Planning (ERP) system on every major cloud provider, including Microsoft Azure, while at the same time Microsoft sells its own ERP solution—Dynamics 365—that directly competes with SAP's version.[66] As of now, it is unclear to what extent the U.S. regulators will seriously investigate these practices of Big Tech companies.

Moving Towards Global AI Regulation?

In 2023, the capabilities of artificial intelligence came to the attention of a global audience far beyond tech circles, thanks largely to ChatGPT and similar AI products. Growing fascination with rapid progress in the field was coupled with alarm at some of the more apocalyptic scenarios that critics thought were possible if the technology was not adequately regulated.[67] The suggestion was made that the EU AI Act could become a global blueprint for classifying risk, enforcing transparency and penalizing tech companies financially for noncompliance. Strong, comprehensive regulation from the EU could "set a powerful example for many governments considering regulation," said Anu Bradford, an expert on the EU and digital regulation, and author of *Digital Empires: The Global Battle to Regulate Technology* (2023). Other countries "may not copy every provision but will likely emulate many aspects of it. AI companies who will have to obey the EU's rules will also likely extend some of those obligations to markets outside the continent," Bradford told the Associated Press. "After all, it is not efficient to re-train separate models for different markets."[68]

Yet it is still not clear whether this global emulation of EU AI regulation will occur, if at all. It is a given that in the worldwide race to develop AI, the EU is lagging behind the United States and China. The largest AI models that would face the most stringent rules might therefore be developed exclusively outside Europe, which could fuel the suspicion of protectionism on the part of the EU. At the same time, voices in Brussels have resounded that the extra administrative costs for these large-scale models are dwarfed by the millions the Big Tech companies were already investing.[69] Meanwhile at the time of writing, the U.S. Congress, after months of hearings and forums focused on the technology, was in the early stages of creating bipartisan legislation addressing artificial intelligence. As the EU AI deal came to be finalized, U.S. Senators indicated that Washington was taking a far lighter approach that focuses on incentivizing developers to build AI in the United States, with lawmakers raising concerns that the EU law could be too draconian.[70]

However, the fact remains that AI is global in nature and can only be kept in check by many nations working together. If one part of the world regulates powerful general-purpose AI (GPAI) systems, such as the one underlying ChatGPT, but another part releases unsecured, "open-source" versions of these tools that rogue actors can weaponize at will, the whole world can still suffer the consequences. Further, the full force of the EU's AI Act will not be felt until 2027. Some parts of it will phase in sooner—and it is designed to be "future proof"—but the regulation may be largely obsolescent by the time it is fully in force due to the rapid evolution of AI, particularly given the rise of generative AI models. This is an even bigger risk if the EU remains alone on legislating AI.

David Evan Harris, a public scholar at UC Berkeley and expert in responsible artificial intelligence and global political development, has listed the most important binding regulations that he thinks global AI safety summits and parallel governance processes need to put in place, thereby building on the EU AI regulation:

- Regarding a remaining gap in the EU AI Act: Affirm the prohibition of uses of AI tools such as "undressing" apps to create non-consensual intimate images (NCII), with liability falling also on the developer of the AI system that created it, and not only on the individual user creating this content. Developers should be prohibited from distributing tools capable of causing

such potentially irreparable harm, especially when children could be both perpetrators and victims;
- Firmly regulate high-risk AI systems, including General Purpose AI systems, requiring thorough risk assessments, testing and mitigations;
- Require companies to secure their high-risk GPAI systems and not release them under "open-source" licenses, unless they are determined to be safe by independent experts;
- Clearly place liability on the developers of GPAI systems as well as their deployers for harms that they cause;
- Require that AI-generated content be "watermarked" in a way that it can be easily detected by lay consumers as well as experts;
- Respect the copyright of creators such as authors and artists, when training AI systems;
- Finally, tax AI companies and use the revenue to protect society from any harms caused by AI, from misinformation to job losses.

The future will tell whether the various stakeholders in regulating AI will follow guidelines such as these to create a safer and more equitable AI landscape globally. Ensuring that AI is developed in ways that serve the public interest is a monumental task that will require participation from citizens and governments around the world. According to Harris, it means that everyone, everywhere, needs to be informed about the risks and benefits of AI, and demand that their elected representatives (assuming they exist locally) take AI's threats seriously. After the start the EU has made, it is imperative that the rest of the world enacts binding legislation that makes AI serve local citizens and their communities.[71]

However, in doing so the massive environmental impact of AI technology should not be ignored. Since AI requires staggering amounts of computing power, and the necessary GPUs (graphics processing units) get very hot (and therefore need cooling), the technology consumes electricity at a colossal rate.[72] One should also not forget the material infrastructures that hold the various digital clouds: vast data centers that use enormous quantities of water and electricity, as well as rare-earth metals.[73] All of this, in turn, results in CO_2 emissions on a gigantic scale.[74]

These environmental costs of AI deployment could even be a critical factor in deciding not to make use of AI where possible, or at least slow down its development. Moreover, the sustainability of AI comes into question because of the insatiable demand not only for energy and natural resources but also for human resources. After all, the generation of much of current AI's output involves the unacknowledged labor (microwork) of an army of exploited people around the world.

11

Proposals to Deprivatize the Internet and Foster Platform Socialism

In resuming the issue of internet policies, it should be reiterated that there are two major strategies for internet reform in the United States as well as in the European Union. The first involves the formulation of new rules about how companies are allowed to behave or merely the enforcement of existing rules. The second one entails the use of antitrust measures; that is, attempts to reduce the market power of big corporations. The latter is the focus of the New Brandeisians in the U.S. and their counterparts in the EU. Both strategies for internet reform have their merits. The rulemakers are correct that the online malls are too lightly regulated. Services mediated by the internet should not be exempt from labor law, civil rights law and any other laws that tech companies have successfully managed to circumvent over the years. The New Brandeisians and other antitrust proponents are correct that rulemaking alone is insufficient without also simultaneously transforming how the internet is owned. Further, they recognize that the current configuration of ownership is not the natural order of things, but rather a contingent state of affairs, constructed through public policy. Political economy is made by humans and can therefore be remade. Yet the political economy preferred by the New Brandeisians is not a radical departure from the present situation. They are still in favor of an internet ruled by markets, albeit one where markets are competitive rather than concentrated. The pursuit of profit would remain the organizing principle, but they envision that profit will be pursued by smaller and more entrepreneurial firms.

More importantly, to what extent would such a restructuring actually address the problems they are supposed to solve? Nick Srnicek, an expert on the digital economy and international politics, stipulates that more competition could very well make things worse: "After all, it's competition—not size—that demands more data, more attention, more engagement and more profits at all costs."[1] In the case of social media, more competition could lead to still more ferocious battles for people's attention and data. Competitive pressure compels companies to seek every possible advantage. The industry's most destructive effects (such as the obsession with user engagement) were first developed by social media firms when they were comparatively leaner and hungrier for data and needed to win market share as quickly as possible. It was only when such companies had developed into a monopoly that they could afford to take a somewhat longer view. For example, by the early 2020s, Facebook spent hundreds of millions of dollars each year on content moderation. It did so completely out of self-interest—to appease advertisers and to prevent public relations debacles—and its efforts clearly fell short. But it seems highly unlikely that Facebook would have spent money on such measures if it were locked in heavy competition with several other firms.[2]

There are also questions about how a company like Google or Facebook could be broken up in any meaningful way. It does not make sense for every country or region to have their own search engine and walled social network. The spread of these services is precisely what makes them socially useful.[3] As an alternative to pro-competition reforms, there have been proposals for regulations of various kinds, modeled on those around airline safety and prescription drugs, for instance. A New Brandeisian might respond that corporate giants are adept at co-opting the regulatory process. Regulation can also reinforce existing concentrations of private power by introducing compliance costs that only large corporations can absorb. This may very well explain why Big Tech firms have repeatedly called for more regulation in recent years—provided they get to decide how they are regulated.[4]

It has become clear that in both the U.S. and the EU, the power and dominance of Big Tech is framed as a problem of unfair competition rather than as an extractive and exploitative business model that is in need of reform. Behind the tough-sounding slogan of "breaking up Big Tech" lies the much weaker idea of "restoring competition in the tech sector" by preventing Amazon and others from selling their own products on their transaction platforms. The limitation of these competition-focused policies is that they do not challenge the fundamental surveillance and commodification model of the tech sector. There is little that would stop new, smaller tech companies from using the same strategies and business models as the current dominant players.[5]

Deprivatization as Another Strategy

Rulemaking and anti-monopoly measures—or some combination of the two—are not the only available options to constrain Big Tech and digital capitalism. There is a third option, that of deprivatization, as Ben Tarnoff suggests. James Muldoon likewise states "We need to shift our focus from 'privacy, data and size' to 'power, ownership and control.'"[6] While the first set of issues are important, they are secondary to a deeper set of concerns about who owns the platforms, who has control and who benefits from the status quo. Technology can either be controlled by private companies and used to generate profit for the few, or it can be directed by communities to benefit the many, which should be the ultimate aim.

Making markets more competitive or more regulated does not deal with the deeper problem, which is the market itself. The online platforms are engineered for profit-making and this is what makes them producers of inequality or so-called "inequality machines."[7] The exploitation of gig and ghost workers; the reinforcement of racism, sexism, and other kinds of oppression; the amplification of right-wing propaganda[8]—these diverse forms of social harm exist primarily because the underlying digital practices of online platforms are profitable. Indeed, these harms can be mitigated somewhat, and larger firms can do this more easily than smaller ones. But here, too, the market sets limits. Facebook, for example, can only spend so much on content moderation before its shareholders revolt. More importantly, its intense preoccupation with engagement, and the symbiotic relationship with the Right this has brought about, is the very foundation of the business model which creates the very problems that content moderation is supposed to address.[9] This situation is expected to worsen even more with the arrival of generative AI models, whereby citizens and voters are likely to face AI-generated content with empty and misleading information that bears no relation to reality.[10] The scale of the challenge is such that even if a company like Meta employed half a million human moderators, it would not be up to the task. And still, even then, section 230 of the 1996 Communications Decency Act would exempt them from legal liability.[11]

Content moderation is a tricky issue anyway. It is unclear how such interventions relate to the right of free speech protected under the First Amendment to the U.S. Constitution and Article 10 in the European Convention on Human Rights (ECHR), which protects the right to freedom of expression, as well as similar legislation in other democratic nations. The question remains regarding the extent to which content moderation unduly infringes on those rights in specific cases. There is also the problem that factually correct information that is controversial in nature, may not please particular moderators, who then label it as misinformation and consequently discard it. This possibility must always be kept in mind when considering (changes in) content moderation practices.

It should be noted here that—as the 2024 elections in the U.S. approached (and those in many other countries as well)—Facebook and other top social-media firms deprioritized content moderation and other user trust and safety protections. This included rolling back platform policies that had previously kept hate, harassment and lies in check on their networks. The declining commitment to content moderation led to a spike in misinformation, as well as hateful and violent content.[12] These social media companies had also laid off critical staff and teams tasked with maintaining platform integrity, according to a study from the non-profit media watchdog Free Press released in December 2023. The report documented 17 major platform policies affecting online integrity that had been eliminated at Alphabet, Meta, and Twitter/X. The report also mentioned that over 40,000 layoffs at these companies posed a threat to the health and safety of their platforms. The study cited a series of remarkable reversals of policies meant to safeguard elections, many of which were reinstated after the insurrection at the U.S. Capitol on January 6, 2021 (following the presidential election in 2020). Taken together with the preferential treatment of VIP users (holding them to lower enforcement standards than other users)—as reflected in the reinstatement of Donald Trump's accounts on Meta, Twitter (X) and YouTube—these developments represented a dangerous relapse. In turn, this had created a toxic online environment that was susceptible to exploitation by anti-democracy forces, white supremacists and other bad actors. According to the report's author,

Nora Benavidez, Big Tech executives such as Mark Zuckerberg and Elon Musk made reckless decisions to maximize their profits and minimize their accountability.[13]

However, even in the extremely unlikely event that online platforms would stop generating the negative effects mentioned above, corporations would still own the internet and provide internet services. Highly consequential decisions would be left in the hands of elites—executives and investors—and these decisions would in turn be bound by the mandates of the market. Most people would have no voice in the matters that critically affect their lives. A privatized internet will always boil down to the rule of the many by the few, and the rule of these few by an imperative that is hard-wired into capitalism: the imperative to accumulate.[14]

As some experts and politicians in the European Union have suggested, in order to address the necessary conditions for a really secure digital landscape, technological affordances should be returned to users through *public* digital services, as outlined in The Public Service Media and Public Service Internet Manifesto, 2021.[15] Commercial parties would be allowed to register for participation, but data would remain users' property. This would mean that the public domain could not be a money-making business enterprise, as it belongs to everybody. In contrast, private tech companies "rob" or "deprive" the public domain ("accumulation by dispossession" in Marxist terms), thus impoverishing and then transforming this domain into a mere transactional space. From this perspective, the obvious path forward is to attack the unequal power distribution and build a more equitable and safer public digital ecosystem, both in Europe and worldwide.[16]

Reversing privatization opens the door to a different kind of internet. In several places community networks are already challenging the legacy of privatization of the internet infrastructure and service provision. Even in the U.S., where the internet service providers lobby holds a powerful influence over the Federal Communications Commission (FCC) and state legislatures, a number of communities have built their own internet networks to counter big telecom's monopoly. By 2021, there were approximately 400 municipal broadband networks serving some 600 communities across the U.S., according to the Institute for Local Self-Reliance (ILSR), a nonprofit that advocates for local economies. Three years later, at least 47 new municipal networks had been added to the ILSR's database (with dozens of other projects still in the planning or pre-construction phase).[17] These networks emerged despite the fact that dark money campaigns, often funded by the big monopoly incumbents, had been popping up across the country in efforts to persuade local officials and residents to reject municipal broadband proposals. This led the American Association for Public Broadband (AAPB) to issue alerts about the misinformation at the center of these campaigns.[18]

The ILSR tally does not include the plethora of other community broadband networks such as member-owned electric cooperatives deploying fiber networks in many hundreds of rural communities across the nation. It also does not include the rising number of Tribal Nations building and operating their own networks to bridge the digital divide in some of the least connected parts of the country.[19] The Internet for All program, part of the Biden Administration's Infrastructure Investment and Jobs Act (2023), amounts to the most ambitious federal effort to date to subsidize broadband. A special $3 billion Tribal Broadband Connectivity Program (TBCP) has been earmarked for expanding high-speed internet access among indigenous nations. Even though much of this federal grant money will likely end up in the pockets of legacy internet providers, some of it is finding its way into the budgets of community-led projects such as the recent Acorn project at Hoopa Valley Reservation, California.[20] Tribes tend to have a rather large uphill battle in winning such funds, but now that money is flowing directly to the tribes, they are expected to have a much better opportunity to build broadband than most small communities.[21]

All of those community-owned internet providers offer fast, affordable internet, compared to the slow and costly services from the big companies in the United States. One may suggest that a similar approach could be applied to the higher layer of the internet that the online malls occupy. But, as Tarnoff points out, while the community network is the main actor at the base in making a democratic network, the situation higher up the stack is markedly different. The distinctiveness of the different online malls (and their specific entanglements with lobbying, data gathering, labor exploitation and financialization) requires a greater range of approaches, which cannot be spelled out in quite as much detail. They still need to be discovered and developed. More experiments are needed; their goal cannot and should not be a one-to-one replacement of each online mall with its deprivatized doppelganger. One cannot simply clone Facebook, place it under public or cooperative ownership, and expect substantially different outcomes. Online malls organize online environments through a particular choice architecture. To create spaces where collective choices can be made, new architectures are required.[22]

Ben Tarnoff proposes a dual strategy. On the one hand, actions can be undertaken to erode the power of the online malls. The anti-monopoly toolkit—breaking up tech giants, banning mergers and acquisitions, and other New Brandeisian methods—still has its merits. On the other hand, a constellation of alternatives can be created that claim the space that online malls currently occupy. The former strategy is designed to open cracks in the digital enclosures. The latter strategy aims at using these openings to build digital public spaces where profit is not the priority and where users govern themselves.[23] Its proponents should take advantage of the current momentum behind antitrust actions by forming tactical alliances with those who want to reduce the power of Big Tech companies. But the antitrust exponents' agenda of strengthening markets and enhancing competition remains at odds with the ultimate goal of creating and sustaining non-market forms of coordinating economic activities.[24]

Specific ideas and innovations put forward by programmers, artists and academics are already in circulation that may be useful in drafting the way forward. For example, media scholar Ethan Zuckerman has come up with a programmatic outline for social digital spaces built with taxpayer money. He envisions a decentralized constellation of such spaces; instead of Facebook, for example, millions of social media communities, each with their own rules and customs, could be created. These communities would be "plural in purpose," hosting different kinds of interactions. "Pool halls, libraries, and churches are all public spaces, but they all have different purposes, norms, and affordances," Zuckerman notes.[25] There is no reason why the proposed online spaces cannot have a similar diversity.

The logic of the online mall is to enclose everything and become as big as possible for a variety of reasons. This size serves the interests of investors; more users mean more rents, more data, more profits. It also brings the benefits of network effects, which neutralizes the competition and enhances the value of the firm. But such a huge scale "makes true participatory governance difficult, if not impossible," argues Zuckerman. "A 'community' of a billion people who have nothing in common but their use of a media platform is not a community in any meaningful sense."[26] This leads to another feature of decentralization; not only does it facilitate greater diversity, it also enables a degree of democracy. Communities in this model are self-governing and self-policing. Policies around appropriate behavior and speech are made by the communities themselves, who then determine who gets a say in making those rules and who belongs to their community. Rather than outsourcing moderation to paid professionals in whichever country can provide services most cheaply, removing content according to the dictates of an opaque, unpublished rulebook, users in this scenario take responsibility for debating which types of speech are appropriate for their communities and how rules and norms will be enforced. At a small enough scale, social media communities can become self-governing. Instead of letting tech executives decide how filtering algorithms work or how content is moderated—behind closed doors and bound by the market—users can make their own choices, and those choices can be guided by considerations other than profit.

Such communities also do not have to be isolated—decentralization is not the same as fragmentation. The internet is made up of distinct networks, but data traffic takes place easily across them as the networks share a common set of protocols. These protocols are open and nonproprietary; any network can join the internet as long as it follows the specified rules.

Mastodon is a good model of an open-source software project that applies this principle to social media. Servers are independently run but interconnect through open protocols to form a federation. They can also connect to non–Mastodon servers within a broader ensemble of federations—the "Fediverse"—so long as those servers use the same protocols.[27] Mastodon resembles X (formerly Twitter), but the Fediverse also offers a range of other services, including those modeled after YouTube, Facebook and Instagram. The interoperability of Mastodon's system means that the advantages of network effects can be preserved without the network residing in the hands of a single entity. E-mail—the internet's original social medium application—works the same way. There are various e-mail services, with distinct features. But users can still exchange messages, thanks to shared protocols. The fact that these protocols are open and nonproprietary is a direct consequence of the internet's public origins.[28]

Mastodon has millions of users, and is certainly not the only experiment of this kind. A number of similar projects such as Bluesky Social, for example, are being pursued across the "decentralized web" community. Still, these alternatives remain relatively limited in reach when compared to the titans of tech. Making them more robust and effective alternatives will require public investment. Two institutions in particular offer ideal vehicles for such investment: public libraries and public media. If each of the more than nine thousand public libraries in the U.S. had a federated social media server and everyone with a library card could have an account, this would open up considerable opportunities. Piggybacking on existing public infrastructure is a good way to make new online spaces more accessible, and tie them to a funding source. Using local libraries also adds some degree of accountability. If one has a complaint one can always approach one's administrator who is an employee at the local library. This is very difficult to do, if not impossible, in the case of Mark Zuckerberg and Facebook. Furthermore, librarians are the original information workers. They retrieve, classify, curate, and contextualize information and they do so not to make a profit, but as a public service. However, the problem is that funding for public libraries has fallen as Google and Facebook have grown. According to Tarnoff, this means that "we have outsourced 'our knowledge needs' to the online malls, which satisfy those needs according to commercial imperatives that inevitably compromise the quality of the knowledge provided."[29] The vision of librarians organizing and curating the internet might be feasible locally. But there seems to be little room for scaling these narrow proposals in today's fast-paced internet.

In turning to the possibility of piggybacking on public media, it is important to recognize that the ethos of journalism as a public service has always been relatively weak within the highly commercialized U.S. media sphere; media deregulation in the 1980s and 1990s weakened it further. In addition, the more recent collapse of local newspapers has led to growing "news deserts." Commercial media's shortcomings have inspired generations of reformers to propagate the alternative of media conceived as a public good. A decade ago, Robert W. McChesney, co-founder of the public interest group Free Press, concluded his eye opening book *Digital Disconnect: How Capitalism Is Turning the Internet Against Democracy* (2013) with bold proposals for more sustainable, independent forms of journalism. These focused on making cooperative and nonprofit media more practical through deliberate public investments similar to those in various European and other Anglophone countries, Japan and South Korea. This would foster larger informed citizen involvement in politics as a quintessential part of progressive moves toward real political democracy in the U.S. McChesney recognized that this requires coalitions of people to form a common front and generate numerical strength—an alliance of copyright activists, independent journalists, community media activists, and privacy advocates alone would not be sufficient. Rather, success would require a broader political movement driven by a general progressive agenda, not one specifically focused on the internet or media. McChesney insisted that such a political movement would likely make progress only if it was aimed at reforming, if not replacing "really existing capitalism" by a new, sustainable economy and much engaged political participation in the form of decentralized and local community control, with the state reinforcing local planning.[30]

In a similar vein, Victor Pickard, a professor of media policy and political economy, promotes a federally funded "permanent trust for public media" that would subsidize local news organizations across the country. He also suggests turning public libraries and post offices in "community media centers" that would help turn local residents into local reporters. Pickard refers to a variety of research findings showing that public media tends to "present a wider range of voices and perspectives, and be more critical of dominant policy positions." These strengths would do much to enhance the quality of information circulating within social media communities, particularly if the latter developed formal relationships with public media organizations and local community media centers.[31]

A kindred proposal by Ethan Zuckerman is to use a tax on targeted ad revenue. This strategy has been suggested by Nobel Prize-winning economist Paul Romer in a 2019 opinion piece in *The New York Times*—along with the Austrian sociologist Christian Fuchs in a 2018 book—to fund a new set of institutions that parallel those set up in the 1960s and 1970s to build public media in the United States.[32] A "Corporation for Digital Public Infrastructure," a parallel to the Corporation for Public Broadcasting, could search and invest in academic research and nonprofit and for-profit experiments to create digital spaces designed "to help us understand our world and participate as citizens."[33] This would mean imagining any number of projects that could result from carefully studying the strengths and weaknesses of civic and social life in digital spaces and building alternatives. However, given the fact that the necessary conditions for broader public investments along the lines described by McChesney are unlikely to materialize any time soon, these proposals are doomed for now, at least in the United States.

Moreover, a social media that is decentralized, self-governing, well-funded, embedded in public libraries and enriched by public media would still have its challenges. Truly decentralized networks would make it impossible to remove content that many people may find offensive. More worrisome is that toxic communities gain more control over the digital spaces they inhabit. A federated social media space means that far-right social networks like 8kun and Gab can create their own spaces online and govern them in the ways they prefer.[34] However, federation also means that other platforms can choose not to connect to those more toxic networks. For example, when the fascist social networking site Gab migrated to Mastodon, the administrators of most of the servers that make up the Fediverse responded by blocking Gab. Because Mastodon is open-source, fascists are free to use it, but their communities can be quarantined.[35] Thus, such toxic communities may have smaller reach and be less likely to spread extremist content beyond those who actively seek it.[36]

Social media is only one type of online mall. To abolish the rest will be harder, requiring more imagination, Tarnoff insists. Nick Srnicek envisions a publicly owned cloud service that ensures privacy, security, energy efficiency and equal access for all. Such a service might be carved out of a corporate provider like Amazon Web Services, which could then be required to donate a portion of its capacity for this purpose through governmental regulation.[37] This would mean a "public lane" for the cloud, serving the digital infrastructural needs of a growing deprivatized sector.

This sector would include a variety of cooperative ventures, not just in social media but across all realms of online life. Here, too, there are ongoing experiments, particularly when it comes to creating worker-owned substitutes for gig economy firms. A well-known example of this is the worker cooperative for home cleaners Up&Go in New York City and more recently in Philadelphia, too, which provides its workers with labor rights and owns a matchmaking platform. Another good example in the U.S. is the Green Taxi Cooperative, the largest taxi company in the Denver metro area, organized by the Communications Workers of America Local 7777.[38] Similar efforts exist around the globe, part of an international community dedicated to "platform cooperativism."[39] The main challenges for worker-run matchmaking platforms for gigs are difficulties in raising capital, managing heterogeneity in collective decision-making, and finding institutional support. A recent analysis of the state of affairs in Europe and North America shows that those challenges can most likely be successfully overcome by platform cooperatives that organize taxi rides and professional jobs (involving, for example, software designers, consultants and artists), while it may prove much more difficult in food delivery, homecare and online micro-tasking.[40]

Regarding public support of platform cooperatives, Tarnoff suggests that various policy instruments, such as grants, loans, contracts, and preferential tax incentives, could be made available. Of course, this all depends on the degree to which politicians and regulators endorse such governmental interventions. One could also imagine municipal regulatory codes that only allow app-base services to be performed by worker-owned firms. This could mean, for instance, that a cooperatively-owned app-based taxi service replaces Uber.[41]

There is also the important question of data. As online malls are organized around the manufacture and monetization of data, a deprivatized sector will need to find ways for dealing with this quintessential feature of platform capitalism. Here one should look beyond the common tendency to see data production in personal terms and recognize its collective character. The data practices of a company like Facebook involve routine violations of individuals' privacy, but Facebook is only interested in individuals insofar as they represent en masse (that is, aggregated) a source of "population-level" insights that can be used to sell ads. User behavior is analyzed in order to sort people into groups that make them legible to advertisers.

Data is made collectively and made valuable collectively. It follows that its governance should also be collective (instead of online malls' control mechanisms); users should have the power to shape the goals and conditions of data production. Toward that end, a number of proposals have been developed for "data trusts" and "data commons" of different kinds.

Furthermore, the ownership of data should be separated from its processing; users would determine under what conditions an online service would have access to their data, and under what conditions more data could be manufactured. This architecture is also well suited to a more decentralized internet, as has been suggested. Users could park their data in a central depot, and then authorize its use by different services as needed—a cooperative social media server, a worker-oriented delivery app or some other kind of digital service.[42] However, users might see such centralized storage and administration of data in itself as problematic from a participatory democratic perspective.

Platform Socialism

In his academic manifesto *Platform Socialism* (2022), the British political scientist, James Muldoon, takes a bolder, comprehensive approach that aims to transcend the prevailing discourse around "fixing" digital technologies, much of which tends to emphasize a constellation of discrete issues—namely "privacy, data, and size," as previously discussed.[43] He argues that focusing on these issues in isolation overlooks the larger systemic questions about decision-making power that underlie all of these issues: who has that power and what do they do with it? As indicated throughout this book, too, the platforms that support so much of people's daily lives and social interactions are controlled by a small group of elites who remain largely unaccountable for their decisions. If users want to fully realize the democratic potential of digital platforms, Muldoon contends, they must disperse this power by reorganizing the digital economy around "social ownership of digital assets and democratic control over the infrastructure and systems that govern [their] daily lives."[44] It will become clear that several of Tarnoff's suggestions mentioned earlier align with this approach. In light of its importance for users' interest in counter-strategies to online platforms embedded in digital capitalism, the theorizing and major arguments in *Platform Socialism* will be considered in more detail here.

Muldoon develops a theory of platform socialism that centers community as the locus of decision-making rather than the state or workers alone. He draws primarily on two early twentieth-century thinkers: the libertarian socialist and labor historian G.D.H. Cole (1889–1959)—whose vision was similar to what was espoused by the Guild Socialists in Britain in the early twentieth century—and the Austrian economist and philosopher Otto Neurath (1882–1945)—who elaborated in the interwar period technocratic socialist ideas for a democratically planned economy.[45]

He also takes inspiration from systems designed in the more recent past, particularly Chile's experiment of using technology to implement democratic control over its economy during Salvador Allende's socialist rule (1970–1973) prior to the military coup led by General Augusto Pinochet and the abrupt introduction of neoliberalism into the country. Allende had taken office in November 1970 and had nationalized key industries with the aim of increasing worker participation and control. He joined forces with British cybernetics theorist Stafford Beer, who was the lead engineer of Cybersyn, a project to assist the democratic management of Chile's economy through a network of telex machines and statistical modeling software. The system would provide real-time updates on factory production to a futuristic operations room that would facilitate comprehension of the data and help predict future economic activity.[46] It must be noted, too, that the engineers were not just technicians. They had a background in economics, a certain geopolitical consciousness about the Cold War, and were earnestly attempting to build an alternative project. The ideology behind it came from the engineers' study of dependency theories and economics. According to the renowned expert on the social and political implications of digital technology, Evgeny Morozov, their engagement was a form of radical engineering.[47]

However, in spite of Beer's good intentions to put forth the maximum effort towards the devolution of power, the resulting network resembled a technocratic approach to organization. This cybernetic management system allowed for workers' input to be filtered upwards, but there was no corresponding political arrangement that could hold decision-makers accountable and subject to democratic control from below.

In drawing lessons from this experiment, Muldoon still supports Cybersyn's unrealized aspirations for democratic transformation. It is striking how much was achieved with a relatively limited technical apparatus under adverse conditions, he asserts. In a rather short period of time, the design team had produced a novel sociotechnical system that represented a different configuration of political power to the command-and-control model of capitalist enterprises. Muldoon points to experimentation as the key to Beer's approach. New inventions had to be prototyped to enable learning from feedback loops so the system could be improved. In this way, systems were not by definition static but fluid, adapting and evolving by learning from data through experience.[48] Thus, a more decentralized version of the Cybersyn project to be developed could now employ digital technology to replace free markets with democratically planned economies. Muldoon points out that such participatory planning today can take advantage of the mass participatory tools enabled through digital technology that could facilitate this process and make coordination easier. He refers to Morozov's writing about the technological possibilities of today's "feedback infrastructure," which makes it possible to bring different parties together to help distribute information about economic needs through non-market forms of coordination.[49] Responding to Friedrich Hayek's influential argument against central planning—that markets offer the best discovery mechanism regarding the necessary information required to make rational economic decisions—Morozov proposes a form of "solidarity as discovery mechanism." This could be accomplished by the design of new social institutions that could coordinate activity more effectively than the market can. Digital platforms make it easier than ever before to examine new ways of distributing resources through democratic methods.[50]

As part of his program for platform socialism, Muldoon makes the case for digital platforms as publicly owned (non-profit) utilities set up in a participatory democratic mold. Platforms provide access to software and apps that can be distinguished from internet infrastructure—the physical hardware and assemblage of data centers, underseas cables, internet exchange points and national and regional networks that form the backbone of the internet. To extend the idea of public utilities to digital platforms would require expanding the concept beyond the idea of brick-and-mortar structures to a less physical and tangible service.

Convincing arguments can be made for taking this path forward. Firstly, certain digital platforms—search engines such as Google and social media such as Facebook or Twitter/X—provide a public service that many people consider essential to their daily lives. Platforms are now widely recognized as necessary services to live a full and rich life in a digital and connected world. Secondly, digital platforms are businesses that compete in marketplaces with tendencies towards natural monopolies. This is not the same as the physical infrastructure of telephone cables or railway lines but the outcomes can be similar. The operation of network effects leads to advantages going to the biggest players. Companies such as Facebook and Google have demonstrated the dangers of predatory monopolistic behavior under such conditions. They provide a service to the public that is difficult for them to turn down and hard to obtain in the same way from other sources.[51]

The argument for certain digital platforms to be public utilities does not rest solely on whether they should be considered natural monopolies. Progressive American reformers of the early twentieth century thought that a much broader range of corporations could be eligible for public regulation and potentially public ownership if their activities were "affected with a public interest."[52] In cases where the concentration of private power could significantly impact upon the common good, democratic governments had a right to assert public control over these facilities.[53] While a number of American legal scholars today have argued for the revitalization of new regulatory powers for the U.S. government over digital platforms, platform socialism would seek to go further in addressing questions of ownership and governance which affect how platforms operate.[54]

As Muldoon argues, the limitation of state regulation is that this does not guarantee greater social empowerment for citizens in how the service is delivered. Public utility regulation adopts a top-down approach of establishing boundaries within which the business can operate and some baselines for service delivery. Absent, however, are more wide-reaching changes in terms of workplace democracy and citizen participation. It also sets a narrow perspective on who should exercise greater decision-making power over these businesses. In the case of big American platforms, a select group of American politicians and lawyers would set the rules for how global businesses affect billions of people across the globe. Adherents of platform socialism agree with the diagnosis of the new regulators, but believe that the remedy lies in a more profound democratization of the platform economy.

Platform socialism as envisioned here also has a more ambitious aspiration of targeting not only the largest companies that would fall within the concept of a public utility but other platform companies as well. It is the structural factors determined by the capitalist economy itself that need to be challenged. Companies such as Uber, Airbnb and Deliveroo could hardly be classified as utilities, but it can very well be argued that the critique of the negative practices of platform businesses should also encompass these companies. Muldoon suggests that new models of social ownership would better enable people to take back control of these services.[55] Such social ownership of digital platforms—which Muldoon discusses under the denominator "guild socialism for the digital economy"[56]—would allow the full potential of technology to become available to the general public. As he depicts this hypothetical: "The transfer of wealth and power from the tech companies to the public would reverse the logic of privatization that has pervaded the growth of technology over the past decades. Platforms should be repurposed so that the services do not extract wealth from communities but provide them with services that are free at the point of use and which generate value for the many."[57] Muldoon gives concrete examples of platform cooperativism and the new municipalist movement, which he considers necessary steps on the path towards platform socialism. Platform cooperatives are digital platforms owned and controlled by the people who take part in them.

By providing equitable work to members/owners and building prosperity in local communities, they offer a more ethical alternative to existing corporate digital platforms. A global network of platform cooperatives has emerged which is organized around the Platform Cooperativism Consortium based at the New School for Social Research.

"New municipalism" is a transnational movement represented by cities across the world who have gathered for the Fearless Cities summits since June 2017.[58] The movement argues that the municipality is a strategic site for experimenting with how citizens can take back power. One of its major goals is a decentralization and devolution of power away from the national state and towards local authorities who are expected to respond better to the specific needs of their citizens. However, this is not simply a parochial localism that eschews addressing global problems and the international dimensions of corporate power. Rather than conflating "local autonomy with greater democracy or justice" and thus repeating "the tendency of researchers and activists to assume something inherent about the local scale," the movement aims to demonstrate the interconnections between local issues that immediately affect citizens and similar issues affecting others across the globe.[59] New municipalists do not simply want to transfer power from one geographic location to another; their aim is to alter the way that power operates. The movement advocates a qualitative shift towards more participatory forms of self-governing community life in developing a transformative and prefigurative politics in municipal settings that potentially extends beyond the city.

These projects of autonomy at the local level combine devolution with a broader agenda of empowering citizens and enabling them to control institutions that exercise control over their lives. Platform socialism can learn from these projects, Muldoon states, provided that it remains cautious against falling into a localist trap by assuming that all problems in question can be solved through devolving power and putting one's faith in small-scale alternatives to global platforms. Case studies from these two traditions provide working examples that show how communities could begin to organize their lives outside of the exploitation and control of corporations, and that they can have efficiency and solidarity, as well as innovation and equality all at the same time. Muldoon thinks that developing alternatives from the local to the international level will ensure a diversity of organizational forms that enable people to solve problems at different points in the system.[60] This is much easier said than done, however. For example, case studies of a new municipalist politics attempting to "jump scales" (in expanding to the wider region or beyond) in Spain have clearly shown the limitations of this strategy. As those movements sought to "scale up" their politics to the regional or national level, they rapidly lost the very qualities and capacities that defined them as transformational. It was found that certain dynamics start developing once one loses the ability to work closely with other activists and more hierarchical and independent structures emerge; discourses become more theoretical and feature less "a politics of proximity" through the given transformation of institutions, while urgency tends to outweigh the trust in collective intelligence.[61] Scaling up to international or "global levels" in relation to city and other local levels would undoubtedly pose much greater challenges.

Muldoon highlights existing success stories like platform co-ops (Up&Go), civic platforms (Barcelona en Comú, Decidim), data commons as in the DECODE pilots (and other commons models like free and open software, Wikipedia,[62] Creative Commons), as well as distributed networks (Mastodon). He also brings up hypothetical examples such as city-owned alternatives to Uber and Airbnb as well as a not-for-profit Google. In their totality, they illustrate the point that platform socialism is not merely utopian; it already exists to some extent in multiple forms. What Muldoon insists upon is for this model to be scaled up, but he does not discuss the problems to be expected mentioned earlier.

He stresses that the widespread adoption of platform socialism does not necessitate large-scale central planning. Instead, he argues, "Democratic platforms should be governed by a principle of subsidiarity—services should be delivered by the most local and proximate level that would be able to undertake the task efficiently, sustainably and in the manner that would maximize its benefits for users."[63] There is no one-size-fits-all model for governing digital platforms because each platform serves different roles and different constituencies. Muldoon envisions the smallest-scale approach in the cases of short-term rental platforms, app-based ride hailing services and food delivery platforms. These are best managed at the municipal level, because they are rooted in the geographic space of existing cities and towns, while health care, child care and social security should preferably be provided at the level of state and national governments. But Muldoon sees a limit to decentralization in the case of social network platforms and internet search engines whose fundamental benefits (including those of convenience) require that they have a global reach and thus global governance.

Platform Socialism gives more general guidance on how various platforms should be managed. At every level, Muldoon argues, platforms should adopt multi-stakeholder, participatory governance structures that include not only employees but also users and community members (or their representatives) in decision-making processes. And again, Muldoon notes that there are existing models that people can draw on for inspiration. On the cooperative stream-to-own music platform Resonate, for example, governance is shared between artists, listeners and workers.[64]

Muldoon's examples enliven the concept of platform socialism and indicate that this may be a realistic goal. Yet the specificity of many of these examples also have the effect of underscoring how overwhelming the "social networking" problem remains. Here the importance of context becomes clear. Civic, mostly municipal, platforms like (as yet mostly hypothetical) city-owned substitutes for Airbnb and Uber or Lyft, and "e-government" platforms like Decidim or Barcelona en Comú, provide services that are tied to a specific geographic locale where the infrastructure required for collective decision-making often already exists. In the case of the hypothetical "RideLondon" app, for example, the change *Platform Socialism* proposes is simply to add a new municipally owned ride hailing platform to the list of existing Transport for London services. But no similar infrastructure currently exists for global digital services like Facebook, Twitter/X and Google.

Muldoon provides some suggestions for new governance structures that these platforms would require. He proposes a new UN-allied agency, a Global Digital

Services Organization (GDSO) that would provide digital services to the world.[65] He brings up the possibility of "a more confrontational" approach to convert Google into a not-for-profit foundation, which shows just how daunting such an effort would be. It would actually mean expropriating its assets and adapting its organizational structure to be used for the common good. This strategy would involve the transfer of capital ownership to an independent, non-partisan foundation that would be administered by the GDSO. Muldoon acknowledges the gigantic effort such a transformation would demand: "It would be achieved after a period of protracted political struggle and the formulation of the demand that access to information is too important and valuable to be left in the hands of a for-profit company."[66] After indicating the organizational and financial hurdles to be overcome, he concludes: "The biggest challenge would be generating the political will to have public organizations fund a foundation to run the service on behalf of humanity."[67]

An enormous, if not insurmountable barrier to such a not-for-profit endeavor is the fact that digital technology is driven to a great extent by venture capitalism in the U.S. As Evgeny Morozov has pointed out, unlike hedge funders or private equity investors, venture capitalists have traditionally held "progressive" credentials. They have styled themselves as the heroes of innovation, and corporate Democrats have done more to promote their progressive image than anyone else. Venture capitalists and Democrats long shared a mutual belief in techno-solutionism, the idea that markets, enhanced by digital technology, could achieve social goods, where government policies had failed. The wave of techno-solutionism began in the early 1980s, as Democrats saw Silicon Valley as the key to boosting environmentalism, worker autonomy and global justice. Venture capitalists, as the financial backers of this new and allegedly benign form of capitalism, figured prominently in this vision. Whenever Republicans pushed for measures favorable to the venture capitalist industry—such as changes in capital gains tax, or the liberalization of pension fund legislation—Democrats have actively advanced the industry's agenda.[68]

This alliance has shaped how the U.S. now finances innovation. Public institutions such as the National Science Foundation and National Institutes of Health fund basic science, while venture capitalists finance the startups that commercialize it, as noted earlier. These startups, in turn, build on intellectual property licensed from recipients of public grants to design apps, gadgets and drugs. And a large part of these profits, naturally, flows back to the venture capitalists who have a stake in these startups.[69] They, along with their political allies, will undoubtedly defend their interests tooth and nail against any attempt (as yet merely hypothetical) to expropriate the digital technology assets in question.

Muldoon further advocates a "fediverse" model of social networking—along the lines of Mastodon—that would "empower users and give them greater autonomy and control over their online publishing and communication" and "guarantee individual autonomy within a larger federal system"[70] In this system, for example, persistent challenges of content moderation would be "democratically decided by actual communities of users ... [who] would make their own decisions about the kinds of speech they would tolerate on their platforms."[71] But here one hits upon a critical issue: in order for actual communities of users to make these decisions, those communities must understand themselves as such. It is not clear that all users of Facebook or Twitter/X (or any of the other global digital platforms for that matter) think of themselves as being in community with one another. The publics associated with them remain inchoate and unorganized. Without those publics thinking of themselves as such, and organizing to make decisions, the ideals of equality, open access, and transparency that Muldoon calls for, are likely to be of little relevance. Collective self-determination requires, above all, a self-conscious collective.

This does not mean that such community life is impossible at a global scale. But the examples of successful digital communities that Muldoon highlights raise further questions about what makes them tick. Mastodon, a decentralized alternative to Twitter/X, seems to perfectly align with the platform socialist model—it is a not-for-profit, user-controlled, open-source platform. By contrast, Reddit, which Muldoon also approvingly mentions, is a for-profit company, and at least until recently, privately owned.[72] While both are examples of democratic communities; the substantive difference in their ownership structure poses a serious challenge for *Platform Socialism*'s approach to social networking. Alongside questions of ownership and control, it appears that there are additional factors to consider when building self-determining digital social networks. These concern questions of design—the ways these platforms structure relationships between their users and whether they incentivize the kind of communal identity on which platform socialism relies.

Problems Inherent to the Designs of Technologies

According to Muldoon's framework, the technologies themselves are presented as mere stakes in the struggle against domination. As sites of political struggle, they are allegedly passive (or neutral) tools that can be harnessed for both capitalist and socialist purposes. The only obstacle to people's democratic freedoms is the private ownership of, and control over, digital technology, but never the technology itself. However, the availability or mere use of digital technology can already markedly constrain human freedom. Automated digital technologies are often presented as a vehicle to empower people to expand their capabilities, but very often they only do so by forcing people to adapt to the technology's operative systems. There are many examples of this. One buys a smartphone to stay connected to friends and colleagues, but ultimately one has to stay on constant alert for incoming messages and charge one's phone often enough. One buys a Roomba (an autonomous robotic vacuum cleaner) for house-cleaning, but ultimately one has to reposition the furniture to accommodate the machine. One creates a profile on a dating app in search of a romantic partner, but has to change one's appearance to fit the algorithm's requirements of what constitutes a legible and attractive profile.[73] And so on; the picture is clear. In such cases, democratizing digital technology will not suffice to attain human autonomy.

The issue here is not the agent controlling digital devices, but the devices themselves, as Gavin Mueller suggests in his insightful history of Luddism and defense of high-tech new Luddism in *Breaking Things at Work* (2021). In Mueller's convincing argument, Luddism does not simply refer to the political uprising and strategic targeting of machines by textile militants (weavers), a specific group of skilled craft workers in early nineteenth-century England. It can be seen in workers' struggles across two centuries in which workers likewise defined and approached machinery as a site for resistance. Along with the rise of the computer, Mueller signals the emergence of the figure of the "hacker," whose very expertise would position him/her superbly to challenge digital capitalism. Acknowledging that the politics of hackers "are complicated," he emphasizes that they are often some of technology's "most critical users, and they regularly deploy their skills to subvert measures by corporations to rationalize and control computer user behavior."[74] He goes on to argue that hackers and some of their projects (like free software) fit within the historical legacy of Luddism. But this is a rather problematic element of Mueller's account. Many hackers will probably not be too pleased to see themselves being described as Luddites, while many self-professed hackers are likely to deride the idea that using bitcoins to buy drugs on the dark web is a Luddite pursuit, as has been suggested. There is also the issue that most hackers tend to be libertarians who are not prone to oppose the essence of the capitalist practices involved. Still, the idea that those most familiar with a technology may know exactly where to strike, certainly resonates with the historic Luddites.

No doubt some types of civic activism, such as the provocative "hacktivism" of Anonymous, for example,[75] are capable of challenging, at least temporarily, the legitimacy of the large-scale power built through data. But the disruption that hacktivism brings to its targets always risks spilling over into a general disorder that damages those who would otherwise support it. As Couldry and Mejias contend, "Hacktivism cannot be a generalizable model for living with data power; if generalized, the result would be a data war whose consequences are incalculable."[76] They recognize other forms of activism, which, admittedly local, might be important exemplars of positive change. These include, among other things, neighborhoods, cities or even larger political entities such as states that are banning Airbnb or Uber, thus putting public values above platform profit and short-term consumer convenience.

There are also instances of direct action, some serious, some tongue-in-cheek or half-way between, that may serve a purpose in drawing attention to the issues at stake. In Los Angeles, opponents of the omnipresent Ring camera doorbells have distributed "Anti-Ring" stickers to be placed over the lenses of the devices. A group of San Franciscans calling themselves Safe Street Rebel have seized traffic cones and placed them on the hoods of the city's self-driving cars, a way of confusing the car's sensors and rendering them inoperable.[77]

It can be argued that, in a strictly historical sense, the term "Luddites" refers to people who are anxious about the interplay of technology and labor markets. New Luddites tend to focus on ground-level concerns—employment especially—because this is where technology enriched by AI appears to cause the most pain, as detailed in Chapter 6. Labor rights go to the very historical core of this movement.

New Luddism is not about forgoing technological innovations per se. Instead, its adherents ask that each new innovation be considered regarding its intrinsic merit, its social fairness and its potential for malicious hidden features. Luddism in this sense is about the idea that just because a technology exists, this does not mean it should be embraced without question. The fact that a new technology has been rolled out does not by definition mean it is a positive advancement. New Luddites remain continually skeptical, especially when technology is being applied in work settings and elsewhere to order social life. They look at technology critically and reject aspects of it that are meant to disempower, deskill or impoverish them.[78]

At the vanguard of the new Luddite movement in the U.S. today is the popular podcast *This Machine Kills*, hosted by tech journalist Edward Ongweso, Jr., and social scientist Jathan Sadowski. *TMK* embraces Luddism as a framework for exploring—and rebuking—the dominance of Big Tech and Silicon Valley. It has become a must-listen podcast for tech critics and insiders alike, and for workers trying to navigate the world that Big Tech has made. The hosts have noticed that their talks about Luddism clearly resonate among their listeners. Sadowski sees a broader movement emerging here: "People are looking for a politics that is ruthlessly critical of Silicon Valley—it's not a primitivism, we don't reject all technology, but we reject the technology that is foisted on us. Because we expect something better."[79] As he explains in a recent interview, "Luddism is founded on a politics of refusal, which in reality just means having the right and ability to say no to things that directly impact upon our life. This should not be treated as an extreme stance, and yet in a culture that fetishizes technology for its own sake, saying no to technology is unthinkable."[80] At least, that was the case until 2023, the year in which ChatGPT, Bard/Gemini and other user-friendly AI systems were embraced by the world. Regarding the development of such AI systems, their serious risks for humankind were highlighted by a variety of people, and some experts in the AI field called for a pause, or even a ban on them altogether.

This begs the pivotal question: to what extent can technologies that are (from a new Luddite perspective) inadequately designed, be modified to be more in tune with a humane way of life? Many of the technological tools have been designed to advance capitalism's role, making it doubtful whether all of them can necessarily be repurposed. As Zachary Loeb succinctly puts it,

> …the underlying question for Luddism remains: are certain technologies irredeemable? Are there technologies that we can remake in a different image, or will those technologies only reshape us in their own image? And if the answer is that these technologies cannot be reshaped, then are there some technologies that we need to break before they can finish breaking us, even if we often find ourselves enjoying some of the benefits of these technologies?[81]

Here one should consider in particular the forms of algorithmic management that utilize apps currently in use in multiple work settings. If not appropriately redesigned before being deployed in a socialist-oriented organization (as imagined here), this kind of system would still have the same detrimental effects as before with regard to surveillance and control of workers' behavior. It is really a question whether such a reshaping is possible at all without dismantling the technology's quintessential features. The resistance practice associated with the inversion of surveillance, called "sousveillance," that one might think could be deployed to achieve this alternative goal, does not fundamentally change the nature of the algorithmic management system. It still assumes a management that is ultimately in control—not the worker community as a whole being democratically in charge—even though the bottom-up practices of surveillance can be used to some extent by workers for collective resistance against workplace exploitation. So, the answer is almost certainly negative, and therefore this management system cannot be repurposed to that end.

If platform socialism truly aims at maximizing freedom, it should expand its reach beyond questions of ownership and governance. Some obstacles to human freedom do not emerge from capitalist power imbalances, but from the behavioral requirements of the digital technology itself. In this sense, people should not only ask how to govern digital technologies, but also whether they want or need all forms of digital technology in the first place, as discussed earlier with regard to smart home devices, for example.[82]

A major challenge for technology criticism today is the simple fact that its exponents are also reliant on these same technologies, and many of them quite like certain aspects of these digital tools and platforms. In this technological climate—where the idea of truly banishing certain technologies seems fantastic and outrageous—feelings of dissatisfaction are often channeled in the direction of appeals to personal responsibility.[83] These can be found, inter alia, in proposals for media literacy about how to better deal with social quantification technologies, in particular the many self-help tips to break one's smartphone and internet addiction that circulate online. This leads, for example, to an individual trying to spend one or two days a week offline; attempting a month-long experimental fast to one's phone's usage, aligning it with more important pursuits in life; using apps to bolster self-control (which is paradoxical, of course); changing the phone settings, such as turning off notifications, removing distraction-based apps from the home screen, using airplane mode, turning on the do not disturb setting, and so forth. And for serious cases, even a range of professional therapies are being offered to treat a user's addiction.[84]

This is the way in which a massive social problem ends up being reduced to telling people that they really just need to implement some tweaks in lifestyle and digital habits. These actions may be useful at an individual level, but they are not a sufficient or practical response to the ways that technology challenges people today. A more appropriate approach is to persuade people that machinery today should still be considered a possible site of political struggle, as it was for Luddites in the past.[85] This means they could be involved in fights (and ideally in decision-making) about the choice, design and construction of digital tools and online platforms that are part of techno-capitalism. This goes well beyond the usual trope of machine breaking. The focus should be on how users can resist the new technology, and how impacted workers do so more specifically in relation to management practices in the labor process.

12

Competing Digital Empires and the Future of Platform Capitalism Globally

The future of digital platforms and the development of new technology is increasingly tied to a global struggle between the U.S. and China. Between China's authoritarian state-led technological ambitions and the United States' free market "industry self-regulation," there is space for a non-aligned movement of countries to promote a democratic alternative for digital platforms and AI. As noted earlier, Europe is taking a third path, focusing on digital sovereignty, regulating digital markets and services as well as the use of AI, and boosting local digital technology. Although Europe is a relatively small player in terms of technology investments, the EU sees itself as a world leader in developing new regulatory approaches that aim to protect privacy, competition and safe data practices.[1]

EU's regulatory model aims for a fairer digital market with a greater redistribution of digital gains, in line with the EU's commitment to the social market economy and the pursuit of more equal wealth distribution in general. In this regard, as the legal expert Anu Bradford has shown in a comprehensive analysis of the EU, U.S. and Chinese systems, there is some common ground with the Chinese regulatory model. In 2021 China enacted a flurry of legislative measures to protect its citizens vis-à-vis profit-seeking Chinese companies and also to restore its own control over an industry that had grown so large that it threatened the power and influence of the state. These measures include the new, stringent Anti-Monopoly Law, the Data Security Law, the China Personal Information Protection Law, and the revisions of the 2016 Cybersecurity Law.[2] Crucially however, what differentiates the European framework from the Chinese one is the absence of citizen protection against state action in the latter. The Chinese digital laws create unrestricted surveillance in the name of a more efficient delivery of public services and safer cities. Similarly, behind the mask of avoiding social unrest, strategic censorship is maintained and any criticism of the Chinese Communist Party (CCP) is severely punished. Additionally, there may be another reason for the Chinese push for a better distribution of digital gains: maintaining social stability, whereby citizens feel their needs are covered and they can improve their conditions—that is, if they align themselves with the regime. This reasoning is apparent in several recent antitrust cases brought by the CCP against domestic tech giants such as Alibaba and Tencent. These companies immediately accepted their responsibility, pledging to do better according to the guide rails laid down by the CCP, and they went beyond the payment of fines to include financial contributions towards the strategic goals set by the Chinese state. This only confirms that the Chinese state extends its authoritarian control not only over its citizens but also over its tech companies, which are expected to actively contribute to the successful implementation of the CCP's political agenda.[3] This applies not only to Chinese corporations, but also to U.S. businesses still operating in China. Given the immense size of the Chinese market and its massive profit potential, some U.S. tech companies seem all too willing to jettison their anti-censorship and "free speech" guidelines. Bradford suggests that the reason that other U.S. multinationals, such as Meta (including Facebook, Instagram and WhatsApp), Google and Amazon no longer operate in China—if their apps were not already banned—was the likelihood of intense public backlash rather than their commitment to liberal democratic values. Some Chinese corporations also face difficulties in operating simultaneously in China and the U.S. due to the conflicting legal obligations in both jurisdictions. This is exemplified by TikTok's and Huawei's struggles in the U.S., which show that despite eschewing legal battles with the Chinese regime, Chinese companies do engage in such battles with the U.S. government.[4]

Digital Technology and Twenty-First-Century Forms of Colonialism

Big Tech is a leading player in new forms of colonialism regarding the Global South. Michael Kwet has argued that just as the railways and maritime trade routes provided the "open veins" of classic colonialism, the digital infrastructure and proprietary services of the Big Tech firms now play a similar role in the plunder of the Global South.[5] This infrastructure is designed to extract value from target countries under the purportedly benevolent goal of "connecting the world." Establishing their infrastructures and services prevent other countries from developing their own alternatives to Big Tech's products, thus keeping populations in the Global South in a position of permanent dependency.

The neocolonial role of international aid has morphed, as the revolving door between the most powerful governments in the world and tech companies manifests itself in global diplomacy. Big Tech CEOs navigate the world as new envoys of digital colonialism, showcasing the power of their enormous technical empires to heads of state and offering deals that appear shiny but are ultimately extractive and deprive emerging economies of a digital future they themselves can govern. This strategy includes building broadband infrastructures, which for some years even meant using high-altitude balloons to create an aerial wireless network connecting remote areas until this practice was halted in 2021.[6] Facebook's "Free Basics" offers a bare-bones version of the internet to users in developing countries (at least 65 in 2017—ranging from Mexico, Colombia, Ghana, Kenya, Pakistan to the Philippines—while reaching 32 countries across the African continent by 2019), who receive free access to a limited number of data-light websites and services.[7] What started as an aggressive public relations campaign around 2013 has continued more discreetly by the five leading U.S. tech companies in developing countries in the Global South. Tech giants have also been providing digital infrastructure to dozens of governments, ranging from cloud services to entire mail and office suites. Amazon and Microsoft have led this process, followed closely by Facebook and Google.

Furthermore, Big Tech has been trying to solidify its position by investing in software in Global South classrooms to entice young people into using its products from an early age. U.S. companies have subsidized tech equipment and software in attempting to foster path dependency and shape the habits, preferences and knowledge base of the next generation of users from childhood. This leads to a *de facto* privatization of public education infrastructure allowing the tech companies to develop further commercial products instead of facilitating an education data commons that would help countries develop public interest digital services. It also enables tech companies to gain valuable data produced by students to develop new services. In this new manifestation of colonialism, data and money flow in one direction, with little or no privacy for digital users, or taxes on the profits of tech companies. Without ownership and control over this data, developing countries cannot create the products they need to become equal participants in the digital economy. A similar state of affairs is occurring with regard to health data, emergency response and even citizen security.[8]

To be sure, this digital imperialism is not exclusively led by the United States. China has increasingly been following the Silicon Valley pattern of behavior and plays a similar role in digital colonialism by building digital infrastructures across the world. Chinese tech companies—all with varying ties to the CCP—have built the physical components of digital infrastructures, provided critical telecommunications and e-commerce services, and supplied surveillance technologies along the Digital Silk Road, a project formally launched in 2015. This is akin to a global initiative in its own right that was originally part of China's massive Belt and Road Initiative (BRI), the biggest infrastructure undertaking in the world. The Chinese digital infrastructure power is felt throughout Africa, Asia, and Latin America and even parts of Europe. This influence has been most tangible in the developing world, even though many developed countries have also imported certain Chinese technologies and in that way, contributed to the global expansion of Chinese technology standards. (The latter include established Western European democracies that have turned to China as their supplier of digital technologies. Chinese companies have built 5G networks in or sold surveillance technologies to France, Germany, the UK, and other European countries.) The implications of the potential for the CCP to use the Chinese tech companies in question to "rewire" the global digital architecture, from physical cables to software, could lead to dominance of the international digital ecosystem by Chinese technology. This could ultimately shift global norms from a free global cyber commons to competing systems of cyber sovereignty or cyber freedom, creating the potential for a further fragmentation of the internet.[9]

Chinese firms are also bringing additional benefits to developing countries by establishing training centers and research and development programs to boost cooperation between scientists and engineers in these countries and their Chinese counterparts, and to transfer technical knowledge locally (but also to China, of course) in areas such as smart cities, artificial intelligence and robotics, and clean energy, among others.[10] China has also gradually assumed control of key positions in relevant international organizations involved in standard setting across technologies, further allowing the Chinese government to entrench its regulatory standards and surveillance practices across the world. Many receiving countries have welcomed Chinese technologies and accompanying regulatory standards as a path toward digital sovereignty and development. For authoritarian governments, an additional motivation has been to gain access to surveillance technologies that they avidly use toward illiberal ends.[11]

The prevailing system of digital colonialism has also become ensconced at the level of international law and trade agreements. Since 2017, more than 70 countries (in 2023, 90 with formal decision-making power), including the U.S., the EU and China, have been in negotiations within the World Trade Organization for a new treaty on e-commerce.[12] Big Tech companies have been looking to solidify their dominance by sponsoring a new framework for keeping data flows as unrestricted as possible and maintaining their position as market leaders. But the WTO negotiations have had to reconcile different member positions on the types or rules to prioritize within an international framework. Earlier in the process, the U.S. pushed for rules to ensure the free flow of data across borders—rules that would of course benefit the expansion of its digital service firms. China and the EU advocated a more cautious approach, seeking to maintain policy space for national security or privacy priorities. China's main push was for trade facilitation and market access benefits, rules that would benefit the flow of traditional trade enabled by digital platforms. Developing countries focused on implementation flexibilities and capacity building benefits. Importantly, at the time of writing, articles on data flows, data localization or location of computing facilities and source code are no longer being considered. These articles on data flow issues were put on the back burner after the U.S. Trade Representative announced in October 2023 that the U.S. was withdrawing support for proposals on these issues. This move meant the shelving of issues considered the most challenging to negotiate, due to the divergent positions of members such as the U.S., the EU and China.[13]

A transnational solidarity movement beyond the confines of the WTO could prevent multinational companies from completely shaping the agenda and entrenching U.S. and Chinese hegemony. From a critical emancipatory perspective, international cooperation is needed around developing local technology that is participatory, public-interest oriented and designed to serve citizens' basic needs.

Decolonial Struggles Against Data Extraction

The broader decolonial approach to data and technology put forward by Couldry and Mejias (outlined in Chapter 1) is highly relevant here. It holds that any struggle against the colonialism advanced through data practices must be global in its framing. This is because the corporate ambitions of data industries are by definition always global just as capitalism and colonialism are. On the other hand, the resistance against data colonialism is always local and must work from, and contend with, local conditions. It is understood that the frame within which such activism unfolds must always account for the global scale of what it is opposing if it is to reach its target—well expressed in the adage "think global, act local." Whatever the local dynamics, data colonialism operates within and takes advantage of a global framing, and it is at this point that its contestation must be launched.

The decolonial approach advocated by Couldry and Mejias acknowledges that decolonial struggles against data extraction must take a dual trajectory: struggles over particular practices of technology *and* struggles over knowledge and rationality, that is, the deep narratives that help frame those specific practices and uses. Regardless of how intense their engagement with the materiality of tools and infrastructures, these decolonial struggles are always acts of imagination, too, constructing alternative visions of data and technology.[14] Sometimes they will build on existing visions; at other times, they will reject all existing alternatives and forge a new direction.[15]

Here it is essential to understand that under data colonialism, what is extracted for value from social life is intimately (not incidentally) linked to the production of knowledge about social life. Resistance to data colonialism draws people inevitably into epistemological questions (about what constitutes knowledge and what its preconditions are) normally delegated to philosophers. What sorts of data should be collected? What sorts of data should be combined with what other sorts of data? Where and when should a limit be posed on algorithmic decision-making? The answers to these questions will shape the social worlds in which people will live in the coming decades.

It is also important that this decolonial approach rejects rather than accommodates prevailing forms of ideology surrounding data. This particularly involves the following responses to crucial features of datafication. The micro-focused targeting of persons by marketing messages is not personalization. The pursuit of continuous automated surveillance does not really bring the benefits it promises, such as the democratization of health or the educational promotion of digital citizenship. It is untrue that limits to connection are always bad or that flows of information must be "seamless." Instead of prioritizing the seamless movement of data, transfers of data must always be done responsibly and be accountable to those affected by the data, otherwise, such transfers should not proceed. It is necessary to reject the implication in metaphorical claims that data create "ecologies" or that data processing is somehow "natural." Even though data accumulated in large quantities across many sources may have a complexity comparable to known ecosystems, this does not mean that such processes are natural and must be left unrestricted, let alone protected and legally sanctioned.[16]

Given the massive global inequalities and asymmetries of cognitive and economic production since at least the beginning of historic colonialism, decolonial struggles against data extractivism (from this perspective) must include, and be defined by, the widest possible range of peoples. The involvement of indigenous and various kinds of marginalized peoples is essential here. Otherwise the practice of struggle will only serve to reinforce the colonialism it claims to challenge. This will require new opportunities to enable those who are typically excluded from debates on the global economy, legal policy, and computing standards to actually lead those debates. According to Couldry and Mejias, the last point means that a new conceptual space must be built that seeks to define and claim techno-social spaces beyond the profit-motivated model of Silicon Valley and the control-motivated model of the Chinese Communist Party, the two epicenters of the new colonial extractivist order.

For this purpose, Couldry and Mejias have themselves been involved in the creation of a Non-Aligned Technologies Movement (NATM), one that mirrors (in some ways) the original collective of non-aligned states, which, during the Cold War, sought to define an alternative to capitalism and communism. NATM held its first international meeting with people from every continent in November 2020. Their bold and all-encompassing approach includes the following components: the boycotting of extractivist technologies and the use of alternative tools; divestment at local and national government levels from Big Tech (by refusing to buy or accept their "free" products); the re-appropriation of data (and the products of data) on behalf of those who generate it; the implementation of taxes and sanctions against Big Tech to repair the damage done by their technologies; the bolstering of public education—in the form of citizen research, literacy campaigns, decolonial thinking—to understand the dangers of data colonialism; the promotion of a broad and diverse culture of non-alignment to reimagine new forms of community without extractivist technologies and their heavy costs, not least to the physical environment; and finally, the building of a solidarity movement that joins non-aligned individuals and communities globally through collective imagination and action.[17] NATM is just one initiative of many aimed at resisting the intensified colonial processes in the field of data and technology.

Anti-digital colonialism activists such as scholars Sareeta Amrute, Nanjala Nyabola, Paola Ricaurte, Abeha Birhane, Michael Kwet and Renata Avila associated with digital rights communities in the Global South (most prominently in Kenya and India) are pointing the way to a more just digital future for everyone. In engaging various publics, they are calling for accountability and pushing for policy and regulatory changes as well as the development of humane new technologies. The annual RightsCon summit is arguably the most significant event of the global digital rights community. However, a pressing concern is that, in its current form, digital rights activism mostly relies on institutional support with Euro-American funding, with a dominance of U.S.-registered nonprofits, and where corporate capture lurks around the corner.[18] At the time of writing, it appears these decolonial efforts have not gained sufficient traction in terms of movement activism to even begin tackling the fundamental problems of data colonialism.

The Prospects of Global Moves Towards Platform Socialism

Needless to say, any serious attempt to put into practice far-reaching proposals regarding the deprivatization of the internet along with the implementation of larger-scale forms of democratic platform socialism, will face an immense uphill battle as it would require an enormous shift in the power relations between social classes. First of all, a broad socio-political movement needs to develop, building upon a growing self-consciousness among publics interested in collective self-determination. This movement should also include collective forms of resistance coming from workers against the power of Big Tech in the current platform economy. If such a movement were to materialize, which seems highly unlikely given the present circumstances, it would draw massive opposition from corporate interests, foremost the Big Tech companies (with those in the U.S. leading the way) and their political allies, particularly those members of the U.S. Congress who receive corporate donations, and from hordes of lobbyists. Such a movement would also encounter fierce opposition from China with its state-capitalist system and tied-in domestic tech companies. In all of this, one must not overlook the power of the state, the armed forces and capital's willingness to fight tooth and claw to preserve their property and profits. And as those on the radical labor left would say, far-reaching ideas for socialist transformation must go hand in glove with a revolutionary consciousness and workers' power—neither one nor the other on its own will do.

In a recent comparison between two reform proposals—which James Muldoon calls "data-owning democracy" and "digital socialism," respectively—he brings up the question about the likelihood of support for such initiatives, given the current dynamics of global politics. He then expresses his pessimism in this regard:

> With numerous examples of democratic backsliding around the world and several Western democracies experiencing dangerous signs of corruption and decay,[19] an argument for the further democratization of the economy and society may seem implausible. How would democratic norms be entrenched in a broader range of economic institutions when they are so poorly actualized in existing political ones? Particularly when considering the enormous shift in the power between social classes that democratic socialism would require, critics are rightly concerned about the political viability of these aspirations.[20]

Indeed, the current global scene appears to be quite inhospitable to such far-ranging democratic socialist initiatives. Rather than reform, there have been initiatives to contain the excesses of platform capitalism, to varying degrees, by means of the existing regulatory models of the digital economy.

The Impact of Competing Regulatory Models on Platform Capitalism Around the Globe

U.S. tech companies have shaped digital economies across the world through their business practices, thus having been key not only in defending the capitalist market-driven principles at home, but also in "universalizing" them through the often-unrestricted influence these companies exercise over the digital lives of internet users abroad. The U.S. government has further paved the way for its companies' global influence by actively promoting its "internet freedom agenda" as a cornerstone of its foreign policy, urging governments around the world to commit to the economic and political "freedoms" that underpin the U.S. regulatory model. However, the outsized influence of U.S. tech companies and their harmful practices have led to a backlash across jurisdictions.

Nowadays, as Anu Bradford points out, U.S. tech companies and the American government are in a more tenuous position than ever before. Over the past years they have been fighting a battle on two fronts, with China as well as with Europe. One of the most notable areas of transatlantic disagreement concerns data protection, where the EU's focus on fundamental rights clashes with America's focus on national security. This disagreement has become a major obstacle for data flows between the EU and the U.S. Other notable conflicts revolve around antitrust policy and digital taxation, two domains where the U.S. government has perceived the EU's attempts to impose obligations on American tech giants as acts of digital protectionism.[21]

More recently, however, U.S. legislators and regulators have shown greater interest in reining in the tech giants. After a long hiatus, the U.S. government is flexing its antitrust muscles, with the Department of Justice and the Federal Trade Commission challenging the business conduct of Google, Facebook (Meta), Amazon and Apple, respectively (see Chapter 9). These developments are bringing the U.S. policy discourse closer to that of the EU as this has evolved over the past decade.

At the same time, many U.S. stakeholders remain skeptical of such change and insist that market-driven principles are deeply embedded in U.S. institutions and prevailing (classical liberal) mindsets. This makes it difficult to reverse the U.S. regulatory model that, despite all its limitations and false promises, continues to be associated with immense wealth and technological progress.[22] Needless to say, this ignores the highly inequitable distribution of all that wealth.

Various commentators believe that any regulatory reform in the U.S. is unlikely at this time. There is no guarantee, Bradford claims, that the political dynamics at play in Congress, including the different concerns that Democrats and Republicans have about Section 230 of the Communications Decency Act (which shields online services as well as internet service providers from legal liability for the "speech" they disseminate) will lead to any substantial legislative reform on content moderation. For similar reasons, bills proposing a federal privacy law will also invariably flounder. While public discontent about tech companies is strong, the political dysfunction of the U.S. Congress might prevent any meaningful legislation in this area from being enacted. It is also unclear if U.S. courts are ready for the renewed antitrust movement, which could lead judges to reject the arguments of antitrust harm that are driving the pending enforcement actions against the leading tech titans. In addition, the concern about the potential pitfalls of regulation remains, including the idea that regulation might curtail further innovation by those companies whose products consumers most appreciate and have come to rely on.[23]

Since late 2020, the Chinese government has taken a turn toward tighter state control of the country's tech industry. As Bradford notices, it abandoned its traditionally lax approach toward tech regulation and has forcefully leveraged its power to crack down on the tech industry in the name of "common prosperity" while facing little resistance from those companies. This reinforces the core tenet of the state-driven regulatory model by ensuring that the Chinese government—not tech companies—reigns supreme over the digital economy there.[24]

China's digital authoritarian governance model infringes individual rights and deprives its citizens of key civil liberties. It also contributes to the political oppression of minorities through far-reaching government surveillance (see last section in Chapter 5). China is also gaining significant global influence by building digital infrastructures across the world as part of its massive Digital Silk Road project. By and large, its state-driven model is faring well, as indicated by the fact that several governments around the world are emulating many of its key censorship and surveillance features.[25] These governments do not share the American and European concerns about the future of liberal democracy. (However, certain illiberal tendencies both in the U.S. and EU member states such as Hungary and Poland should be acknowledged here, too.) The number of countries embracing China's governance view is also rising as the world tilts more towards authoritarianism. In the past decade, there has been a decline in global internet freedom, as governments were increasingly curtailing online speech, arresting internet users for nonviolent political, social, or religious speech, and suspending access to various social media platforms, or to the internet itself. A growing number of governments are also obtaining spyware, facial recognition, and other data-extraction technologies from private vendors, equipping themselves with tools for authoritarian control in violation of individual rights. This trend bodes well for the global appeal of Chinese digital authoritarianism, while increasing the demand for Chinese-made surveillance technologies or censorship techniques that help governments weave that approach into the very fabric of their societies.[26] Again, it should be recognized that both in the U.S. and the EU, governments, national security and law enforcement agencies are also acquiring and deploying digital technologies to monitor and surveil their citizenries, sometimes too in violation of individual rights. As outlined in Chapter 1, in relation to Foster and McChesney's political-economic conception of surveillance capitalism, the U.S. National Security Agency (NSA) engages in extensive surveillance practices, often in collaboration with U.S. tech companies. Even if the NSA targets foreign individuals, including terrorists and others deemed a threat to U.S. national security, it also collects extensive incidental data on its own citizens in the process.[27] Further, Chapter 8 detailed behavioral control practices in relation to city surveillance and predictive policing in the U.S. and elsewhere.

The Chinese state-driven model also appeals to many developing, authoritarian countries because it combines political control with tremendous technological success, which has fueled China's rapid economic growth. Although the U.S. has traditionally been seen as the technological superpower, China is quickly catching up and even surpassing the U.S. in several domains. For example, Chinese tech companies currently lead the world in terms of smartphone and telecommunications equipment. China is well on its way to becoming the world's premier AI superpower, potentially surpassing the U.S. in the next decade. For many developing countries hesitant to embrace the American market-driven model that is centered on economic and political freedoms, these facts show that political freedom is not necessary for technological and economic progress, which provides these countries with an additional reason to embrace the Chinese regulatory model.

According to Bradford, the U.S. government's strategy regarding its tech war with China may have further elevated the Chinese statist regulatory model, however unintentionally.[28] Citing national security concerns, the U.S. has moved to restrict exports of strategic technologies, such as advanced semiconductors, to China.[29] The U.S. has also banned Chinese investors from critical digital infrastructures, such as 5G networks. China has been responding in kind, imposing extensive export and investment restrictions on U.S. companies. This ongoing rivalry has also fueled a subsidy race as both the U.S. and China are turning to industrial policy to shore up their capabilities in critical technologies. However, deeply intertwined supply chains and commercial pressures in both countries are likely to prevent a full decoupling of U.S. and Chinese technological assets. Other countries, including those in the EU, are also turning to industrial policy in the midst of growing U.S.-China tensions and unraveling global supply chains. As a result, the tech war risks entrenching techno-nationalism as a global norm.[30]

As Bradford indicates, the European regulatory model rests on three pillars—fundamental rights, democracy, and fairness—all of which require government protection in the modern digital era. These key elements are considered essential in upholding a more equitable and human-centric digital economy. In practice, this has entailed leveraging European antitrust, employment and tax laws to redistribute power away from platforms, vesting more power instead with internet users and consumers, platform workers, smaller businesses, and the public at large. This regulatory philosophy is more in tune with today's political environment where the ideological underpinnings of neoliberalism and the manifestations of capitalist excess have increasingly come under attack. Another normative argument in favor of the EU model stems from the growing discontent toward the U.S.'s uncompromising commitment to free speech, even when that speech harms individuals and destabilizes societies. One can be a staunch proponent of free speech yet still argue that the U.S. government has gotten more than it bargained for with Section 230 of the CDA. That provision created an online world in which techno-libertarianism has run amok, turning the American regulatory model against the country's own democratic institutions and wreaking havoc around the world.[31]

Even though the EU shares the U.S. commitment to protecting free speech, it is prepared to restrict that fundamental right in the name of other fundamental rights and important public policies, be it human dignity, personal privacy, public safety, or democracy. The EU's willingness to intervene in internet freedoms by restricting hateful or dangerous content is increasingly seen as necessary in today's society, lending further support to the EU model.

The European regulatory model also protects internet users' privacy more than the extensive surveillance capitalism enabled by the U.S. market-driven model, where tech companies track internet users' every move online and acquire a trove of personal data that they then monetize through targeted advertising. The perverse impact of targeting users can go even deeper. For example, Facebook's algorithms have intentionally targeted the vulnerabilities of young users, with the risk of harming their mental health and well-being. At worst, privacy-infringing data extraction can compromise decisional privacy by subverting individual choice, liberty and self-governance. Voter behavior can also be manipulated whenever one gains access to internet users' personal data and deploys that data for psychographic profiling that enables micro-targeted political advertising. The EU has wielded international influence through its digital regulations that have spread round the world. By adopting laws such as the General Data Protection Regulation (GDPR), the EU is apt to shape the global business practices of leading tech companies, which often extend these EU regulations across their global business operations in an effort to standardize their products and services worldwide—a phenomenon known as the "Brussels effect."[32] While the GDPR may be the poster child of the EU's global regulatory influence, antitrust law, regulation of online content, and rules for emerging technologies such as AI likewise have the potential to gain global influence. Moreover, European digital regulations have not only been incorporated into tech companies' global business practices, but are often infused in legislation by foreign governments.[33]

In many domains, the EU has been successful in articulating policy goals that resonate globally and adopting stringent regulations that are being emulated across jurisdictions. However, the EU has seen less success in translating those policy goals and regulations into concrete changes in the marketplace. Tech companies' market dominance remains intact, the internet continues to be flooded with hate speech and other harmful content, and internet users' personal data is still being exploited by companies and governments alike.[34]

Today's largest tech companies rival some countries in their size and influence, and their tremendous economic, political and social clout make it difficult for governments to police their business practices. As Bradford explains, the tech industry has almost unlimited resources to spend on lobbying against regulations and defending themselves in legal battles against various governments. The tech giants' size and resources are not the only reasons why governments' battles against them are challenging. There is no realistic way that some agency in Brussels (or anywhere else) could review the enormous amount of content that is disseminated by those companies to ensure that nothing illegal ends up online. Finally, some of the EU's enforcement challenges can be traced back to its commitment to democracy and fundamental rights; this involves a delicate balancing act between numerous competing rights. Thus, digital regulation poses difficult trade-offs that complicate any enforcement task. It is naturally easier for an authoritarian government to "effectively" enforce online content rules when they do not need to balance freedom of speech considerations with other grounds for content removal.[35]

The Chinese government can effectuate its regulatory model in the absence of democratic rulemaking constraints. It faces little resistance in enforcing those regulations and tolerates little dissent from tech companies, all of which know that compliance is their only option. This applies to Chinese and foreign companies alike. The relative success of the Chinese regulatory model in obtaining compliance from tech companies stands in stark relief against the repeated difficulties faced by European and American regulators in holding these companies accountable. Regulators in both the EU and the U.S. often have to contend with lengthy legal battles as tech companies contest, rather than relent to, the regulatory actions targeting them. These features make the Chinese state-driven model attractive for other governments that are reluctant to be drawn into the kind of legal battles that the U.S. and the EU are struggling to win.[36]

The EU's commitment to the rule of law and due process explains, at least in part, the shortcomings in antitrust enforcement. The big delays in the EU's enforcement of antitrust laws against tech companies stem from the EU's adherence to due process in its investigations. A dramatic and speedy crackdown on the tech industry, comparable to what has recently taken place in China, is not achievable in Europe. Instead, proposed laws must go through a legislative process that consists of numerous democratic checks and balances across several institutions—the Commission, the Council of Ministers, the European Parliament and the legislative institutions across all twenty-seven member states—to ensure that all interests are considered before any regulations are enacted. And in enforcing these laws, the EU acknowledges the tech companies' right to be heard and offers two layers of appeals following any adverse decision the Commission makes against them.[37]

The EU has recently adopted new digital regulations that are explicitly designed to address some of the known deficiencies of its existing policies. To avoid the lengthy delays associated with its past antitrust investigations (focusing on the tech industry), the EU adopted the Digital Markets Act (DMA), as outlined in Chapter 9, which bans *ex ante* (beforehand) a set of anticompetitive practices by digital gatekeepers—including Amazon, Apple, Google, Meta and Microsoft—without the need to show that these companies harm competition.

Prohibiting a set of business practices outright allows the Commission to circumvent lengthy investigations, which have hindered its efforts to intervene in digital markets in a timely manner. Mandating a set of practices (such as interoperability or data portability) and banning a set of practices (such as self-preferencing or using data across different services) will give European regulators stronger tools to enhance competition.[38]

Similarly, the EU's Digital Services Act (DSA), also discussed in Chapter 9, introduced binding obligations on tech companies in relation to the content that is hosted on their platforms. The DSA calls for platform transparency and accountability, and strengthens these mandatory assignments. While it does not impose a general monitoring obligation in terms of the content they host, the DSA sets clear limits on platforms' freedoms, for example, by banning targeting advertising directed at minors and forbidding manipulative designs that distort autonomous and informed decisions or choice. These specific provisions accompany a host of general rules on algorithmic transparency and due process regarding the users' ability to contest any content moderation decisions. The largest platforms will be subject to additional rules, including independent audits and an obligation to grant access to authorities and researchers to investigate their business models.[39]

The EU's latest digital regulations, epitomized by the EU AI Act (discussed in Chapter 10), likewise imposes binding obligations consisting of a two-tier system of guardrails for general-purpose AI (GPAI) systems with stricter obligations for the use of "high-impact" GPAIs involving "systemic risk." The Act also prohibits *ex ante* a variety of monitoring, data collection, review, categorization and surveillance practices. These involve, among other things: cognitive behavioral manipulation; facial recognition data mining; emotion recognition in the workplace and educational institutions; public and private social scoring; biometric categorization; and the use of AI to exploit people's vulnerabilities due to age, disability, and social or economic situation. In addition, the AI Act establishes a series of safeguards for the use of remote biometric identification systems in public spaces for the purpose of law enforcement. In effect, police and national security bodies in the EU will be prohibited from using real-time biometric data driven by AI in most circumstances if they do not have judicial authorization. Of course, all these prohibitions *ex ante* do not address the shortcomings of the EU AI Act mentioned earlier that weaken its impact in certain respects. But they are likely to prevent lengthy delays in judicial investigations associated with future AI regulation court cases.

The EU's regulatory model is becoming more influential in the wider world of liberal democracies. At the same time, the continuing appeal of the Chinese regulatory model limits the EU's ability to embed its standards and values outside of this orbit. The growing demand for China's model enhances the likelihood that the U.S. will increasingly align itself with the EU. This transatlantic proclivity toward the European rights-driven model may pave the way for a bipolar world order, in which the U.S. and the EU would jointly lead a coalition of "techno-democracies" to challenge digital authoritarian norms and principles embraced by China and its allies.[40]

Any proposed collaboration among techno-democracies, however, raises the question regarding which countries should join such a group and what its agenda should be. While specifics vary across the proposals in circulation, Australia, Canada, Japan, South Korea, and the UK are often cited as the leading techno-democracies alongside the U.S. and the EU.[41] China and Russia are seen as the most prominent techno-autocracies, but there are several others, including Iran, Pakistan and Saudi Arabia, that suppress internet freedoms.[42] But there is no clear delineation between techno-democracies and techno-autocracies; many countries are not perfect democracies or full autocracies but are often characterized as "hybrid regimes" or "flawed democracies." Some critics might include the leading liberal democracy, the United States, in the latter category, given the flaws inherent to its non-parliamentary, presidential system: a winner-take-all electoral system (currently allied with hyper partisan gerrymandering at the state level) rather than proportional representation; two U.S. Senators per state irrespective of population size; its elitist, cumbersome Electoral College instead of a direct popular vote for presidential elections; and the profound structural inequality of the Supreme Court as the ultimate judicial arbiter.[43] It should also be mentioned that, at the time of writing, both the U.S. and the EU are facing stark challenges to their own democratic institutions at home.

The idea of closer cooperation among techno-democracies has gained momentum, in part because of strong political support by the current U.S. government. In his political speeches, President Biden often portrays the world as a contest between democracy and autocracy, and his foreign policy doctrine is aimed at countering

China's ambitions and global influence. He has made cooperation with democracies a hallmark of his foreign policy, and even hosted a "Summit for Democracy" in December 2021, which sought to "set forth an affirmative agenda for democratic renewal."[44] In April 2022, these efforts culminated in the U.S. government announcing a partnership with sixty countries who signed a "Declaration for the Future of the Internet," which pledged to harness digital technologies to "promote connectivity, democracy, peace, the rule of law, sustainable development, and the enjoyment of human rights and fundamental freedoms."[45] Despite these efforts to build a more cohesive coalition of techno-democracies, it is unclear how effective such a coalition would be in shaping the global digital order toward the principles articulated in the 2022 Declaration. The group of techno-democracies is heterogeneous, and existing disagreements may hamper any meaningful collaboration among signatory countries.

Bradford suggests that even in a bipolar digital world marked by continuing conflict, stabilizing forces are likely to keep those conflicts under control—and the existing digital order functioning. For example, the U.S. and China retain market opportunities for their tech companies in each other's markets despite their simultaneous attempts to limit each other's access to key technologies as investment opportunities. After all, to win its tech race against China, U.S. tech companies need to be able to fund their innovations, which they often do by means of profits generated in China. In the same way China wants its tech companies to grow, which may compel them to gain access to U.S. capital markets.[46]

Another force driving de-escalation in these battles is that none of the entities (the U.S., China, and the EU) in the conflicts is uniform or internally coherent; there are competing priorities within each of them. The pro-market voices in the EU often moderate the EU's regulatory impulses, making them more acceptable to the U.S. Similarly, the U.S.'s pro-market forces are tempered by growing internal dissent calling for tighter regulations, paving the way for continuing transatlantic alignment. The Chinese government refrains from total censorship, choosing not to enforce a total ban on virtual private networks (VPNs), hence allowing Chinese internet users to remain partially connected to the outer world. These balancing impulses within each jurisdiction moderate the extremes and pave the way for a world characterized by limited cooperation, managed conflict, or bearable coexistence. These forces of restraint also explain why the continuing disputes are unlikely to lead to full technological decoupling. The internet has already lost (in part) its global character. China is excluding many foreign providers from its market and blocking access to numerous foreign websites. The U.S. is limiting the ability of some American companies' software—such as Google apps—to operate in Chinese-built Huawei phones. However, any such technological decoupling is likely to be partial and incomplete at best. The U.S., China and the EU are all providing a different layer for the operation of the internet and the digital economy more broadly. As a result, the three regulatory models will continue to overlap in several—if not most—countries that continue to rely on Chinese digital infrastructure, U.S. tech companies, and European regulations to govern that infrastructure and those companies. The complementary roles that each of the digital empires plays in the world's tech ecosystem will thus sustain at least a minimal foundation for a global, interconnected digital economy.[47]

However, there is something very important going on, which is the breakup of the U.S.-dominated form of economic globalization. The system involved western companies outsourcing manufacturing to China and other low-cost destinations that, for a while, delivered cheap goods, which kept inflation low and made life easy for central banks conducting their monetary policies. These days are now over. The U.S. has tilted towards nationalistic protectionism regarding its domestic businesses and jobs by limiting imports of Chinese products through significantly increased tariffs on electric vehicles (100%); semiconductors and solar cells (50%); EV batteries, components and parts, and critical minerals (25%); steel and aluminum, ship-to-shore cranes, as well as certain medical supplies and personal protective equipment (all 25%).[48] At the same time, the U.S. is restricting exports of strategic technologies to China, and subsidizing its own manufacturing sector. The latter encompasses a series of measures—such as the Inflation Reduction Act and the Chips and Science Act—to boost U.S. industry in high-tech sectors and the development of "green technology" in particular. The EU is expected to take similar protectionist measures, as it has conducted investigations into possible state subsidies regarding certain imports from China (electric vehicles and solar panels, and specific products of iron or steel).[49] Its first manifestation is a planned tariff of up to 38 percent on Chinese electric vehicles (on top of the 10 percent it already charges on all car imports), provisionally coming into force in July 2024.[50] The EU is subsidizing strategic domestic manufacturing, too, most notably the building of a home-grown battery industry and chips industry.[51] In response, the Chinese government is threatening to impose higher import tariffs up to 25 percent on products such as European and U.S.-made cars, French cognac, and pork products from Europe,[52] while it is investing heavily in its own chip manufacturing sector.[53]

There are also the deleterious impacts on the global economy of ongoing military conflicts and wars, particularly in Eastern Europe and the Middle East. And growing U.S.-China tensions centered on the South China Sea (which includes the Taiwan Strait) could have serious implications for the global balance of power and influence. Beijing considers the greater part of the South China Sea—one of the most strategically and economically important waterways in the world—to be an indisputable part of its territory; exercising full sovereignty over this area is a key component of President Xi Jinping's "China Dream." Next to China, the Philippines, Vietnam, Malaysia, Brunei and Taiwan all have claims over areas within the South China Sea region, many of which overlap. The U.S. is not a claimant, but maintains that the waterway is crucial to its national interests (both economic and in terms of security) and often conducts "freedom of navigation" operations through the area in a message to all parties. For its part, the U.S. asserts that this waterway needs to stay free and open if it is to deter Chinese aggression, live up to its regional alliance commitments and prevent Beijing from displacing the U.S. as the dominant power in the Indo-Pacific.[54]

This is not the first time such things have happened. The pre-first world war era of globalization fell apart as a result of war, a pandemic, inflation and protectionism, which is a stark reminder of what might lie ahead. Little by little, history appears to be repeating itself, as signaled by some observers.[55]

Given the current state of affairs, the best that proponents of a more equitable platform economy can hope for—against all odds—is a global shift to the EU's regulatory model that is associated with greater economic fairness and a more human-centric digital society. This would restrain the excesses of platform capitalism, but with protections against authoritarian state action. For that purpose the U.S. regulatory model will have to change over to the European model (as has been suggested) and subsequently, the U.S. and the EU, alongside other liberal democracies, will need to unite and offer undecided countries an affordable alternative to the Chinese state-driven digital infrastructures. How realistic this change scenario regarding platform capitalism is, remains uncertain. One thing should be clear, however. Even if such a global shift were to materialize, it would still provide little to no leeway for a broader adoption of forms of democratic platform socialism across the world.

Chapter Notes

Chapter 1

1. Shoshana Zuboff, *The Age of Surveillance Capitalism: The Fight for a Human Future at the New Frontier of Power* (New York: Public Affairs, 2019).
2. Zuboff already pointed to the phenomenon in a series of articles, the first of which was published in the German newspaper *Frankfurter Allgemeine Zeitung* in the summer of 2013.
3. Evgeny Morozov, "Capitalism's New Clothes," *The Baffler*, February 4, 2019, https://thebaffler.com/latest/capitalisms-new-clothes-morozov.
4. Frank Webster and Kevin Robins, *Information Technology: A Luddite Analysis* (Norwood, NJ: Ablex, 1986), 328–43.
5. John B. Foster and Robert W. McChesney, "Surveillance Capitalism. Monopoly-Finance Capital, the Military-Industrial Complex, and the Digital Age," *Monthly Review* 66, no. 3 (2014), https://monthlyreview.org/2014/07/01/surveillance-capitalism/.
6. Zuboff, *Age of Surveillance Capitalism*, 8.
7. Brett Christophers, "Capitalism Has Always Been 'Rogue,'" *Jacobin* (March 2020), https://jacobin.com/2020/03/surveillance-capitalism-shoshana-zuboff-review/.
8. Mark Whitehead et al., *Neuroliberalism: Behavioural Government in the 21st Century* (New York: Routledge, 2018).
9. Richard T. Thaler and Cass R. Sunstein, *Nudge: Improving Decisions About Health, Wealth, and Happiness* (New York: Penguin, 2008), 6.
10. Matthias Jesse and Dietmar Jannach, "Digital Nudging with Recommender Systems: Survey and Future Directions," *Computers in Human Behavior Reports* 3 (2021): 1–14, https://doi.org/10.1016/j.chbr.2020.100052.
11. Zuboff, *Age of Surveillance Capitalism*, 68–82.
12. Ibid., 46–37.
13. An article published in the *Proceedings of the National Academy of Sciences* in June 2014 reported on this experiment that demonstrated that emotional states can be transferred to others by emotional contagion. Adam D.I. Kramer, Jamie E. Guillory, and Jeffrey T. Hancock, "Experimental Evidence of Massive-Scale Emotional Contagion through Social Networks," *Proceedings of the National Academy of Sciences of the United States of America* 111, no. 24 (2014): 8788–90, https://doi.org/10.1073/pnas.1320040111; see also Brett Frischmann and Evan Selinger, *Re-Engineering Humanity* (Cambridge: Cambridge University Press, 2019), 117–18.
14. Zuboff, *Age of Surveillance Capitalism*, 293, 299, 309–29, 431–41.
15. Morozov mentions that the books, *Reinventing Capitalism in the Age of Big Data* (New York: Basic Books, 2018) by Viktor Mayer-Schönberger and Thomas Range, and *The World After Capital* by Albert Wenger (2016), https://worldaftercapital.org/products/the-world-after-capital-epub, have a similar format: "data capitalism" and "knowledge-age" post-capitalism, respectively, versus the previous stage of capitalism. Evgeny Morozov, "Digital Socialism? The Calculation Debate in the Age of Big Data," *New Left Review*, March-June 2019, https://newleftreview.org/issues/ii116/articles/evgeny-morozov-digital-socialism.
16. Morozov, "Capitalism's New Clothes."
17. Matthew Crain, *Profit over Privacy: How Surveillance Advertising Conquered the Internet* (Minneapolis: University of Minnesota Press, 2021), 10. See also Jathan Sadowski, *Too Smart: How Digital Capitalism Is Extracting Data, Controlling Our Lives, and Taking over the World* (Cambridge: MIT Press, 2020), 50.
18. Christophers, "Capitalism Has Always Been 'Rogue.'"
19. Oscar H. Gandy, *The Panoptic Sort: A Political Economy of Personal Information* (Oxford: Oxford University Press, [1993] 2021).
20. Qtd. in Ben Lee, "Putting the 'Capitalism' in Surveillance Capitalism," *Current Affairs*, May 2021, https://www.currentaffairs.org/2021/05/putting-the-capitalism-in-surveillance-capitalism.
21. Ibid.
22. Morozov argues that Zuboff's theorizing puts her directly within the intellectual tradition of "managerial capitalism" linked to Alfred Chandler, and to a wider functionalist tradition in sociology associated with Talcott Parsons. He attributes this partly to an unacknowledged mindset: Zuboff's failure to understand the extent to

which her critique of surveillance capitalism is actually a critique of capitalism, pure and simple. Morozov, "Capitalism's New Clothes." See also Blayne Haggart, "Evaluating Scholarship, or Why I Won't Be Teaching Shoshana Zuboff's the Age of Surveillance Capitalism," *Blayne Haggart's Orangespace* (February 15, 2019). https://blaynehaggart.com/2019/02/15/evaluating-scholarship-or-why-i-wont-be-teaching-shoshana-zuboffs-the-age-of-surveillance-capitalism/.

23. Zuboff, *Age of Surveillance Capitalism*, 309.

24. Morozov, "Capitalism's New Clothes."; "Apple's Independent Repair Provider Program Expands Globally," Apple news release, March 29, 2021, https://www.apple.com/newsroom/2021/03/apples-independent-repair-provider-program-expands-globally/.

25. Zuboff, *Age of Surveillance Capitalism*, 46–47.

26. Both the iBeacon service and the iPhone's built-in Wallet app enable push notifications from marketers. Nick Couldry and Ulises A. Mejias, *The Costs of Connectivity: How Data Is Colonizing Human Life and Appropriating It for Capitalism* (Stanford: Stanford University Press, 2019), 10.

27. Research bureau Evercore ISI has estimated Apple's advertising turnover in 2022 at $5bn a year and expected a six time increase over the next four years, to $30bn. At this rate, Apple is on its way to joining the largest advertising networks: Google, Facebook and Amazon.The latter company earned over $31bn in advertisements in 2021. Marc Hijink, "Apple Doet Een Greep Naar De Advertentiedollars Van De iPhone," *NRC*, September 8, 2022, https://www.nrc.nl/nieuws/2022/09/08/apple-doet-een-greep-naar-de-advertentiedollars-van-de-iphone-a4141255.

28. While Zuboff's analysis suggests that the platform business itself contributes nothing to the value of the behavioral data, several critics have pointed to contributions to value addition, such as the company's investment in services that attract users, the development of matching algorithms or the act of retailing ad opportunities to advertisers. Mueller also points to a mutual economic benefit that underlies the user-platform interaction in question: "Consumers trade the tracking of their online behavior for free information services: search engines, maps and navigation services, email services, informational and entertainment content, document storage, and millions of connections to products, other individuals and groups." Milton Mueller, "A Critique of the 'Surveillance Capitalism' Thesis: Toward a Digital Political Economy," *SSRN*, August 2, 2022, https://dx.doi.org/10.2139/ssrn.4178467. But obviously, this relationship entails a lopsided balance of power, especially an asymmetrical loss of privacy, whereby users are scrutinized more than ever, while the institutions conducting surveillance remain "stubbornly opaque." Mark Andrejevic, *Ispy: Surveillance and Power in the Interactive Era* (Lawrence, Kansas: University Press of Kansas, 2007), 7. Of course, the benefits (free information services) to the users could conceivably be facilitated by non-surveillance capitalist means, but at present they are not. Mark Whitehead, "Review of Shoshana Zuboff, *The Age of Surveillance Capitalism: The Fight for a Human Future at the New Frontier of Power*. New York: Public Affairs, 2019," *Antipode online*, October 2, 2019, https://antipodeonline.org/wp-content/uploads/2019/10/Book-review_Whitehead-on-Zuboff.pdf.

29. Morozov, "Capitalism's New Clothes," 46.

30. James Muldoon, *Platform Socialism: How to Reclaim Our Digital Future from Big Tech* (London: Pluto Press, 2022), 24.

31. Morozov, "Capitalism's New Clothes."

32. Eisenhower 1946, as qtd. in Foster and McChesney, "Surveillance Capitalism," 3. This 1946 Memorandum was published as Appendix A in Seymour Melman, *Pentagon Capitalism* (New York: McGraw Hill, 1971), 231–34.

33. Military Keynesianism is an economic policy based on the position that government should raise military spending to boost economic growth. It is a fiscal stimulus policy as was advocated by the economist John Maynard Keynes. But where Keynes advocated increasing public spending on socially useful items (infrastructure in particular), additional public spending is allocated to the arms industry, the area of defense over which the executive exercises greater discretionary power. In the United States this theory was applied during and after the Second World War, during the presidencies of Franklin Delano Roosevelt and Harry Truman, the latter with a sweeping secret national security program outlined in 1950 in the document NSC-68. The influence of military Keynesianism on U.S. economic policy choices allegedly lasted until the Vietnam War. Yet, even in the midst of its official abandonment, during the Reagan period, the U.S. in reality continued to deploy military Keynesianism. After the dot-com bubble initiated a recession, in early 2001, the terrorist attacks of 9/11 triggered yet another episode of military Keynesianism, while there were also new resurgences in 2011 and 2021. James M. Cypher, "The Political Economy of Systemic U.S. Militarism," *Monthly Review* 73, no. 11 (2022), https://monthlyreview.org/2022/04/01/the-political-economy-of-systemic-u-s-militarism-2/.

34. Foster and McChesney, "Surveillance Capitalism," 17.

35. Julian Vigo, "The World Google Controls and Surveillance Capitalism," *Counterpunch*, December 17, 2018, https://www.counterpunch.org/2018/12/17/the-world-google-controls-and-surveillance-capitalism/.

36. Foster and McChesney, "Surveillance Capitalism," 21.

37. *Ibid.*, 17–27.

38. John B. Foster and Brett Clark, "Notes from the Editors," *Monthly Review* 70, no. 2 (2018), https://monthlyreview.org/category/2018/volume-70-issue-02-june/.

39. Couldry and Mejias, *Costs of Connection*, x.

40. *Ibid.*, 4 (italics in original).

41. David Harvey, *A Brief History of Neoliberalism* (Oxford: Oxford University Press, 2005), 159–65.

42. Couldry and Mejias, *Costs of Connection*, 9.

43. Vicki Husueh, "Cultivating and Challenging the Common: Lockean Property, Indigenous Traditionalisms, and the Problem of Exclusion," *Contemporary Political Theory* 5 (2006): 193–214, https://doi.org/10.1057/palgrave.cpt.9300233.

44. Julie E. Cohen, "The Biopolitical Public Domain: The Legal Construction of the Surveillance Economy," *Philosophy & Technology* 31, no. 2 (2018): 213–33, https://doi.org/10.1007/s13347-017-0258-2.

45. This edict also informed indigenous people that if they willingly subjected to the new order, their lives, property and religious beliefs would be respected and left alone. Sensing that not many natives would choose this option, however, the *Requerimiento* spelled out the terrible treatments that awaited those who refused: forceful invasion of their territory and waging wars against them in all possible ways and manners; taking their wives and children, enslaving them and disposing of them as the Church's Highness might command; taking away their goods and "do all the mischief and damage that we can." It is ironic, of course, that voluntary submission was not even a real option, because the document was read in Spanish to non-Spanish speaking peoples. Even if they would have obtained a limited translation, they would most likely not have understood the legal and theological concepts by which they were suddenly and forcefully dispossessed of their property, or even understood the concept of property in the same way as the invaders did. Couldry and Mejias, *Costs of Connection*, 92–93.

46. Zuboff, *Age of Surveillance Capitalism*, 176–80.

47. Nick Couldry and Ulises A. Mejias, "The Decolonial Turn in Data and Technology Research, What Is at Stake and Where Is It Heading?" *Information, Communication & Society* 26, no. 4 (2023): 790–91, https://www.tandfonline.com/doi/full/10.1080/1369118X.2021.1986102.

48. Carl J. Griffin, "Enclosure as Internal Colonisation: The Subaltern Commons, *Terra Nullius* and the Settling of England's Waste," *Transactions of the RHS* 1 (December 2023): 95, 119, https://doi.org/10.1017/S0080440123000014.

49. Ibid., 95–96, 105.

50. E.P. Thompson, *Customs in Common: Studies in Traditional Popular Culture* (New York: Free Press, 1991), 163–75.

51. Griffin. "Enclosure," 97–98.

52. World Economic Forum, *Personal Data: The Emergence of a New Data Asset Class* (Geneva, CH: WE Forum, February 17, 2011), https://www.weforum.org/publications/personal-data-emergence-new-asset-class/.

53. David Hart, "On the Origins of Google," *National Science Foundation*, August 17, 2004, http://www.nsf.gov/discoveries/disc_summ.jsp?cntn_id=100660&org=NSF.

54. Qtd. in Zuboff, *Age of Surveillance Capitalism*, 179.

55. Couldry and Mejias, *Costs of Connection*, 93–94; José van Dijck, *The Culture of Connectivity: A Critical History of Social Media* (Oxford: Oxford University Press, 2013).

56. Trebor Scholz, ed., *Digital Labor* (New York: Routledge, 2013); Christian Fuchs and Klaus Unterberger, eds., *The Public Service Media and Public Service Internet Manifesto* (London: University of Westminster Press, 2021).

57. Nick Srnicek, *Platform Capitalism* (Malden, MA: Polity Press, 2017), 41–42.

58. Couldry and Mejias, *Costs of Connectivity*, 101–02.

59. Ibid., 5

60. Ibid., xii. This is the core difference between Couldry and Mejias' argument and other influential critiques of what is happening with data, such as Zuboff's.

61. Ibid., 19–20, 189.

62. In this conception, at colonialism's core are not particular methods (its methods were always complex and variable), but is a consistent vision of the world's resources, economic and cognitive, that undergirds universal claims to those resources by a particular few. While colonial practices always have a material basis—a land grab of particular types of resources—the underlying rationale remains the same: the specific cosmic vision of a particular ethnic group understood as universal rationality. Anibal Quijano, "Coloniality and Modernity/Rationality," *Cultural Studies* 2, no. 2–3 (2007): 177, https://www.tandfonline.com/doi/abs/10.1080/09502380601164353. This is how the provincial discourse of Big Data today claims privileged application everywhere. The data colonialism thesis foregrounds this continuity with the *epistemic* violence of earlier colonialism. Paola Ricaurte, "Data Epistemologies, the Coloniality of Power, and Resistance," *Television & New Media* 20, no. 4 (2019): 350–65, https://doi.org/10.1177/1527476419831640.

63. Couldry and Mejias, "Decolonial Turn," 11–12.

64. Couldry and Mejias, *Costs of Connectivity*, xiii, 85.

65. Ibid., xiii.

66. Ibid., 190–91.

67. FLIA, "Notice of the State Council Issuing the New Generation of Artificial Intelligence Development Plan," State Council Document No. 35, July 2017, https://flia.org/notice-state-council-issuing-new-generation-artificial-intelligence-development-plan/; DigiChina, "State Council Notice Concerning Issuance of the Planning Outline for the Construction of a Social Credit System (2014–2020)," *DigiChina*, June 14, 2014, https://digichina.stanford.edu/work/planning-outline-for-the-construction-of-a-social-credit-system-2014-2020/.

68. Cohen, "Biopolitical Public Domain."

69. Couldry and Mejias, *Costs of Connectivity*, 11, 55, 100. 134.

70. *Ibid.*, 13.
71. Safiya Umoja Noble, *Algorithms of Oppression* (New York: New York University Press, 2018), 2; Ruha Benjamin, *Race after Technology: Abolitionist Tools for the New Jim Code* (London: Polity Books, 2019), 5–6.
72. Noble, *Algorithms*, 186.
73. Couldry and Mejias, *Costs of Connection*, 191.
74. Ricaurte, "Data Epistemologies."
75. Couldry and Mejias, *Costs of Connection*, 10.
76. *Ibid.*, xiii.
77. *Ibid.*, xiv.
78. *Ibid.*, 191–92 (italics in original).
79. *Ibid.*, 33.

Chapter 2

1. Ben Tarnoff, *Internet for the People: The Fight for Our Digital Future* (London: Verso, 2022),7–8; Janet Abbate, *Inventing the Internet* (Boston: MIT Press, 2000), 7–81; Johnny Ryan, *A History of the Internet and the Digital Future* (London: Reaktion Books, 2010), 23–40.
2. Tim Wu, *The Master Switch: The Rise and Fall of Information Empires* (New York: Vintage, 2010).
3. John Naughton, *What You Really Need to Know About the Internet. From Gutenberg to Zuckerberg* (London: Quercus, 2012), 45–46.
4. Susan Landau, *Surveillance or Security? The Risks Posed by New Wiretapping Technologies* (Cambridge: MIT Press, 2010), 18.
5. Ryan, *History of the Internet*, 16–17; Abbate, *Inventing the Internet*, 135, 195; Robert W. McChesney, *Digital Disconnect: How Capitalism Is Turning the Internet against Democracy* (New York: The New Press, 2013), 99–100.
6. Tarnoff, *Internet for People*, 8–9; Abbate, *Inventing the Internet*, 70–71; Katie Hafner and Matthew Lyon, *Where the Wizards Stay up Late: The Origins of the Internet* (New York: Simon & Schuster, [1996] 2006]), 233–34.
7. Yasha Levine, *Surveillance Valley: The Secret Military History of the Internet* (London: Icon Books Ltd, 2019), 7–8.
8. The internet was invented and grew as part of a long line of government projects, mainly militarily, dating back to the virtual reality of naval warfare in World War I. For this background, see Scott Malcomson, *Splinternet: How Geopolitics and Commerce Are Fragmenting the World Wide Web* (New York: OR Books, 2016), 13–58.
9. Tarnoff, *Internet for People*, 9–10; Abbate, *Inventing the Internet*, 113–45; Ryan, *History of the Internet*, 31–44; Hafner and Lyon, *Wizards*, 219–37.
10. Vint Cerf and Robert Kahn, "A Protocol for Packet Network Intercommunication," *IEEE Transactions on Communications* 22, no. 5 (1974): 637–48.
11. Tarnoff, *Internet for People*, 11.
12. Hafner and Lyon, *Wizards*, 240–46, 248–49; Ryan, *History of the Internet*, 88–91; Abbate, *Inventing the Internet*, 186–88, 195.
13. Tarnoff, *Internet for People*, 13.
14. Rajiv C. Shah and Kesan Jay P., "The Privatization of the Internet's Backbone Network," *Journal of Broadcasting & Electronic Media* 51, no. 1 (2007): 93–94, https://www.tandfonline.com/doi/abs/10.1080/0883815070130807.
15. Thomas Streeter, "'That Deep Romantic Chasm': Libertarianism, Neoliberalism and the Computer Culture," in *Communication, Citizenship and Social Policy: Rethinking the Limits of the Welfare State*, ed. Andrew Calabrese and Jean-Claude Burgelman (Boulder: Rowman & Littlefield, 1999), 49–64.
16. McChesney, *Digital Disconnect*, 97, 101.
17. Qtd. in Wu, *Master Switch*, 276.
18. John Markoff, *What the Doormouse Said ... How the Sixties Counterculture Shaped the Personal Computer* (New York: Viking Penguin, 2005).
19. McChesney, *Digital Disconnect*, 101–02.
20. Heather Brooke, *The Revolution Will Be Digitised: Dispatches from the Information War* (London: Heinemann, 2011), 24; Johan Soderberg, *Hacking Capitalism: The Free and Open Software Movement* (New York: Routledge, 2008); Howard Rheingold, *The Virtual Community:Homesteading on the Electronic Frontier* (Reading, MA: Addison Wesley, 1993).
21. Esther Dyson, *Release 2.0: A Design for Living in the Digital Age* (New York: Broadway Books, 1997); Kevin Kelly, *New Rules for the New Economy:10 Radical Strategies for a Connected World* (New York: Viking, 1998); Rheingold, *The Virtual Community*.
22. Fred Turner, *From Counterculture to Cyberculture: Stewart Brand, the Whole Earth Network, and the Rise of Digital Utopianism* (Chicago: University of Chicago Press, 2008), 3.
23. *Ibid.*, 3–4.
24. *Ibid.*, 4.
25. *Ibid.*, 5.
26. Gregory Bateson, *Steps to an Ecology of Mind* (San Francisco: Chandler Pub Co., 1972).
27. Streeter, "Deep Romantic Chasm."
28. Ronald S. Burt, "The Network Entrepreneur," in *Entrepreneurship: The Social Science View*, ed. Richard Swedberg (Oxford: Oxford University Press, 2000), 281–307.
29. Turner, *From Counterculture*, 5.
30. *Ibid.*, 8.
31. *Ibid.*, 248.
32. The Global Business Network grew out of the Learning Conferences, network events organized by Stewart Brand and attended by corporate executives, sponsored by Shell, AT&T and Volvo.
33. Peter Schwartz and Stewart Brand, *The 1989 GBN Scenario Book:Decades of Restructuring* (Emeryville: Global Business Network, 1989).
34. Paulina Borsook, an insider of this technolibertarian world, wrote the book *Cyberselfish*. It was based on an essay that appeared in *Mother Jones* in 1996 and traces the origins of technolibertarianism. See Paulina Borsook, *Cyberselfish: A*

Critical Romp through the Terribly Libertarian Culture of High Tech (New York: Public Affairs, 2000). Borsook characterized the culture of the digital technology community as predominantly anarchic market-based libertarian, anti-government, and anti-regulation. She criticized the lack of philanthropy in digital technology circles (yet overlooked the tax avoidance practices of Silicon Valley companies), and questioned how an industry given birth through large government funding could be so vehemently anti-government. The book also includes Borsook's experiences as a woman at *Wired* magazine and in Silicon Valley.

35. Turner, *From Counterculture*, 248–49.
36. Tarnoff, *Internet for People*, 15.
37. Ibid., 15–16.
38. Josie Fischels, "A Look Back at the Very First Website Ever Launched, 30 Years Later," *NPR*, August 6, 2021, https://www.npr.org/2021/08/06/1025554426/a-look-back-at-the-very-first-website-ever-launched-30-years-later.
39. Tarnoff, *Internet for People*, 16–17.
40. Ibid., 17.
41. Ibid., 19.
42. McChesney, *Digital Disconnect*, 104.
43. Ibid., 107.
44. Al From, *Democrats and the Return to Power* (New York: Palgrave Macmillan, 2013); Jennifer Holt, *Empires of Entertainment: Media Industries and the Politics of Deregulation, 1980–1996* (New Brunswick: Rutgers University Press, 2011); Robert Horowitz, *Irony of Regulatory Reform: The Deregulations of American Telecommunications* (Oxford: Oxford University Press, 1989).
45. Robert Brenner, *The Boom and the Bubble* (London: Verso, 2001).
46. Crain, *Profit over Privacy*, 23–25, 29–30, 52.
47. David Hesmondhalgh, *The Cultural Industries* (Los Angeles: Sage, 2013), 125.
48. Crain, *Profit over Privacy*, 33.
49. Tarnoff, *Internet for People*, 20.
50. John Markoff, "Building the Superhighway," *The New York Times*, January 24, 1993, https://www.nytimes.com/1993/01/24/business/building-the-electronic-superhighway.html.
51. McChesney, *Digital Disconnect*, 117.
52. Given the fact that even Gore as a centrist politician ("New Democrat") in a powerful position could not hold a modest public grip on the new internet, there was of course little room for more radical initiatives and less influential voices. Tarnoff mentions one interesting bold initiative: a bill, introduced by Senator Daniel Inouye (D-Hawaii) in 1994, that would have telecom companies reserve up to 20% of their capacity for "public uses." This capacity would be considered "public property"—the telecoms would have no control over it. It would be used to offer free access to qualifying organizations, such as libraries, nonprofits, and educational institutions, so long as they provided "educational, informational, cultural, civic, or charitable services directly to the public without charge for such services." Such organizations would also receive funding to support their ability to provide these services. The idea came from the Telecommunications Policy Roundtable, a coalition of unionized workers, consumer activists, computer professionals, and others who offered a counterpoint to the deregulatory craze of the era by demanding a "public lane on the information highway." A major source of inspiration was public media: the Public Broadcasting Act of 1967, which had created the Corporation for Public Broadcasting to promote noncommercial broadcasting on radio and television. Inouye's bill went nowhere; telecom lobbying ensured the public lane would never be realized. The Telecommunications Policy Roundtable made a valient effort, but could not muster the kind of mass mobilization that was needed to overcome industry opposition. See Tarnoff, *Internet for People*, 21–22.
53. Ibid., xiii.

Chapter 3

1. Joseph Turow, *The Daily You: How the New Advertising Industry Is Defining Your Identity and Your Worth* (New Haven: Yale University Press, 2011), 38, 40.
2. Lee, "Putting the 'Capitalism' in Surveillance Capitalism."
3. See Appendix A: Advertising and Mixed Motives to Sergey Brin and Lawrence Page, "The Anatomy of a Large-Scale Hypertextual Web Search Engine," *Computer Networks and ISDN Systems* 30, no. 1–7: 18, https://doi.org/10.1016/S0169-7552(98)00110-X.
4. Ibid.
5. Zuboff, *Age of Surveillance Capitalism*, 71–82.
6. White House, *The Framework for Global Electronic Commerce* (Washington, DC, 1997), https://clintonwhitehouse4.archives.gov/WH/New/Commerce/.
7. Crain, *Profit over Privacy*, 9.
8. Ibid., 11, 13–14.
9. Crain, *Profit over Privacy*, 19–21; Edward Lee Lamoureux, *Privacy Surveillance and the New Media You* (New York: Peter Lang, 2016).
10. Crain, *Profit over Privacy*, 21–22.
11. Shane Greenstein, "Commercialization of the Internet: The Interaction of Public Policy and Private Choices or Why Introducing the Market Worked So Well," in *Innovation Policy and the Economy*, ed. Adam B. Jaffe, Josh Lerner, and Scott Stern (Cambridge: MIT Press, 2001), 151–61.
12. Crain, *Profit over Privacy*, 57, 59.
13. Ibid., 61.
14. Ibid., 15, 57.
15. Ibid., 66.
16. Ibid., 57–58
17. Ibid., 71–72.
18. David M. Kristol, "Cookies: Standards, Privacy, and Politics," *ACM Transactions on Internet Technology* 1, no. 2 (2001): 155–98, https://dl.acm.org/doi/10.1145/502152.502153.
19. Crain, *Profit over Privacy*, 15, 75–76, 78.

20. *Ibid.*, 94–95.
21. Srnicek, *Platform Capitalism*, 103.
22. Florian Cord and Simon Schleusener, "Looking Backward at the Present, 2020–1990: Deleuze's 'Postscript on Control Societies,'" *Coils of the Serpent: Journal for the Study of Contemporary Power* 6 (2020): 1–12, https://ul.qucosa.de/api/qucosa%3A72917/attachment/ATT-0/.
23. Crain, *Profit over Privacy*, 96–98.
24. David Lyon, *Surveillance and Social Sorting: Privacy, Risk and Digital Discrimination* (New York: Routledge, 2002).
25. Crain, *Profit over Privacy*, 101–03.
26. Andrejevic, *iSpy*, 2–4.
27. Kevin Robins and Frank Webster, *Times of the Technoculture: From the Information Society to the Virtual Life* (London: Routledge, 1999).7; E.P. Thompson, *The Making of the English Working Class* (London: Penguin, [1968] 1979), 237–43.
28. Crain, *Profit over Privacy*, 106.
29. Marx used this term to refer to the dispossession of peasants from their land and their entrance into labor markets as landless owners. In *Capital*, Marx described how the old English institution of communal property was gradually eroded with help from what he calls a "parliamentary form of robbery": the Enclosure Acts, which granted landlords private ownership over land previously belonging to the people in common. For centuries, English peasants depended on rights over common and waste lands for planting crops, grazing animals, gathering wood and foraging for food. Enclosures occurred during two main periods in the sixteenth and late eighteenth to early nineteenth centuries, with over 5,300 enclosure bills enacted between 1604 and 1914 relating to 6.8 million acres—over a fifth of the total area of England. Basically, the periods of enclosure were about the conversion of public goods in such a way that they could be redistributed as private assets. Muldoon, *Platform Socialism*, 40–41; UK Parliament, "Enclosing the Land," accessed July 8, 2024, https://www.parliament.uk/about/living-heritage/transformingsociety/towncountry/landscape/overview/enclosingland/.
30. Harvey, *Brief History of Neoliberalism*, 159–65.
31. Toby Lester, "The Reinvention of Privacy," *Atlantic Monthly* 287, no. 3 (March 2001): 28, https://www.theatlantic.com/magazine/archive/2001/03/the-reinvention-of-privacy/302140/.
32. Andrejevic, *iSpy*, 4.
33. AWS, "Healthcare.Gov Case Study—Amazon Web Services (AWS)," https://aws.amazon.com/solutions/case-studies/healthcare-gov/.
34. Jathan Sadowski, "The Internet of Landlords: Digital Platforms and New Mechanisms of Rentier Capitalism," *Antipode* 52, no. 2 (2020): 562–80, https://doi.org/10.1111/anti.12595.
35. Tarnoff, *Internet for People*, 86–87; Paul Langley and Andrew Leyshon, "Platform Capitalism: The Intermediation and Capitalization of Digital Economic Circulation," *Finance and Society* 3, no. 1 (2017): 11–31, https://doi.org/10.2218/finsoc.v3i1.1936.

36. Vincent Mosco, *The Political Economy of Communication* (London: Sage, 1996); Dan Schiller, *How to Think About Information* (Urbana: University of Illinois Press, 2007).
37. Crain, *Profit over Privacy*, 42.
38. Hesmondhalgh, *Cultural Industries*, 40, 99.
39. Andrejevic, *iSpy*, 201, 3.
40. Cass Sunstein, *Republic.Com* (Princeton: Princeton University Press, 2002), 37.
41. Andrejevic, *iSpy*, 202.
42. *Ibid.*, 49.
43. Webster and Robins, *Information Technology*, 312.
44. James R. Beniger, *The Control Revolution: Technological and Economic Origins of the Information Society* (Cambridge: Harvard University Press, 1986).
45. Webster and Robins, *Information Technology*, 328–43.
46. Frank Webster and Kevin Robins, "'I'll Be Watching You': Comment on Sewell and Wilkinson," *Sociology* 27, no. 2 (1993): 247, https://doi.org/10.1177/0038038593027002004.
47. Andrejevic, *iSpy*, 74.
48. Robins and Webster, *Times of the Technoculture*, 98–99.
49. Stuart Ewen and Elizabeth Ewen, *Channels of Desire* (New York: McGraw-Hill, 1982).
50. Stuart Ewen, *Captains of Consciousness: Advertising and the Social Roots of the Consumer Culture* (New York: McGraw-Hill, 1976).
51. *Ibid.*, 33.
52. Andrejevic, *iSpy*, 79.
53. Robert W. McChesney, *The Problem of the Media: U.S. Communication Politics in the 21st Century* (New York: Monthly Review Press, 2004), 32.
54. Andrejevic, *iSpy*, 80.
55. Roland Marchand, *Advertising the American Dream: Making Way for Modernity, 1920–1940* (Berkeley: University of California Press, 1985), 25.
56. Robins and Webster. *Times of the Technoculture*, 100–01.
57. Andrejevic, *iSpy*, 83.
58. *Ibid.*, 55–56, 71.
59. Crain, *Profit over Privacy*, 113–14.
60. Colin J. Bennett, *The Privacy Advocates: Resisting the Spread of Surveillance* (Cambridge: MIT Press, 2008). Other participating organizations included the Electronic Frontier Organization, Junk Busters, Privacy International, Privacy Rights Clearinghouse, Consumer Federation of America and the National Parent Teacher Association.
61. Crain, *Profit over Privacy*, 115.
62. Kathryn C. Montgomery, *Generation Digital: Politics, Commerce, and Childhood in the Age of the Internet* (Cambridge: MIT Press, 2007), 95.
63. Crain, *Profit over Privacy*, 121.
64. Montgomery, *Generation Digital*, 102.
65. Crain, *Profit over Privacy*, 122.
66. *Ibid.*, 126–27.

67. Evan Hansen, "Doubleclick under Email Attack for Consumer Profiling Plans," *CNET*, February 2, 2000, https://www.cnet.com/tech/services-and-software/doubleclick-under-email-attack-for-consumer-profiling-plans/.
68. Crain, *Profit over Privacy*, 128.
69. Ibid., 129–30.
70. Ibid., 132–33.
71. Joseph Turow, *Niche Envy:Marketing Discrimination in the Digital Age* (Cambridge: MIT Press, 2006), 63.
72. Frischmann and Selinger, *Re-engineering Humanity*, 60, 288–90.
73. As qtd. in *ibid.*, 289; Nancy S. Kim, *Wrap Contracts: Foundations and Ramifications* (New York: Oxford University Press, 2013).
74. For a more detailed analysis of the current legal and technical architecture of electronic contracting and the Taylorism involved in it, see Frischmann and Selinger, *Re-engineering Humanity*, 60–80.
75. Tim Wilkinson-Ryan, "The Perverse Consequences of Disclosing Standard Terms," *Cornell Law Review* 103, no. 1 (2020): 117–75, https://www.cornelllawreview.org/2020/07/28/the-perverse-consequences-of-disclosing-standard-terms/.
76. Andrejevic, *iSpy*, 17.
77. Crain, *Profit over Privacy*, 136.
78. Ibid., 137.
79. Ken Auletta, *Googled: The End of the World as We Know It* (New York: Penguin, 2010), 61.
80. Crain, *Profit over Privacy*, 137–38.
81. Google's initial business plan anticipated three revenue streams: licensing search technology to other sites; selling a hardware device that enabled other companies to perform internal searches; and advertising. But initially, Google expected to earn most of its revenue through licensing. See Steven Levy, *In the Plex: How Google Thinks, Works, and Shapes Our Lives* (New York: Simon & Schuster, 2011), 77–84.
82. *Ibid.*, 85–93; Tim Hwang, *Subprime Attention Crisis: Advertising and the Time Bomb at the Heart of the Internet* (New York: FSG Originals x Logic, 2020), 36–41.
83. Crain, *Profit over Privacy*, 138.
84. Douglas Rushkoff, *Throwing Rocks at the Google Bus: How Google Became the Enemy of Prosperity* (New York: Portfolio/Penguin, 2016).
85. Levy, *In the Plex*, 103–08.
86. Crain, *Profit over Privacy*, 139.
87. Ibid.: 140.
88. Auletta, *Googled*, 174.
89. Due to low user engagement and disclosed software design flaws that potentially allowed outside developers access to personal information of its users, Google Plus was shut down in April 2019.
90. Crain, *Profit over Privacy*, 141–43.
91. Stacey Lynn Schulman, "Hyperlinks and Marketing Design," in *The Hyperlinked Society*, ed. Joseph Turow and Lokman Tsui (Ann Arbor: University of Michigan Press, 2011), 145.
92. Crain, *Profit over Privacy*, 145–46.

Chapter 4

1. Steve Vallas and Juliet Schor, "What Do Platforms Do? Understanding the Gig Economy," *Annual Review of Sociology* 46 (2020): 273–94, https://www.annualreviews.org/doi/abs/10.1146/annurev-soc-121919-054857. For an early and influential typology, see Srnicek, *Platform Capitalism*, 49–88. Scrinek distinguishes five different types of platforms: advertising platforms (e.g., Google, Facebook), cloud platforms (e.g., AWS, Salesforce), industrial platforms (e.g., General Electric, Siemens), product platforms (e.g., Rolls Royce, Spotify), and lean platforms (e.g., Uber, Airbnb).
2. Muldoon, *Platform Socialism*, 14.
3. Brett Christophers, *Rentier Capitalism* (London: Verso, 2020), 41.
4. Muldoon, *Platform Socialism*, 15.
5. Christophers, *Rentier Capitalism*, 50.
6. Muldoon, *Platform Socialism*, 15–16.
7. Alex Rosenblat and Luke Stark, "Algorithmic Labor and Information Asymmetries: A Case Study of Uber's Drivers," *International Journal of Communication* 10 (2016): 3758–84, https://papers.ssrn.com/sol3/papers.cfm?abstract_id=2686227.
8. Niels van Doorn, "Platform Labor: On the Gendered and Racialized Exploitation of Low-Income Service Work in the 'On-Demand' Economy," *Information, Communication, Society* 20, no. 6 (2017): 898–914, https://doi.org/10.1080/1369118X.2017.1294194.
9. Juliet B. Schor, *After the Big Gig: How the Sharing Economy Got Hijacked and How to Win It Back* (Oakland: University of California Press, 2020), 121.
10. Muldoon, *Platform Socialism*, 17.
11. Tim O'Reilly, "What Is Web 2.0: Design Patterns and Business Models for the Next Generation of Software," *Communications and Strategies*, no. 1 (First Quarter 2007): 17–37, https://papers.ssrn.com/sol3/papers.cfm?abstract_id=1008839. For an analysis of Web.2.0 and the rise of the "platform," see Anne Helmond, "The Platformization of the Web: Making Web Data Platform Ready," *Social Media + Society* 1, no. 2 (2015): 1–11, https://doi.org/10.1177/2056305115603080. For Web 2.0 and the rise of the social media, see van Dijck, *Culture of Connectivity*.
12. Tarnoff, *Internet for People*, 94.
13. Muldoon, *Platform Socialism*, 27.
14. Ibid., 41.
15. Bianca Bosker, "Facebook's Mark Zuckerberg 2005 Interview Reveals Ceo's Doubts," *Huffington Post*, August 11, 2011, https://www.huffpost.com/entry/facebook-mark-zuckerberg-2005-interview_n_924628.
16. Qtd. in David Kirkpatrick, *The Facebook Effect: The Inside Story of the Company That Is Connecting the World* (New York: Simon & Schuster, 2010), 42.
17. Muldoon, *Platform Socialism*, 28–29.
18. Mark Zuckerberg, "Building Global Community," *Facebook.com*, February 16, 2017,

https://www.facebook.com/zuck/posts/10154544292806634.

19. Muldoon, *Platform Socialism*, 29.

20. Mark Zuckerberg, "Harvard Commencement Address," *The Harvard Gazette*, May 25, 2017, https://news.harvard.edu/gazette/story/2017/05/mark-zuckerbergs-speech-as-written-for-harvards-class-of-2017/.

21. Robert Putnam, *Bowling Alone: The Collapse and Revival of American Community* (New York: Simon & Schuster, 2000).

22. Harvey, *History of Neoliberalism*.

23. Muldoon, *Platform Socialism*, 30–31.

24. Zuckerberg, "Building Global Community."

25. Muldoon, *Platform Socialism*, 32.

26. On the exploitation of user-generated content by technology companies, see Christian Fuchs, "Labor in Informational Capitalism and on the Internet," *The Information Society* 26, no. 3 (2010): 176–96, https://doi.org/10.1080/01972241003712215; Adam Arvidson and Eleanor Colleoni, "Value in Informational Capitalism and on the Internet," *The Information Society* 28, no. 3 (2012): 135–50, https://doi.org/10.1080/01972243.2012.669449.

27. Muldoon, *Platform Capitalism*, 33.

28. Catherine Price, "You Have One Life: Do You Really Want to Spend It Looking at Your Phone?" *The Guardian*, January 2, 2024, https://www.theguardian.com/lifeandstyle/2024/jan/02/smartphones-attention-economy-reclaim-free-time.

29. Tarnoff, *Internet for People*, 95.

30. Philip E. Agre, "Surveillance and Capture: Two Models of Privacy," *The Information Society* 10, no. 2 (1994): 117, https://www.tandfonline.com/doi/abs/10.1080/01972243.1994.9960162.

31. Tarnoff, *Internet for People*, 95–96.

32. *Ibid*., 98; John C. Wu, "Anatomy of a Dot-Com," *Supply Chain Management Review* 5, no. 6 (2001): 42–51, https://kupdf.net/download/operations-management_5afdf9f5e2b6f57070550a90_pdf.

33. Tarnoff, *Internet for People*, 99; Brad Stone, *The Everything Store: Jeff Bezos and the Age of Amazon* (New York: Little, Brown, 2013), 78, 100–35.

34. Tarnoff, *Internet for People*, 100; Stone, *Everything Store*, 115–16, 134.

35. Keyvan Kashkool, *The Making of a Modern Market: Ebay.Com* (Doctoral dissertation, University of California, Berkeley, 2010), 75–82, https://escholarship.org/uc/item/16t905b5; Stone, *Everything Store*, 263–65, 134.

36. Tarnoff, *Internet for People*, 101.

37. *Ibid*., 101–02; Stone, *Everything Store*, 193–96.

38. Linda M. Kahn, "Sources of Tech Platform Power," *Georgetown Law & Technology Review* 2, no. 2 (2018): 325–34, https://georgetownlawtechreview.org/sources-of-tech-platform-power/GLTR-07-2018/.

39. Tarnoff, *Internet for People*, 102.

40. Phil Jones, *Work without the Worker: Labor in the Age of Platform Capitalism* (London: Verso, 2021), 77.

41. Callum Jones, "Amazon Profits Surge on Strong Trading Season and Cloud Computing Growth," *The Guardian*, February 1, 2024, https://www.theguardian.com/technology/2024/feb/01/amazon-earnings-q4.

42. Tarnoff, *Internet for People*, 103.

43. Tung-Hui Hu, *A Prehistory of the Cloud* (Cambridge: MIT Press, 2015), 37–71; Devin Kennedy, "The People's Utility," *Logic(s)*, August 1, 2018, https://logicmag.io/failure/the-people's-utility/.

44. Tarnoff, *Internet for People*, 104–05.

45. *Ibid*., 108.

46. Audrey Kurenkow, "A Brief History of Neural Nets and Deep Learning," *Skynet Today*, September 27, 2020, https://www.skynettoday.com/overviews/neural-net-history; Alex Hanna et al., "Lines of Sight," *Logic(s)*, December 20, 2020, https://logicmag.io/commons/lines-of-sight/.

47. Marion Fourcade and Kieran Healy, "Seeing Like a Market," *Socio-Economic Review* 15, no. 1 (2017): 9–29, https://doi.org/10.1093/ser/mww033.

48. Tarnoff, *Internet for People*, 109.

49. Jones, *Work without the Worker*, 77.

50. The term "algorithmic management" was initially coined in 2015 by Min Kyung Lee, Daniel Kusbit, Evan Metsky and Laura Dabbish to describe the managerial role by algorithms on the Uber and Lyft platforms. See Min Kyung Lee et al., "Working with Machines: The Impact of Algorithmic and Data-Driven Management on Human Workers," *Proceedings of the 33rd Annual ACM Conference on Human Factors in Computing Systems* (April 18, 2015): 1603–12, https://doi.org/10.1145/2702123.2702548.

51. Rosenblat and Stark, "Algorithmic Labor."

52. Krishnan Vasudevan and Ngai Keung Chan, "Gamification and Work Games. Examining Consent and Resistance among Uber Drivers," *New Media & Society* 24, no. 4 (2022): 869–70, https://doi.org/10.1177/14614448221079028.

53. Sarah Mason, "Chasing the Pink," *Logic(s)*, January 1, 2019, https://logicmag.io/play/chasing-the-pink/.

54. Vaseduvan and Chun, "Gamification," 869.

55. Alex Rosenblat, *Uberland: How Algorithms Are Rewriting the Rules of Work* (Berkeley: University of California Press, 2018), 138–42.

56. Jones, *Work without the Worker*, 57–58. For more details about the role of microworkers in algorithmic management systems, see last section of Chapter 6.

57. California passed Proposition 22 in November 2020, allowing companies like Uber to classify their workers as independent contractors in the gig economy. The Uber-backed ballot measure became the most expensive in California history, with over $200m spent on campaigning in support. The constitutionality of Proposition 22 was immediately challenged legally, but a California appeals court ruled in March 2023 that Prop 22 was largely constitutional and would remain in effect. See Dan Blystone, "The History of Uber: How the Controversial Ride-Sharing Company Came to Dominate Its Market Worldwide," *Investopedia*, April

18, 2023, https://www.investopedia.com/articles/personal-finance/111015/story-uber.asp.

58. Veena Dubal, "A Brief History of the Gig," *Logic(s)*, May 4, 2020, https://logicmag.io/security/a-brief-history-of-the-gig/; Tarnoff, *Internet of the People*, 115–16.

59. David Weil, *The Fissured Workplace: Why Work Became So Bad for So Many and What Can Be Done to Improve It* (Cambridge: Harvard University Press, 2014).

60. Joan Greenbaum, *Windows on the Workplace: Technology, Jobs, and the Organization of Office Work* (New York: Monthly Review Press, 2004), 92–94.

61. Dubal, "Brief History of Gig."

62. Within the advanced capitalist world, different countries have responded in different ways to this new service. A comparative analysis of Uber's arrival and reception in the United States, Germany and Sweden showed three very different responses to the disruptive new actor, ranging from welcome embrace and regulatory adjustments to complete rejection and legal bans. Conflicts over Uber centered on different issues in these three countries. The specific regulatory flashpoints that Uber triggered, mobilized different actors, led to the formation of different coalitions and shaped the terms on which conflicts over Uber were framed and fought. See Kathleen Thelen, "Regulating Uber: The Politics of the Platform Economy in Europe and the United States," *Perspectives on Politics* 16, no. 4 (2018): 938–53. This section about Uber concentrates on the U.S. state of affairs.

63. Tarnoff, *Internet for People*, 128.

64. Milo van Bokkum, "Voor Het Eerst Sinds De Oprichting in 2009 Maakt Uber Winst," *NRC*, August 1, 2023, https://www.nrc.nl/nieuws/2023/08/01/voor-het-eerst-sinds-de-oprichting-in-2009-maakt-uber-winst-a4171026.

65. Jasper Jolly and Graeme Wearden, "Landmark Moment as Uber Unveils Its First Annual Profit as Limited Company," *The Guardian*, February 7, 2024, https://www.theguardian.com/technology/2024/feb/07/landmark-moment-as-uber-unveils-first-annual-profit-as-limited-company.

66. Quantitative easing (QE) is a monetary policy action where a central bank purchases predetermined amounts of government bonds or other financial assets in order to stimulate economic activity. Quantitative easing is a novel form of monetary policy that came to be widely implemented after the financial crisis of 2007–2008.

67. Taylor Tepper and Michael Adams, "Federal Funds Rate History 1990 to 2024," *Forbes Advisor*, May 10, 2024, https://www.forbes.com/advisor/investing/fed-funds-rate-history/.

68. Srnicek, *Platform Capitalism*, 30.

69. Carole Cadwalladr, "'Capitalism Is Dead. Now We Have Something Much Worse': Yanis Varoufakis on Extremism, Starmer and the Tyranny of Big Tech," *The Guardian*, September 24, 2023, https://www.theguardian.com/world/2023/sep/24/yanis-varoufakis-technofeudalism-capitalism-ukraine-interview.

70. Tarnoff, *Internet for People*, 119–21.

71. Niels van Doorn and Adam Badger, "Platform Capitalism's Hidden Abode: Producing Data Assets," *Antipode* 52, no. 5 (2020): 1477, https://doi.org/10.1111/anti.12641.

72. Tarnoff, *Internet for People*, 121–22.

73. Blystone, "History of Uber."

74. Rebecca Aydin, "How 3 Guys Turned Renting Air Mattresses in Their Apartment into a $31 Billion Company, Airbnb," *Business Insider*, September 19, 2019, https://www.businessinsider.in/How-3-guys-turned-renting-an-air-mattress-in-their-apartment-into-a-25-billion-company/articleshow/51114238.cms.

75. Sri Rahaju Hijrah Hati et al., "A Decade of Systematic Literature Review on Airbnb, the Sharing Economy from a Multiple Stakeholder Perspective," *Heliyon* 7 (2021), https://www.semanticscholar.org/paper/A-decade-of-systematic-literature-review-on-Airbnb:-Hati-Balqiah/652ed1dc827b5b48182c98a18cf392b7ad8e8dbb.

76. Zainab Hussain and Joshua Franklin, "Valuation Surges Past $100 Billion in Biggest U.S. IPO of 2020," *Reuters*, December 10, 2020, https://www.reuters.com/article/airbnb-ipo-idUSKBN28K261.

77. For instance, Airbnb's net loss for 2019 was $674m; for 2020 $4,585bn (its highest loss ever); and for 2021 $352m. John Hughes, "Airbnb SWOT Analysis," *Business Chronicler*, accessed January 31, 2024, https://businesschronicler.com/swot/airbnb-swot-analysis/; "Airbnb Q4 2022 and Full-Year Financial Results," Airbnb.com, February 14, 2023, https://news.airbnb.com/airbnb-q4-2022-and-full-year-financial-results/.

78. Hati et al., "A decade of systematic literature review on Airbnb."

79. Muldoon, *Platform Socialism*, 44.

80. Douglas Atkin, *The Culting of Brands: When Customers Become True Believers* (New York: Portfolio, 2004).

81. Muldoon, *Platform Socialism*, 44–45.

82. *Ibid.*, 46.

83. For an excellent exposé of the more general "grassroots" countermobilization against regulation, protest or controversy by elite consultants working for corporations and powerful interest groups, see Edward T. Walker, *Grassroots for Hire* (Cambridge: Cambridge University Press, 2014).

84. Niels van Doorn, "A New Institution on the Block: On Platform Urbanism and Airbnb Citizenship," *New Media & Society* 22, no. 10 (2020): 1808–26, https://doi.org/10.1177/1461444819884377.

85. It is telling that as part of its strategy to mitigate the fierce resistance from the taxi industry and government regulators during its expansion, Uber hired David Plouffe, a high-profile political and corporate strategist, who worked on Barack Obama's 2008 presidential campaign. Blystone, "History of Uber."

86. Hahrie Han and Elizabeth McKenna,

Groundbreakers: How Obama's 2.2 Million Volunteers Transformed Campaigning in America (Oxford: Oxford University Press, 2015).

87. Muldoon, *Platform Socialism*, 49.

88. Luke Yates, "Understanding the Airbnb 'Movement': How Platform-Sponsored Grassroots Lobbying Is Changing Politics," *Rosa Luxemburg Stiftung News*, October 22, 2021, https://www.rosalux.de/en/news/id/45224/understanding-the-airbnb-movement.

89. Mike Miller, "Alinsky for the Left: The Politics of Community Organizing," *Dissent*, Winter 2010, https://www.dissentmagazine.org/article/alinsky-for-the-left-the-politics-of-community-organizing/; Luke Yates, "Understanding the Airbnb 'Movement.'"

90. Muldoon, *Platform Socialism*, 48, 50.

91. Josh Bivens, *The Economic Costs and Benefits of Airbnb. No Reason for Local Policymakers to Let Airbnb Bypass Tax for Regulatory Obligations* (Economic Policy Institute, 2019), https://files.epi.org/pdf/157766.pdf.

92. "Inside Airbnb," Airbnb.com, accessed May 12, 2024, http://insideairbnb.com/.

93. Anna Minton, "New York Is Breaking Free of Airbnb's Clutches. This Is How the Rest of the World Can Follow Suit," *The Guardian*, September 23, 2023, https://www.theguardian.com/commentisfree/2023/sep/27/new-york-airbnb-renters-cities-law-ban-properties.

Chapter 5

1. Tarnoff, *Internet for People*, 109–12.

2. Veena Pureswaran and Paul Brody, *Device Democracy: Saving the Internet of Things* (IBM Executive Report Electronics Industry, 2014), https://www.ibm.com/downloads/cas/Y5ONA8EV.

3. Couldry and Mejias, *Costs of Connection*, xx.

4. Naomi Klein, *This Changes Everything: Capitalism Vs. the Climate* (New York: Simon & Schuster, 2015), 170.

5. Couldry and Mejias, *Costs of Connection*, 137.

6. Tarnoff, *Internet for People*, 112–13.

7. Sadowski, *Too Smart*, 5–6; Langdon Winner, *The Whale and the Reactor: A Search for Limits in an Age of High Technology* (Chicago: University of Chicago Press, 1986).

8. Sadowski, *Too Smart*, 33–34.

9. Ibid., 9–11.

10. Ibid., 11–12.

11. Ibid., 32–33.

12. Jones, *Work without the Worker*, 33.

13. Mel van Elteren, *Managerial Control of American Workers: Methods and Technology from the 1880s to Today* (Jefferson, NC: McFarland, 2017), 144–208, 224–35.

14. Sadowski, *Too Smart*, 32–33.

15. Ibid., 39–40.

16. Michel Foucault, *The Birth of Biopolitics: Lectures at the College of France, 1978–1979* (New York: Picador, 2008).

17. Surveillance technologies like the panopticon were first tried in India and then imported to the United Kingdom. Fingerprinting was also first applied in South Asia in the mid-nineteenth century as a means of controlling prisoners, pensioners and contractors, but the practice reached England only later. Along with other forms of identification that were introduced in the colonies, the aim of fingerprinting was not primarily to identify unknown subjects (who had not been fingerprinted yet) but to extend police power over groups who were already under suspicion and already registered in the system. Couldry and Mejias, *Costs of Connection*, 99.

18. Michel Foucault, *Discipline and Punish: The Birth of the Prison* (New York: Vintage Books, 1995).

19. Couldry and Mejias remind the reader that even before the panopticon, methods such as taxation and the confession of sins were used during historical colonialism to ensure that colonial subjects felt an obligation to behave as though everything (from their material possessions to their innermost thoughts and desires) was open to scrutiny and auditing, the results of which had to be reported to the secular or religious authorities involved. Couldry and Mejias, *Costs of Connection*, 100; Ahmad H. Sai'di, "Colonialism and Surveillance," in *Routledge Handbook of Surveillance Studies*, ed. Kirstie Ball, Kevin Haggerty, and David Lyon (New York: Routledge, 2012), 151–58.

20. Sadowski, *Too Smart*, 40–41.

21. Benjamin H. Bratton, *The Stack: On Software and Sovereignty* (Cambridge: MIT Press, 2016).

22. Couldry and Mejias, *Costs of Connection*, 100.

23. Sadowski, *Too Smart*, 41–42, 46–48.

24. Andrejevic, *iSpy*, 106–107.

25. Ibid., 108.

26. Robins and Webster, *Times of Technoculture*, 117.

27. Kevin Robins and Frank Webster, "Capitalism: Information, Technology, and Everyday Life," in *The Political Economy of Information*, ed. Vincent Mosco and Janet Wasko (Madison: University of Wisconsin Press, 1998), 52.

28. Sadowski, *Too Smart*, 46–47.

29. Margaret Jane Radin, *Boilerplate: The Fine Print, Vanishing Rights, and the Rule of Law* (Princeton: Princeton University Press, 2013), 27.

30. Ian Ayres and Alan Schwartz suggest that optimism bias—specifically misplaced optimism about terms and conditions—may affect consumers. Ian Ayres and Alan Schwartz, "The No-Reading Problem in Consumer Contract Law," *Stanford Law Review* 66, no. 3 (2014): 545–610. Oren Bar-Gill claims: "Myopic consumers care more about the present and not enough about the future…. Myopia is common. People are impatient, preferring immediate benefits even at the expense of future costs." Oren Bar-Gill, *Seduction by Contract: Law, Economics, and Psychology in Consumer Markets* (Oxford: Oxford University Press, 2012),

85–87. Nancy S. Kim discusses various cognitive biases that affect contracting behavior. Kim, *Wrap Contracts*, 85–87. There are also the findings about how people respond to information presented by machines, such as "automation bias" and "automation complacency" detailed by Nicholas Carr. See Nicholas Carr, *The Glass Cage: How Our Computers Are Changing Us* (New York: W.W. Norton, 2015).

31. Frischmann and Selinger, *Re-engineering Humanity*, 62–63, 67.

32. Ibid., 64–65.

33. Joseph Weizenbaum, *Computer Power and Human Reason: From Judgment to Calculation* (New York: W.H. Freeman & Co., 1976), 252.

34. Weizenbaum saw a shift in the pattern of the co-evolution between humans and their social and technological tools. There seemed to be an all too convenient marriage between tools and means. The tools—computers, systems analysis, science, instrumental reason—work together synergistically to define reality, just as the light under a lamp-post defines the territory where a drunkard might look for his lost keys. See Frischmann and Selinger, *Re-engineering Humanity*, 51.

35. Ibid., 11.

36. Ibid., 78.

37. The label "subprime" tends to be associated with mortgages and the toxic assets that contributed to the 2008 financial debacle, but subprime loans are not just for homes.

38. Michael Corkery and Jessica Silver-Greenberg, "Miss a Payment? Good Luck Moving That Car," *The New York Times*, September 24, 2014, https://archive.nytimes.com/dealbook.nytimes.com/2014/09/24/miss-a-payment-good-luck-moving-that-car/.

39. Sadowksi, *Too Smart*, 76–78.

40. Michael Sainato, "Ford Seeks to Remotely Repossess Cars after Missed Payments in U.S. Patent," *The Guardian*, March 3, 2023, https://www.theguardian.com/business/2023/mar/03/ford-reposses-patent-remote-lock.

41. Virginia Eubanks, *Automating Inequality: How High-Tech Tools Profile, Police, and Punish the Poor* (New York: St. Martin's Press, 2018).

42. Progressive Corporation, *Linking Driving Behavior to Automobile Accidents and Insurance Rates: An Analysis of Five Billion Miles Driven* (Mayfield, OH: Progressive Corporation, 2012), https://goodtimesweb.org/documentation/2012/snapshot_report_final_070812.pdf.

43. Sadowski, *Too Smart*, 78–79.

44. Gina Neff and Dawn Nafus, *Self-Tracking* (Cambridge: MIT Press, 2016); Kate Crawford, Jessica Lingel, and Tero Karppi, "Our Metrics, Ourselves: A Hundred Years of Self-Tracking from the Weight Scale to the Worst Device," *European Journal of Cultural Studies* 18, no. 4–5 (2015): 479–96, https://doi.org/10.1177/1367549415584857.

45. Deborah Lupton, *The Quantified Self: A Sociology of Self-Tracking* (Cambridge: Cambridge University Press, 2016).

46. Tamar Sharon and Dorien Zandbergen, "From Data Fetishism to Quantifying Selves: Self-Tracking Practices and the Other Values of Data," *New Media & Society* 19, no. 11 (2016): 1695–709, https://doi.org/10.1177/1461444816636090.

47. Couldry and Mejias. *Costs of Connection*, 169–71.

48. Mark Andrejevic, "The Big Data Divide," *International Journal of Communication* 8 (2014): 1674, https://ijoc.org/index.php/ijoc/article/viewFile/2161/1163.

49. Sadowski, *Too Smart*, 104 (italics in original).

50. Deborah Lupton, ed., *Self-Tracking, Health and Medicine: Sociological Perspectives* (London: Routledge, 2018).

51. Sarah Brayne, *Predict and Surveil: Data, Discretion, and the Future of Policing* (New York: Oxford University Press, 2021), 24–25. It is hard to fully understand the scope of the data brokerage industry. Even the Federal Trade Commission cannot find out exactly where data brokers obtain their information, because brokerages cite trade secrecy as an excuse to withhold that information.

52. Sadowski, *Too Smart*, 82–83.

53. Cathy O'Neil, *Weapons of Math Destruction: How Big Data Increases Inequality and Threatens Democracy* (New York: Crown Publishers, 2016).

54. Marcia Stepanek, "Weblining," *Bloomberg*, April 3, 2000, https://www.bloomberg.com/news/articles/2000-04-02/weblining.

55. Crain, *Profit over Privacy*, 104–05; Sadowski, *Too Smart*, 84.

56. Couldry and Mejias, *Costs of Connection*, 56.

57. Will Knight, "China's AI Awakening," *MIT Technology Review*, October 10, 2017, https://www.technologyreview.com/2017/10/10/148284/chinas-ai-awakening/.

58. Sadowski, *Too Smart*, 85–86.

59. Mara Hvistendahl, "Inside China's Vast New Experiment in Social Ranking," *Wired*, December 14, 2017, https://www.wired.com/story/age-of-social-credit/.

60. Rachel Botsman, "Big Data Meets Big Brother as China Moves to Rate Its Citizens," *Wired*, October 12, 2017, https://www.wired.co.uk/article/chinese-government-social-credit-score-privacy-invasion.

61. Mark Kear, "Playing the Credit Score Game: Algorithms, 'Positive' Data and the Personification of Financial Objects," *Economy and Society* 46, no. 3–4 (2017): 346–68, https://doi.org/10.1080/03085147.2017.1412642.

62. Hvistendahl, "Inside China's Vast New Experiment."

63. Sadowski, *Too Smart*, 88.

64. Diana Fu and Rui Hou, "Rating Citizens with China's Social Credit System," in *CPC Futures: The New Era of Socialism with Chinese Characteristics*, ed. Frank N. Pieke and Bert Hofman (Singapore: East Asian Institute, National University of Singapore, 2022), 78–85, https://epress.nus.edu.sg/cpcfutures/9789811852060-10.pdf.

65. Vincent Brussee, *Social Credit: The Warring States of China's Emerging Data Empire* (Singapore: Palgrave Macmillan, 2023), 12.

66. K.W. Tan, "As China Shuts out World, Internet Access from Abroad Gets Harder Too," *Freedom House*, June 23, 2022, https://freedomhouse.org/article/china-shuts-out-world-internet-access-abroad-gets-harder-too.

67. Zuboff, *Age of Surveillance Capitalism*, 393, 468.

68. Sadowski, *Too Smart*, 88–89.

69. The *Golden Shield Project*, which began in 1998 and was completed in 2008, is the Chinese fundamental nation-wide network-security construction project. It resulted in a constellation that includes a security management information system, a criminal information system, an exit and entry administration information system, a supervisor information system and a traffic management information system, among others. The Golden Shield Project also manages the Bureau of Public Information and Network Supervision, which is widely believed to operate a subproject called the Great Firewall of China, which is a censorship and surveillance project that blocks data from foreign countries that may be unlawful in the PRC. It is operated by the Ministry of Public Security. Angela Romano, "Asia," in *Public Sentinel: News Media and Governance Reform*, ed. Pippa Norris (New York: World Bank Publications, 2020), 360. *Sharp Eyes* is a project which aims to surveil one hundred percent of public space using surveillance in China by 2020, according to the 13th Five Year Plan released in 2016. Although it is highly questionable whether such bold targets as outlined in the plan had been achieved by then, the 14th Five Year Plan (released in 2021) continues with the project, instructing public security agencies to "closely guard against, and crack down on, the infiltration, sabotage, subversion and separatist activities by hostile forces." Dave Gershgorn, "China's Sharp Eyes' Program Aims to Surveil," *One Zero*, March 2, 2021, https://onezero.medium.com/chinas-sharp-eyes-program-aims-to-surveil-100-of-public-space-ddc22d63e015.

70. Johana Bhuiyan, "How Chinese Firm Linked to Repression of Uyghurs Aids Israeli Surveillance in West Bank," *The Guardian*, November 11, 2023, https://www.theguardian.com/technology/2023/nov/11/west-bank-palestinians-surveillance-cameras-hikvision.

71. Paul Mozur, "Inside China's Dystopian Dreams: A.I., Shame and Lots of Cameras," *The New York Times*, July 8, 2018, https://www.nytimes.com/2018/07/08/business/china-surveillance-technology.html; Phoebe Zhang, "Cities in China Most Monitored in the World, Report Finds," *South China Morning Post*, August 19, 2019, https://www.scmp.com/news/china/society/article/3023455/report-finds-cities-china-most-monitored-world.

72. "China: Police 'Big Data' Systems Violate Privacy, Target Dissent," *Human Rights Watch*, November 19, 2017, https://www.hrw.org/news/2017/11/19/china-police-big-data-systems-violate-pr.ivacy-target-dissent.

73. Couldry and Mejias, *Costs of Connection*, 57.

Chapter 6

1. Nelson Lichtenstein, *The Retail Revolution: How Wal-Mart Created a Brave New World of Business* (New York: Metropolitan Books, 2009); Nelson Lichtenstein, "Wal-Mart's Authoritarian Culture," *The New York Times*, June 21, 2011, https://www.nytimes.com/2011/06/22/opinion/22Lichtenstein.html?_r=1.

2. Simon Head, *Mindless: Why Smarter Machines Are Making Dumber Humans* (New York: Basic Books, 2014), 31–33.

3. The data-driven micromanagement of the workers in Amazon's warehouses may also include the method of "voice picking" (short for "voice-directed order picking"), which works to channel surveillance directly through the worker's body. The worker is instructed through a headset that relays automated verbal comments—issued not by a person but by the warehouse management system—while simultaneously watching the worker's movements. The system uses voice recognition to understand what the worker is saying. Couldry and Mejias, *Costs of Connection*, 6.

4. Sadowski, *Too Smart*, 93; Chris Baraniuk, "How Algorithms Run Amazon's Warehouses," *BBC*, August 1, 2015, https://www.bbc.com/future/article/20150818-how-algorithms-run-amazons-warehouses; Colin Lecher, "Amazon Automation Tracks and Fires Warehouse Workers for Productivity," *The Verge*, April 25, 2019, https://www.theverge.com/2019/4/25/18516004/amazon-warehouse-fulfillment-centers-productivity-firing-terminations; Lauren Kaori Gurley, "Internal Documents Show Amazon's Dystopian System for Tracking Workers Every Minute of Their Shifts," *Vice*, June 2, 2022, https://www.vice.com/en/article/5dgn73/internal-documents-show-amazons-dystopian-system-for-tracking-workers-every-minute-of-their-shifts.

5. Head, *Mindless*, 30–31.

6. Sadowski, *Too Smart*, 94.

7. Mac McClelland, "I Was a Warehouse Wage Slave," *Mother Jones*, March/April, 2012, https://www.motherjones.com/politics/2012/02/mac-mcclelland-free-online-shipping-warehouses-labor/.

8. Sadowski, *Too Smart*, 95, 103; Thuy Ong, "Amazon Patents Wristband That Track Warehouse Employees' Hands in Real Time," *The Verge*, February 1, 2018, https://www.theverge.com/2018/2/1/16958918/amazon-patents-trackable-wristband-warehouse-employees.

9. John Markoff, *Machines of Loving Grace: The Quest for Common Ground between Humans and Robots* (New York: Ecco, 2015), 206–7; Martin Ford, *Rise of the Robots: Technology and the Threat of a Jobless Future* (New York: Basic Books, 2015), 16–17.

10. Qtd. in Callum Jones, "Fears of Employee Displacement as Amazon Brings Robots into Warehouses," *The Guardian*, October 19, 2023, https://www.theguardian.com/technology/2023/oct/18/amazon-robot-warehouses-digit-workers. Separately, Amazon announced at the same event that it was deploying a robotic system called Sequoia at one of its Houston warehouses in an effort to speed up deliveries. The system was designed to help identify and store inventory 75% more quickly and reduce the processing time of orders by as much as 25%. The implications for the existing workers were not specified, however.

11. Jorn Boewe and Johannes Schulten, *The Long Struggle of the Amazon Employees: Laboratories of Resistance* (New York: Rosa Luxemburg Stiftung, 2017), 13.

12. Karen E.C. Levy, "The Contexts of Control: Information, Power, and Truck-Driving Work," *The Information Society* 31, no. 2 (2015): 166–74, https://doi.org/10.1080/01972243.2015.998105.

13. Christopher Haubursin, "Automation Is Coming for Truckers. But First, They're Being Watched," *Vox*, November 20, 2017, https://www.vox.com/videos/2017/11/20/16670266/trucking-eld-surveillance.

14. Levy, "Contexts of Control."

15. Haubursin, "Automation Is Coming."

16. Couldry and Mejias, *Costs of Connection*, 65; Jessica Bruder, "We're Watching You Work. Labor Is Fighting Employer's Techno-Utopian Dream of a Perfectly Efficient—and Totally Surveilled—Workforce," *The Nation*, June 15, 2015, 28–29.

17. Mark Schremmer, "UPS Receives Five-Year Renewal on ELD Exemption," *Landline*, October 26, 2022, https://landline.media/ups-receives-five-year-renewal-on-eld-exemption/.

18. Bruder, "We're Watching You Work," 30.

19. John Klyce, "Safety Vs. Privacy Issues Arise as Fedex Express Installs Driver-Facing Cameras in Vehicles," *Memphis Business Journal*, October 18, 2021, https://www.bizjournals.com/memphis/news/2021/10/18/fedex-express-installing-driver-cameras-delivery.html.

20. Jodi Kantor, "Working Anything but 9 to 5 Scheduling: Technology Leaves Low-Income Parents with Hours of Chaos," *The New York Times*, August 13, 2014, https://www.nytimes.com/interactive/2014/08/13/us/starbucks-workers-scheduling-hours.html.

21. Sadowski, *Too Smart*, 97.

22. van Elteren, *Managerial Control*, 138–42, 160–62.

23. Matthew Cantor, "Idle No More: How Mouse Jigglers Are Taking on Nosy Bosses," *The Guardian*, March 6, 2023, https://www.theguardian.com/technology/2023/mar/05/idle-no-more-how-automatic-mouse-jigglers-are-taking-on-nosy-bosses.

24. van Elteren, *Managerial Control*, 158–67, 171–82, 224–35.

25. *Ibid.*, 160, 261–62.

26. Brayne, *Predict and Surveil*, 77.

27. *Ibid.*, 15.

28. Erum Salam, "Uber and Lyft to Pay out $328m to New York Ride-Share Drivers," *The Guardian*, November 2, 2023, https://www.theguardian.com/technology/2023/nov/02/uber-lyft-settlement-new-york-driver-lawsuit.

29. See Drivers Union, https://www.driversunionwa.org/; Massachusetts Drivers United, https://massdriversunited.org/about/; Rideshare Drivers United, https://www.drivers-united.org/; Gig Workers Rising, https://www.coworker.org/partnerships/gig-workers-rising.

30. Jones, *Work without the Worker*, 60; Jamie Woodcock, *The Fight against Platform Capitalism: An Inquiry into the Global Sruggles of the Gig Economy* (London: University of Westminster Press, 2021), 28.

31. Timo Nijssen, "Zonder Blauwe Hesjes, Maar Met Instagram Wil De FNV Fietskoeriers Verenigen," *NRC*, February 8, 2024, https://www.nrc.nl/nieuws/2024/02/08/zonder-blauwe-hesjes-maar-met-instagram-wil-de-fnv-fietskoeriers-verenigen-a4189603.

32. Maria E. Cecchinato, Sandy J.J. Gould, and Frederick Harry Pitts, "Self-Tracking and Surveillance at Work: Insights from Human-Computer Interaction and Social Science," in *Augmented Exploitation: Artificial Intelligence, Automation and Work*, ed. Phoebe V. Moore and Jamie Woodcock (London: Pluto Press, 2021), 127–28.

33. Steve Mann, Jason Nolan, and Barry Wellman, "Sousveillance: Inventing and Using Wearable Computing Devices for Data Collection in Surveillance Environments," *Surveillance and Society* 1, no. 3 (2003): 331–55, https://ojs.library.queensu.ca/index.php/surveillance-and-society/article/view/3344.

34. Cecchinato et al., "Self-Tracking and Surveillance," 132–33.

35. Joanna Bronowicka and Mirela Ivanova, "The Algorithmic Boss: Guessing, Gaming, Reframing and Contesting Rules in App-Based Management," in *Augmented Exploitation: Artificial Intelligence, Automation and Work*, ed. Phoebe V. Moore and Jamie Woodcock (London: Pluto Press, 2021), 149–62.

36. Ursula Huws, "Logged Labour: A New Paradigm of Work Organisation?" *Work Organisation, Labour and Globalisation* 10, no. 1 (2016): 7–26, https://doi.org/10.13169/workorgalaboglob.10.1.0007.

37. Bronowicka and Ivanova, "Algorithmic Boss," 159.

38. Michael Burawoy, *Manufacturing Consent: Changes in the Labor Process under Monopoly Capital* (Chicago: University of Chicago Press, 1979), 52, 57, 60, 85, 89.

39. Vasudevan and Chan, "Gamification and Work Games," 874–81.

40. Gavin Mueller, *Breaking Things at Work: The Luddites Are Right About Why You Hate Your Job* (London: Verso, 2021), 117–19; Mary L. Gray and Suri Siddarth, *Ghost Work: How to Stop Silicon*

Valley from Building a New Global Underclass (Boston: Houghton Mifflin, 2019).

41. Moritz Altenried, "The Platform as Factory: Crowdwork and the Hidden Labour Behind Artificial Intelligence," *Capital and Class* 44, no. 2 (2020): 145–58, https://doi.org/10.1177/0309816819899410.

42. Jones, *Work without the Worker*, 2.

43. Tarnoff, *Internet for People*, 118–19, 131–32.

44. Jones, *Work without the Worker*, 3.

45. Ibid., 32.

46. Ibid., 72–73.

47. Ibid., 74–76.

48. Mike Davis, *Planet of Slums* (New York: Verso, 2007), 174.

49. Astra Taylor, "The Automation Charade," *Logic(s)*, August 1, 2018, https://logicmag.io/failure/the-automation-charade/; Woodcock, *Fight against Platform Capitalism*, 56.

50. Jones, *Work without the Worker*, 7.

51. Ibid., 36–37; Mark Graham and Jamie Woodcock, *The Gig Economy: A Critical Introduction* (London: Polity Press, 2019), 54.

52. *Guardian* editorial, "The Guardian View on Microworking: Younger, Educated Workers Left Powerless," *The Guardian*, August 21, 2022, https://www.theguardian.com/commentisfree/2022/aug/21/the-guardian-view-on-microworking-younger-educated-workers-left-powerless.

53. Jones, *Work without the Worker*, 5.

54. Phil Jones and James Muldoon, *Rise and Grind: Microwork and Hustle Culture in the UK* (Thinktank Autonomy, June 2022), https://autonomy.work/wp-content/uploads/2022/06/riseandgrind11.pdf.

55. Janine Berg et al., *Digital Labour Platforms and the Future of Work: Toward Decent Work in the Online World* (International Labour Organization, 2018), https://www.ilo.org/global/publications/books/WCMS_645337/lang—en/index.htm.

56. Jeremias Prassl, *Humans as a Service: The Promise and Perils of Work in the Gig Economy* (Oxford: Oxford University Press, 2018).

57. Jones, *Work without the Worker*, 54–55.

58. Lehdonvirta, Vili, "From Millions of Tasks to Thousands of Jobs: Bringing Digital Work to Developing Countries," *blogs.worldbank.org*, January 31, 2012, https://blogs.worldbank.org/psd/from-millions-of-tasks-to-thousands-of-jobs-bringing-digital-work-to-developing-countries.

59. Jones, *Work without the Worker*, 60–61.

60. Davis, *Planet of Slums*, 181.

61. Jones, *Work without the Worker*, 61.

62. Ibid., 16–17.

63. Ibid., 47.

64. Alexander J. Quinn et al., "Crowdflow: Integrating Machine Learning with Mechanical Turk for Speed-Cost-Quality Flexibility" (2020), https://www.semanticscholar.org/paper/CrowdFlow-%3A-Integrating-Machine-Learning-with-Turk-Quinn-Bederson/53db354fdcfa8fe8d22c1d5ae84b9e86e6f57abe.

65. Alex J. Wood et al., "Good Gig, Bad Gig: Autonomy and Algorithmic Control in the Global Gig Economy," *Work, Employment and Society* 33, no. 1 (2019): 56–75, https://journals.sagepub.com/doi/pdf/10.1177/0950017018785616.

66. Jones, *Work without the Worker*, 49.

67. Lilly Irani and M. Six Silberman, "Turkopticon: Interrupting Worker Invisibility in Amazon Mechanical Turk," *Proceedings of the SIGCHI Conference on Human Factors in Computing Systems* (2013), 611–20, http://crowdsourcing-class.org/readings/downloads/ethics/turkopticon.pdf.

68. Gray and Siddarth, *Ghost Work*, 85–91.

69. Berg et al., *Digital Labour Platforms*, 74.

70. Gray and Siddarth, *Ghost Work*, 124–25; Jones, *Work without the Worker*, 53.

71. Frank Pasquale, *The Black Box Society: The Secret Algorithms That Control Money and Information* (Cambridge: Harvard University Press, 2016), 3–4.

72. Jones, *Work without the Worker*, 64–68, 70.

73. Ibid., 84–85.

74. Niloufar Salehi et al., "We Are Dynamo: Overcoming Stalling and Friction in Collective Action for Crowd Workers," *CH' 15: Proceedings of the 13rd Annual ACM Conference on Human Factors in Computing Systems* (2015), 1621–30, https://dl.acm.org/doi/10.1145/2702123.2702508.

75. Irani and Silberman, "Turkopticon," 611–15.

76. Jones, *Work without the Worker*, 92–93.

77. Thompson, *Making of English Working Class*, 604.

78. Jones, *Work without the Worker*, 93.

Chapter 7

1. Jathan Sadowski, "When Data Is Capital: Datafication, Accumulation, Extraction," *Big Data & Society* 6, no. 1 (2019), https://doi.org/10.1177/2053951718820549.

2. Sadowski, *Too Smart*, 117–18.

3. Justin McGuirk, "Honeywell, I'm Home! The Internet of Things and the New Domestic Landscape," *e-flux*, https://www.e-flux.com/journal/64/60855/honeywell-i-m-home-the-internet-of-things-and-the-new-domestic-landscape/.

4. Sadowski, *Too Smart*, 118.

5. Adam Hayes, "Smart Home: Definition, How They Work, Pros and Cons," *Investopedia*, September 14, 2022, https://www.investopedia.com/terms/s/smart-home.asp.

6. Kim J. Kaaz et al., "Understanding User Perceptions of Privacy, and Configuration Challenges in Home Automation," in *2017 IEEE Symposium on Visual Languages and Human-Centric Computing (VL/HCC)* (Raleigh, NC: IEEE, 2017), 297–301.

7. Measures to prevent, or at least mitigate, such attacks that have been suggested, include setting up smart appliances and devices with a strong password, using encryption when available, and only connecting trusted devices to one's network. Hayes, "Smart Home." This defeats the purpose of attaining efficiency to some extent, of course.

8. Adrian Kingsley-Hughes, "The Android 'Toxic Hellstew' Survival Guide," *ZDnet*, June 9, 2014, https://www.zdnet.com/article/the-android-toxic-hellstew-survival-guide/; Daniel R. Thomas, Alistair R. Beresford, and Andrew Rice, "Securing Metrics for the Android Ecosystem," in *Proceedings of the 5th Annual ACM CCS Workshop on Security and Privacy in Smartphones and Mobile Devices—SPSM '15* (Cambridge: Computer Laboratory, University of Cambridge, 2015), https://www.cl.cam.ac.uk/~drt24/papers/spsm-scoring.pdf; Stacey Higginbotham, "Why Insurance Companies Want to Subsidize Your Smart Home," *MIT Technological Review*, October 12, 2016, https://www.technologyreview.com/2016/10/12/157045/why-insurance-companies-want-to-subsidize-your-smart-home/; Sadowski, *Too Smart*, 120–21.

9. Qtd. in Allyson Chiu, "She Installed a Ring Camera in Her Children's Room for 'Peace of Mind.' A Hacker Accessed It and Harassed Her 8-Year-Old Daughter," *The Washington Post*, December 12, 2019, https://www.washingtonpost.com/nation/2019/12/12/she-installed-ring-camera-her-childrens-room-peace-mind-hacker-accessed-it-harassed-her-year-old-daughter/.

10. By March 2020, Ring supposedly had suspended the use of most third-party analytics services in Ring apps and was working to provide greater ability to opt out in its then-new Control Center. This would let users manage privacy and security options such as two-factor authentication, sharing information with third parties for personalized advertising, and managing any shared users. Davey Winder, "How to Stop Your Smart Home Spying on You," *The Guardian*, March 8, 2020, https://www.theguardian.com/technology/2020/mar/08/how-to-stop-your-smart-home-spying-on-you-lightbulbs-doorbell-ring-google-assistant-alexa-privacy.

11. Kari Paul, "Dozens Sue Amazon's Ring after Camera Hack Leads to Threats and Racial Slurs," *The Guardian*, December 23, 2020, https://www.theguardian.com/technology/2020/dec/23/amazon-ring-camera-hack-lawsuit-threats.

12. Lauren Bridges, "Ring Is the Largest Civilian Surveillance Network the U.S. Has Ever Seen," *The Guardian*, May 18, 2021, https://www.theguardian.com/commentisfree/2021/may/18/amazon-ring-largest-civilian-surveillance-network-us?ICID=ref_fark.

13. Johana Bhuiyan, "Amazon Ring Says U.S. Police Will Now Need Warrant to Access User Footage," *The Guardian*, January 24, 2024, https://www.theguardian.com/technology/2024/jan/24/police-warrant-amazon-ring-footage.

14. Bridges, "Ring Largest Civilian Surveillance Network."

15. The company had previously come under fire for its broader privacy policies around access to users' footage. In May 2023, Amazon had already entered into a 20-year $5.8m settlement with the FTC that required the company to disclose to its customers how much access it had to their data. Bhuiyan, "Amazon Ring."

16. It is interesting how smart televisions collect real-time data about what people watch—or even their remarks in front of the TV—and then use that information to send targeted messages to other devices in people's homes. Sapna Maheshwari, "Smart TV in Millions of U.S. Homes Track More Than What's on Tonight," *The New York Times*, July 5, 2018, https://www.nytimes.com/2018/07/05/business/media/tv-viewer-tracking.html.

17. Qtd. in Herb Weisbaum, "The Downside of Connected Tech: Are the Smart Devices in Your Home Spying on You?" *NBC*, December 16, 2019, https://www.nbcnews.com/better/lifestyle/downside-connected-tech-are-smart-devices-your-home-spying-you-ncna1101906.

18. For some time, the annual Mozilla "privacy not included" guide (available online) has been listing the privacy options for many smart home devices. If one has already invested in a particular technology that does not have the wanted privacy settings, often a few configuration tweaks can still help balancing device performance with data privacy, experts have claimed. Winder, "How to Stop."

19. Derek Kravits and Marshall Allen, "Your Medical Devices Are Not Keeping Your Health Data to Yourselves," *ProPublica*, November 21, 2018, https://www.propublica.org/article/your-medical-devices-are-not-keeping-your-health-data-to-themselves.

20. Marshall Allen, "You Snooze, You Lose: Insurers Make the Old Adage Literally True," *ProPublica*, https://www.propublica.org/article/you-snooze-you-lose-insurers-make-the-old-adage-literally-true.

21. Sadowski, *Too Smart*, 120–21; Higginbotham, Why Insurance Companies."

22. Rik Myslewski, "The Internet of Things Helps Insurance Firms Reward, Punish," *The Register*, May 23, 2014, https://www.theregister.com/2014/05/23/the_internet_of_things_helps_insurance_firms_reward_punish/.

23. John Rappoport, "How Private Insurers Regulate Public Police," *Harvard Law Review* 130, no. 6 (2017): 1539–614, https://harvardlawreview.org/print/vol-130/how-private-insurers-regulate-public-police/.

24. Sadowski, *Too Smart*, 121–22; Tom Baker and Jonathan Simon, "Embracing Risk," in *Embracing Risk: The Changing Culture of Insurance and Responsibility*, ed. Tom Baker and Jonathan Simon (Chicago: University of Chicago Press, 2002), 13.

25. Scott R. Peppet, "Unraveling Privacy: The Personal Prospectus and the Threat of a Full-Disclosure Future," *Northwestern University Law Review* 105, no. 3 (2011): 1153–204, https://scholarlycommons.law.northwestern.edu/cgi/viewcontent.cgi?article=1157&context=nulr.

26. Sadowski, *Too Smart*, 125.

27. The tragedy of the commons is a metaphoric label for a concept used in economics, ecology and other sciences. The concept holds that if numerous independent individuals should enjoy unfettered access to a finite, valuable resource, for example a

pasture, they will tend to over-use it, and may end up by destroying its value altogether. To exercise voluntary restraint is not a rational choice for any one individual—if he/she did, the others would merely replace him/her—yet the predictable result is a tragedy for all.

28. Frischmann and Selinger, *Re-engineering Humanity*, 133.

29. Writer Mark Wilson illustrates this by the example of an expensive smart toaster (which runs on AI and deep learning): "When *you* cook salmon wrong, you learn about cooking it right. When June [this toaster] cooks salmon wrong, its findings are uploaded, aggregated, and averaged into a June-database that you hope will allow all June ovens to get it right the next time. Good thing the firmware updates are installed automatically." Qtd. in Frischmann and Selinger, *Re-engineering Humanity*, 129–30. The other side of the coin is, of course, that such toasters along with other smart kitchen appliances, are likely to disincentivize people from cultivating culinary skills.

30. Frischmann and Selinger, *Re-engineering Humanity*, 130.

31. Weisbaum, "Downside of Connected Tech."

32. Evgeny Morozov and Francesca Bria, *Rethinking the Smart City: Democratizing the Smart City* (New York: Rosa Luxemburg Stiftung, New York Office, 2018), 2,

33. Mani Dhingra, "What's So Smart About Smart Cities?" *RTE Brainstorm*, March 22, 2023. https://www.rte.ie/brainstorm/2023/0321/1364457-whats-so-smart-about-smart-cities/.

34. Stephen Goldsmith, "As the Chorus of Dumb City Advocates Increases, How Do We Define the Truly Smart City?" *Data-Smart City Solutions*, September 16, 2021, https://datasmart.hks.harvard.edu/chorus-dumb-city-advocates-increases-how-do-we-define-truly-smart-city.

35. Duncan McLaren and Julian Agyeman, *Sharing Cities: A Case for Truly Smart and Sustainable Cities* (Cambridge: MIT Press, 2015); Sam Musa, "Smart Cities—a Road Map for Development," *IEEE Potentials* 37, no. 2 (2018): 19–23, https://doi.org/10.1109/MPOT.2016.2566099.

36. Morozov and Bria, *Rethinking Smart City*, 8.

37. Townsend sees smartphones enabling an open-source approach to urban planning that calls upon the creative intelligence of the people who actually live there, as opposed to the wholesale standardization of smart cities through the global sales teams of a few corporations. Anthony M. Townsend, *Smart Cities: Big Data, Civic Hackers, and the Quest for a New Utopia* (New York: W.W. Norton, 2013). Antoine Picon sees a promising development in the "smart mobs" enabled by social media, who can quickly move to protest injustices and inequities perpetuated by governments. Antoine Picon, *Smart Cities: A Spatialised Intelligence* (West Sussex: Wiley, 2015).

38. Ayona Datta, "Three Big Challenges for Smart Cities and How to Solve Them," *The Conversation*, June 9, 2016, https://theconversation.com/three-big-challenges-for-smart-cities-and-how-to-solve-them-59191.

39. Morozov and Bria, *Rethinking Smart City*, 6, 8.

40. Jennifer Clark, "What Cities Need Now: Smart Cities Haven't Brought the Tangible Improvements That Many Hoped They Would. What Comes Next?" *MIT Technology Review*, April 28, 2021, https://www.technologyreview.com/2021/04/28/1023104/smart-cities-urban-technology-pandemic-covid/.

41. Picon, *Smart Cities*, 75.

42. Morozov and Bria, *Rethinking Smart City*, 7. In 2008, Illinois had adopted the State's Biometric Information Privacy Act (BIPA). The law does not completely ban facial recognition, but it tightly regulates *corporate use* of biometrics by mandating consent and disclosure. For law enforcement use, a new Illinois law adopted in June 2023 banned the use of facial recognition on drones, except in cases of preventing terrorist attacks or immediate threats to life. Scarlett Evans, "Illinois Bans Drones from Using Facial Recognition," *IOT World Today*, June 29, 2023, https://www.iotworldtoday.com/security/illinois-bans-drones-from-using-facial-recognition-#close-modal.

43. Morozov and Bria, *Rethinking Smart City*, 8.

44. Clark, "What Cities Need."

45. *Ibid*.

46. Morozov and Bria, *Rethinking Smart City*, 2.

47. Richard Sennett, "No One Likes a City That's Too Smart," *The Guardian*, December 14, 2012, https://www.theguardian.com/commentisfree/2012/dec/04/smart-city-rio-songdo-masdar.

48. Adam Greenfield, *Against the Smart City: A Pamphlet* (New York: Do projects, 2013).

49. Rob Kitchin, "Making Sense of Smart Cities: Addressing Present Shortcomings," *Cambridge Journal of Regions, Economy and Society* 8, no. 1 (2015): 131–36, https://doi.org/10.1093/cjres/rsu027; Marianna Cavada, Chris Rogers, and Dexter Hunt, "Smart Cities: Contradicting Definitions and Unclear Measures," *The 4th World Sustainability Forum 2014—Conference Proceedings Paper*, 1–12, https://sciforum.net/manuscripts/2454/manuscript.pdf.

50. It is not urban planners who have been driving the smart cities trend, but the tech sector, which has very different norms and goals. For example, test beds and experimentation are common in tech but do not sit well with cities. At best, cities customize complicated networks of old and new sociotechnical systems to work in a particular place for communities with different cultures, interests and priorities. For the tech sector, however, such local variation challenges the whole notion of an urban operating system that can scale. Clark, "What Cities Need."

51. Jane Jacobs, *The Death and Life of Great American Cities* (New York: Random House, 1961).

52. Robert Caro, *The Power Broker: Robert Moses and the Fall of New York* (New York: Vintage Books, 1975).

53. Morozov and Bria, *Rethinking Smart City*, 3.

54. Patrick Russell, "Smart Way Forward: A Review of S*mart Cities: A Spatialised Intelligence*," *Planning Forum*, 17 (2016):109–12, https://sites.utexas.edu/planningforum/files/2021/02/PlanningForumVol17-1.pdf.

55. Morozov and Bria, *Rethinking Smart City*, 11.

56. *Ibid.*, 9–10.

57. Datta, "Three Big Challenges."

58. Shiuh-Shen Chien and Max D. Woodworth, "China's Urban Speed Machine: The Politics of Speed and Time in a Period of Rapid Urban Growth," *International Journal of Urban and Regional Research* 42, no. 4 (2018): 723–37, https://onlinelibrary.wiley.com/doi/10.1111/1468-2427.12610.

59. Frederico Caprotti and Robert Cowley, "Varieties of Smart Urbanism in the UK: Discursive Logics, the State and Local Urban Context," *Transactions of the Institute of British Geographers* 44, no. 3 (2019): 587–601, http://dx.doi.org/10.1111/tran.12284.

60. Maria Skou and Nicklas Echsner-Rasmussen, "Smart Cities around the World," *Geoforum Perpektiv* 14, no. 25 (2016): 61–67, https://doi.org/10.5278/ojs.perspektiv.v14i25.1289.

61. Rob Kitchin, "Conceptualising Smart Cities," *Urban Research & Practice* 5, no. 1 (2022): 157, https://doi.org/DOI:10.1080/17535069.2022.2031143.

62. Caprotti and Cowley, "Varieties of Smart Urbanism," 587.

63. Morozov and Bria, *Rethinking Smart City*, 9.

64. *Ibid.*

65. Clark, "What Cities Need."

66. Seeta P. Gangadharan and Jedszej Niklas, "Decentering Technology in Discourse on Discrimination," *Information, Communication & Society* 22, no. 7 (2021): 882–99, https://doi.org/10.1080/1369118X.2019.1593484.

67. Morozov defines technological solutionism as an endemic ideology that recasts complex social phenomena like politics, public health, education, and law enforcement "either as neatly defined problems with definite, computable solutions or as transparent and self-evident processes that can be easily optimized—if only the right algorithms are in place...." Evgeny Morozov, *To Save Everything, Click Here: The Folly of Technological Solutionism* (New York: Public Affairs, 2013), 5.

68. Clark, "What Cities Need."

69. Rob Kitchin, "Decentering Smart Cities," in *Equality in the City: Imaginaries of the Smart Future*, ed. Susan Flynn (Bristol: Intellect, 2021), 260–66.

70. Kitchin, "Conceptualising Smart Cities."

71. Sommer Mathis and Alexandra Kanik, "Why You'll Be Hearing a Lot Less About 'Smart Cities,'" *City Monitor*, February 18, 2021, https://citymonitor.ai/government/why-youll-be-hearing-a-lot-less-about-smart-cities.

72. Andrew J. Hawkins, "Sidewalk Labs Shuts Down Toronto Smart City Project," *The Verge*, May 7, 2020, https://www.theverge.com/2020/5/7/21250594/alphabet-sidewalk-labs-toronto-quayside-shutting-down.

73. For an overview of the state of affairs in 2018, see Morozov and Bria, *Rethinking Smart City*, 26–42.

74. Barcelona Urban Platform, "Commitment to Boosting Innovation in Barcelona," June 21, 2021, https://ajuntament.barcelona.cat/digital/sites/default/files/commitment_to_boosting_innovation_in_barcelona_bithabitat.pdf.

75. Ajuntament de Barcelona, "Decidim Barcelona." https://ajuntament.barcelona.cat/digital/en/technology-accessible-everyone/accessible-and-participatory/accessible-and-participatory-5.

76. Muldoon, *Platform Socialism*, 109.

Chapter 8

1. Steven Graham, *Cities under Siege: The New Military Urbanism* (London: Verso Books, 2011).

2. ACLU, *War Comes Home:The Excessive Militarization of American Police* (New York: American Civil Liberties Union, 2015), https://www.aclu.org/report/war-comes-home-excessive-militarization-american-police.

3. Louise Amoore, "Algorithmic War: Everyday Geographies of the War on Terror," *Antipode* 41, no. 1 (2009): 49–69, https://doi.org/10.1111/j.1467-8330.2008.00655.x.

4. Bruce Schneier, "Mission Creep: When Everything Is Terrorism," *The Atlantic*, July 16, 2013, https://www.theatlantic.com/politics/archive/2013/07/mission-creep-when-everything-is-terrorism/277844/.

5. Brayne, *Predict and Surveil*, 19.

6. Sadowski, *Too Smart*, 136–37.

7. NYCLU, "Stop and Frisk Data in the De Blasio Era (2019)," *New York Civil Liberties Union*, March 14, 2019, https://www.nyclu.org/en/publications/stop-and-frisk-de-blasio-era-2019.

8. Sadowski, *Too Smart*, 137.

9. Andrew Guthrie Ferguson, "High-Tech Surveillance Amplifies Police Bias and Overreach," *The Conversation*, June 12, 2020, https://theconversation.com/high-tech-surveillance-amplifies-police-bias-and-overreach-140225.

10. Ali Winston, "Palantir Has Secretly Been Using New Orleans to Test Its Predictive Policing Technology," *The Verge*, February 27, 2018, https://www.theverge.com/2018/2/27/17054740/palantir-predictive-policing-tool-new-orleans-nopd.

11. Brayne, *Predict and Surveil*, 7.

12. Peter Waldman, Lizette Chapman, and Jordan Roberson, "Palantir Knows Everything About You,"*Bloomberg*, April 19, 2018, https://www.bloomberg.com/features/2018-palantir-peter-thiel/?leadSource=uverify%20wall.

13. Jay Stanley, "New Orleans Program Offers Lessons in Pitfalls of Predictive Policing," *ACLU*, March 15, 2018, https://www.aclu.org/

news/privacy-technology/new-orleans-program-offers-lessons-pitfalls-predictive-policing.

14. Sadowski, *Too Smart*, 134; Ali Winston and Ingrid Burrington, "A Pioneer in Predictive Policing Is Starting a Troubling New Project," *The Verge*, April 2, 2015, https://www.theverge.com/2018/4/26/17285058/predictive-policing-predpol-pentagon-ai-racial-bias.

15. Andrew Guthrie Ferguson, *The Rise of Big Data Policing: Surveillance, Race, and the Future of Law Enforcement* (New York: New York University Press, 2017), 40–42.

16. Qtd. in Jonathan Bullington and Emily Lane, "How a Tech Firm Brought Data and Worry to New Orleans Crime Fighting," *NOLA.com | The Times Picayune*, March 1, 2018, https://www.nola.com/news/crime_police/how-a-tech-firm-brought-data-and-worry-to-new-orleans-crime-fighting/article_33b8bf05-722f-5163-9a0c-774aa69b6645.html.

17. Michael Isaac Stein, "New Orleans City Council Bans Facial Recognition, Predictive Policing and Other Surveillance Tech," *The Lens*, December 18, 2020, https://thelensnola.org/2020/12/18/new-orleans-city-council-approves-ban-on-facial-recognition-predictive-policing-and-other-surveillance-tech/.

18. Sarah Brayne, "Big Data Surveillance: The Case of Policing," *American Sociological Review* 82, no. 5 (2017): 977–1008, https://doi.org/10.1177/0003122417725865; Sadowski, *Too Smart*, 138.

19. Brayne, *Predict and Surveil*, 15.

20. Sadowski, *Too Smart*, 139.

21. Craig D. Uchida et al., "The Los Angeles Smart Policing Initiative: Reducing Gun-Related Violence through Operation Laser," BJA Bureau of Justice Assistance, October 1, 2012, https://bja.ojp.gov/sites/g/files/xyckuh186/files/media/document/LosAngelesSPI.pdf.

22. Ferguson, *Rise of Big Data Policing*, 102.

23. Ibid., 84–85; Justin Jouvenal, "The New Way Police Are Surveilling You: Calculating Your Threat Score," *The Washington Post*, January 10, 2016, https://www.washingtonpost.com/local/public-safety/the-new-way-police-are-surveilling-you-calculating-your-threat-score/2016/01/10/e42bccac-8e15-11e5-baf4-bdf37355da0c_story.html.

24. Sadowski, *Too Smart*, 140–41.

25. The broken windows theory states that visible signs of crime, antisocial behavior and civil disorder, such as broken windows (hence the name of the theory), vandalism, loitering, public drinking and drug use, jaywalking, and transportation fare evasion, create an environment that promotes even more crime and disorder. James Q. Wilson and George L. Kelling, "Broken Windows," *Atlantic Monthly* 249, no. 3 (1982): 29–38. The theory suggests that policing methods that target those minor crimes help to create an atmosphere of order and lawfulness. It was popularized in the 1990s by New York City Police Commissioner William Bratton and Mayor Rudy Giuliani, whose aggressive policing policies were influenced by the theory. The theory has drawn several criticisms. Specifically, many scholars point to the fact that there is no clear causal relationship between lack of order and crime. Rather, crime going down when order is being restored is a coincidental correlation. Additionally, the theory has contributed to racial and class bias in police practices, especially in the form of stop-and-frisk methods. Charlotte Ruhl, "Broken Windows Theory of Criminiology," *Simply Psychology*, February 13, 2024, https://www.simplypsychology.org/broken-windows-theory.html.

26. Sadowski, *Too Smart*, 141–42.

27. Emmanuel Didier, "Globalization of Quantitative Policing: Between Management and Statactivism," *Annual Review of Sociology* 44 (2018): 515–35, https://doi.org/10.1146/annurev-soc-060116-053308.

28. Brayne, *Predict and Surveil*, 21.

29. Didier, "Globalization of Quantitative Policing." 519.

30. "Juking the stats" is the unethical practice of manipulating or falsifying data to present a misleading picture of reality, often to make performance appear better than it actually is. This term was popularized by the TV show, *The Wire*, which depicted police officers misreporting crime statistics to make it seem like they were more effective at fighting crime.

31. Brayne, *Predict and Surveil*, 14.

32. Ibid., 18.

33. Ibid., 20. As police commissioner again in New York City in 2014, Bratton embraced an even more robust data driven system of surveillance with the NYPD. Ferguson, *Rise of Big Data Policing*, 29.

34. Brayne, "Big Data Surveillance," 989.

35. Sadowski. *Too Smart*, 143–44.

36. Brayne, *Predict and Surveil*, 45.

37. Sadowski. *Too Smart*, 145.

38. Ibid., 146–47.

39. Brayne, "Big Data Surveillance," 992.

40. Brayne, *Predict and Surveil*, 14.

41. Ibid., 16.

42. Rob Kitchin, *Getting Smarter About Smart Cities: Improving Data Privacy and Data Security* (Dublin, Ireland: Data Protection Unit, Department of the Taoiseach, 2016), 36, https://www.academia.edu/27367984/Getting_smarter_about_smart_cities_Improving_data_privacy_and_data_security.

43. John Gilliom and Torin Monahan, *SuperVision: An Introduction to the Surveillance Society* (Chicago: University of Chicago Press, 2012); Bruce Schneier, *Data and Goliath: The Hidden Battles to Collect Your Data and Control Your World* (New York: W.W. Norton, 2015).

44. Sadowksi, *Too Smart*, 148.

45. Ibid., 148–49.

46. The federal government's response to 9/11, which included congressional acts and new federal data sharing programs, shaped the legacy of counterterrorism on local policing. Such initiatives

include: the Information Sharing Environment (ISE), established by the Intelligence Reform and Terrorism Prevention Act of 2004; Secure Communities, a deportation program that relies on integrated databases and partnerships among federal, state and local law enforcement agencies; and Centers for Excellence—Department of Homeland Security-sponsored consortiums of universities conducting research on homeland security issues. They also included the fusion centers elucidated in this chapter. Brayne, *Predict and Surveil*, 21.

47. *Ibid.*, 22; Pasquale, *Black Box Society*, 45–46.

48. Richard V. Ericson and Kevin D. Haggerty, *Policing the Risk Society* (Oxford: Oxford University Press, 1997); Brayne, *Predict and Surveil*, 24.

49. *Ibid.*, 25, 54; Chris J. Hoofnagle, "Big Brother's Little Helpers: How Choicepoint and Other Commercial Data Brokers Collect and Package Your Data for Law Enforcement," *North Carolina Journal of International Law* 29, no. 4 (2003): 595–637, https://scholarship.law.unc.edu/ncilj/vol29/iss4/1.

50. Torin Monahan and Priscilla M. Regan, "Zones of Opacity: Data Fusion in Post 9/11 Security Organizations," *Canadian Journal of Law and Society* 27, no. 3 (2012): 301–2, 307, http://dx.doi.org/10.1017/S0829320100010528.

51. Sadowski, *Too Smart*, 149; Brayne, *Predict and Surveil*, 9.

52. *Ibid.*, 53.

53. Sarah Brayne, "Surveillance and System Avoidance: Criminal Justice Contact and Institutional Attachment," *American Sociological Review* 79, no. 3 (2014): 367, https://doi.org/10.1177/0003122414530398; Brayne, *Predict and Surveil*, 114–15.

54. Sadowski, *Too Smart*, 150.

55. Kevin D. Haggerty and Richard V. Ericson, "The Surveillant Assemblage," *British Journal of Sociology* 51, no. 4 (2000): 619, https://doi.org/10.1080/00071310020015280.

56. Brayne, *Predict and Surveil*, 115.

57. Thomas H. Davenport, "How Big Data Is Helping the NYPD Solve Crimes Faster," *Fortune*, July 17, 2016, https://fortune.com/2016/07/17/big-data-nypd-situational-awareness/.

58. Ferguson, *Rise of Big Data Policing*, 86.

59. R. Joshua Scannell, "Both a Cyborg and a Goddess: Deep Managerial Time and Informatic Governance," in *Object-Oriented Feminism*, ed. Katherine Behar (Minneapolis: University of Minnesota Press, 2016), 256.

60. E.S. Levine et al., "The New York City Police Department's Domain Awareness System," *Interfaces* 47, no. 1 (2017): 70–84, https://dl.acm.org/doi/10.1287/inte.2016.0860.

61. Sadowksi, *Too Smart*, 152.

62. John Fasman, *We See It All: Liberty and Justice in an Age of Perpetual Surveillance* (New York: Public Affairs, 2021), 12.

63. Ferguson, *Rise of Big Data Policing*, 90.

64. Alfred Ng, "Amazon's Helping Police Build a Surveillance Network with Ring Doorbells," *CNET*, June 5, 2019, https://www.cnet.com/home/smart-home/features/amazons-helping-police-build-a-surveillance-network-with-ring-doorbells/.

65. Beryl Lipton, "Neigborhood Watch Out: Cops Are Incorporating Private Cameras into Their Real-Time Surveillance Works," *Electronic Frontier Foundation*, May 11, 2023, https://www.eff.org/deeplinks/2023/05/neighborhood-watch-out-cops-are-incorporating-private-cameras-their-real-time.

66. *Ibid.*

67. James Vincent, "Artificial Intelligence Is Going to Supercharge Surveillance," *The Verge*, January 23, 2018, https://www.theverge.com/2018/1/23/16907238/artificial-intelligence-surveillance-cameras-security.

68. Ferguson, "High-Tech Surveillance Amplifies."

69. Dave Davies, "Surveillance and Local Police: How Technology Is Evolving Faster Than Regulation," *NPR*, January 27, 2021, https://www.npr.org/2021/01/27/961103187/surveillance-and-local-police-how-technology-is-evolving-faster-than-regulation.

70. Fasman, *We See It All*, 21–22.

71. *Ibid.*, 9–10.

72. Davies, "Surveillance and Local Police."

73. MacArthur Justice Center, "Shotspotter Generated over 40,000 Dead-End Police Deployments in Chicago in 21 Months, According to New Study," May 3, 2021, https://www.macarthurjustice.org/shotspotter-generated-over-40000-dead-end-police-deployments-in-chicago-in-21-months-according-to-new-study/.

74. Johana Bhuiyan, "'Ready for Some Help?': How a Controversial Technology Firm Courted Portland Police," *The Guardian*, May 3, 2023, https://www.theguardian.com/us-news/2023/may/03/oregon-police-gunshot-detection-shotspotter.

75. Fasman, *We See It All*, 109. In addition, this part of the Bill of Rights sets requirements for issuing warrants. These must be issued by a judge or magistrate, justified by probable cause, supported by oath or affirmation, and must specify the place to be searched and the persons or things to be seized.

76. Fasman, *We See It All*, 25; Davies, "Surveillance and Local Police."

77. *Ibid.* Arrest patterns born from racial bias are used to determine where to send police to ferret out further crimes. In her 2016 book *Weapons of Math Destruction*, Cathy O'Neil calls this "a pernicious feedback loop" in which "the policing itself spawns new data, which justifies more policing," as qtd. in Fasman, *We See It All*, 28.

78. Will Douglas Heaven, "Predictive Policing Algorithms Are Racist. They Need to Be Dismantled," *MIT Technology Review*, July 17, 2020, https://www.technologyreview.com/2020/07/17/1005396/predictive-policing-algorithms-racist-dismantled-machine-learning-bias-criminal-justice/.

79. Jathan Sadowski, "'Anyway, the Dashboard Is Dead': On Trying to Build Urban Informatics,"

New Media & Society 26, no. 1 (2021), 313–28, https://doi.org/10.1177/14614448211058455.

80. Ferguson, *Rise of Big Data Policing*, 184.

81. Xerxes Minocher and Caelyn Randall, "Predictable Policing: New Technology, Old Bias, and Future Resistance in Big Data Surveillance," *Convergence* 26, no. 5–6 (2020): 1108–24, https://doi.org/10.1177/1354856520933838.

82. Amnesty International, "USA: NYPD Ordered to Hand over Documents Detailing Surveillance of Black Lives Matter Protests Following Lawsuit," *Amnesty International*, August 1, 2022, https://www.amnesty.org/en/latest/news/2022/08/usa-nypd-black-lives-matter-protests-surveillance/.

83. Qtd. in Ed Shanahan, "Robot Dog's Work in Garage Collapse Should Quiet Critics, Adams Suggests," *The New York Times*, April 25, 2023 (updated April 28, 2023), https://www.nytimes.com/2023/04/25/nyregion/digidog-robot-nypd.html.

84. Gloria Oladipo, "Robot Dogs Deployed in New York Building Collapse Revive Surveillance Fears," *The Guardian*, April 26, 2023, https://www.theguardian.com/technology/2023/apr/26/robot-dogs-new-york-building-collapse-surveillance.

85. Nina Lakhani, "Investment Fund Links to Atlanta Police and 'Cop City' Project Revealed," *The Guardian*, March 22, 2023, https://www.theguardian.com/us-news/2023/mar/22/investment-fund-links-atlanta-police-cop-city-project.

86. Bill Chapell, "What's at Stake in Atlanta's 'Cop City' Protests," *NPR*, March 7, 2023, https://www.npr.org/2023/03/07/1161343394/atlanta-cop-city-protests-explained.

87. Akin Olla, "Why Is Georgia Prosecuting Leftwing Activists with the Same Law as Trump?" *The Guardian*, September 20, 2023, https://www.theguardian.com/commentisfree/2023/sep/20/georgia-prosecuting-cop-city-activists-rico-trump. Rico statutes target the unique structure of the mafia, a hyper-centralized organization with an insulated leadership that cannot be caught in street-level crimes. The laws allow for different crimes to be linked together and used to prosecute an entire organization at the same time. The Stop Cop City movement is the opposite; it is neither centralized nor a criminal organization.

88. Alleen Brown, "Federal Agencies Pushed Extreme View of Cop Ciy Protesters, Records Show," *The Guardian*, December 6, 2023, https://www.theguardian.com/us-news/2023/dec/06/cop-city-atlanta-georgia-environment-protesters-terrorism.

89. Timothy Pratt, "Outcry as Atlanta Refuses to Handle Petitions over 'Cop City' Police Campus," *The Guardian*, September 11, 2023, https://www.theguardian.com/us-news/2023/sep/11/atlanta-petition-police-cop-city-georgia.

90. Activists burned construction equipment at least twice in January 2024, while another pair of activists tied themselves to an Atlanta construction site linked to Brasfield & Gorrie, a company working on Cop City. One activist's social media account explained the dual approach as follows: "There is not two movements, one that dismantles earth-destroying machinery by night, another that coordinates nonviolent direct action by day. There is one movement, dedicated to victory, using methods mostly suited for the building of serious and capable networks of resistance," as qtd. in Timothy Pratt, "Georgia Police and FBI Conduct Swat-Style Raids on 'Cop City' Activists' Homes," *The Guardian*, February 10, 2024, https://www.theguardian.com/us-news/2024/feb/10/georgia-police-fbi-raids-cop-city-activists-atlanta.

91. By late April 2024, this opposition also included student-led protests at Emory University against Israel's war in Gaza as well as Cop City, which sought transparency about and divestment from both Israel and Cop City, seeing the two as connected. Atlanta cops were being trained by the Israel Defense Forces as part of the Georgia International Law Enforcement Exchange (Gilee), a non-profit program that sends U.S. police officers to Israel and vice versa, for training and other activities characterized by a marked militarization. Timothy Pratt, "'Our Struggles Are Connected': Atlanta Protesters Link Cop City to Gaza War," *The Guardian*, May 13, 2024, https://www.theguardian.com/us-news/article/2024/may/13/cop-city-emory-atlanta-israel.

92. Timothy Pratt, "Atlanta Police Surveil People Opposing 'Cop City': 'There's This Constant Stalking Feeling,'" *The Guardian*, May 29, 2024, https://www.theguardian.com/us-news/article/2024/may/29/atlanta-police-cop-city-surveillance.

93. Stanley, "New Orleans Offers Lessons."

94. Fasman, *We See It All*, 212–13.

95. *Ibid.*, 213.

96. *Ibid.*, 220.

97. Kashmir Hill, *Your Face Belongs to Us: The Secret Startup Dismantling Your Privacy* (New York: Simon & Schuster, 2023); Charles Arthur, "Your Face Belongs to Us by Kashmir Hill Review—Nowhere to Hide," *The Guardian*, September 21, 2023, https://www.theguardian.com/books/2023/sep/21/your-face-belongs-to-us-by-kashmir-hill-review-nowhere-to-hide.

98. Stijn Bronzwaer, "Clearview AI: Het Bedrijf Dat Met Gezichtsherkenning Verder Gaat Waar Zelfs Google En Facebook Stoppen," *NRC*, October 5, 2023, https://www.nrc.nl/nieuws/2023/10/05/clearview-ai-het-bedrijf-dat-met-gezichtsherkenning-verder-gaat-waar-zelfs-google-en-facebook-stoppen-a4176272.

99. ACLU, "In Big Win, Settlement Ensures Clearview AI Complies with Groundbreaking Illinois Biometric Privacy Law," ACLU news release, May 9, 2022, https://www.aclu.org/press-releases/big-win-settlement-ensures-clearview-ai-complies-with-groundbreaking-illinois.

100. Colleen McCue, *Data Mining and Predictive Analysis* (Oxford: Butterworth-Heinemann, 2007), 48.

Chapter 9

1. Couldry and Mejias, *Costs of Connection*, 194.
2. Crain, *Profit over Privacy*, 147. In taking this stance, Crain explicitly follows Gabriel Weinberg, the CEO of search engine DuckDuckGo who points to politics, advocating for strong privacy laws to force the digital advertising industry to return to contextual advertising without consumer surveillance rather than enhanced transparency in data collection, greater market competition, more responsible executives, or industry regulation. Gabriel Weinberg, "What If We All Just Sold Non-Creepy Advertising?" *The New York Times*, June 19, 2019, https://www.nytimes.com/2019/06/19/opinion/facebook-google-privacy.html.
3. Part of this law was foreshadowed by the action of the Spanish Data Protection Agency which, in August 2011, concluded that citizens had the right to request the removal of links and ordered Google to stop indexing the information and to remove existing links to its original sources. Ninety Spaniards had fought for this aspect of their informational privacy under the banner of "the right to be forgotten." Zuboff, *Age of Surveillance Capitalism*, 57–60.
4. It refers to Article 8(1) of the EU Charter of Fundamental Rights and Article 16(1) of the Treaty on the Functioning of the European Union (TFEU), which say that everyone has the right to the protection of personal data concerning themselves. General Data Protection Regulation (GDPR). "Recital 1: Data Protection as a Fundamental Right," 2018, https://gdpr-info.eu/recitals/no-1/.
5. Frischmann and Selinger, *Re-engineering Humanity*, 378n74.
6. Couldry and Mejias, *Costs of Connection*, 180.
7. Zuboff. *Age of Surveillance Capitalism*, 481.
8. Couldry and Mejias, *Costs of Connection*, 180–81.
9. Muldoon, *Platform Socialism*, 63–64.
10. Couldry and Mejias, *Costs of Connection*, 181.
11. Ben Lovejoy, "GDPR-Style Privacy Law in California Acting as a 'Blueprint' for Other States," *9to5mac.com*, December 27, 2019, https://9to5mac.com/2019/12/27/gdpr-style-privacy-law/.
12. U.S. State Privacy Legislation Tracker, https://iapp.org/resources/article/us-state-privacy-legislation-tracker/, accessed February 25, 2024.
13. Tarnoff, *Internet for People*, 149–50.
14. David Streitfeld and Edward Wyatt, "U.S. Escalates Google's Case by Hiring Noted Outside Lawyer," *The New York Times*, April 27, 2012, https://www.nytimes.com/2012/04/27/technology/google-antitrust-inquiry-advances.html.
15. McChesney, *Digital Disconnet*, 142–43.
16. Ibid., 143.
17. In addition, Google then faced a €25bn lawsuit in the UK and EU over anti-competitive conduct regarding digital advertising. Dan Milmo, "Google Fails to Overturn €4bn Fine over Android Bundling," *The Guardian*, September 14, 2022, https://www.theguardian.com/technology/2022/sep/14/google-fails-to-overturn-4bn-fine-over-android-bundling; Dan Milmo, "Google Faces $25bn Lawsuit in UK and EU over Digital Advertising," *The Guardian*, September 13, 2022, https://www.theguardian.com/technology/2022/sep/13/google-lawsuit-uk-eu-digital-advertising.
18. While still a law student, Kahn argued in a now-famous 2017 article in the *Yale Law Journal* that U.S. antitrust law was fundamentally broken, allowing for the rise of tech behemoths like Apple, Google and Amazon. She was also on the staff of the House Judiciary Committee that wrote the report issued in 2020 that advocated reining in the four tech giants. Diane Bartz, "Amazon Faces Landmark Monopoly Lawsuit by FTC," *MSN*, September 26, 2023, https://www.msn.com/en-us/news/us/amazon-faces-landmark-monopoly-lawsuit-by-ftc/ar-AA1hmZF7; *The Guardian* staff and agencies, "'Different Set of Rules': How FTC Head Lina Khan Is Fighting Tech Giants Such as Amazon," *The Guardian*, September 27, 2023, https://www.theguardian.com/technology/2023/sep/27/ftc-head-lina-khans-fight-against-amazon-has-been-years-in-the-making.
19. Tarnoff, *Internet for People*, 150–51.
20. Steven Pearlstein, "Here's the Inside Story of How Congress Failed to Rein in Big Tech," *The Washington Post*, July 6, 2023, https://www.washingtonpost.com/opinions/2023/07/06/congress-facebook-google-amazon-apple-regulation-failure/.
21. When testifying as a witness during a separate antitrust trial against Google on November 13, 2023, Google CEO Sundar Pichai confirmed that the company paid Apple a 36% revenue share (between $18bn and $20bn every year) to remain Safari's default search engine. Nick Robins-Early, "Sundar Pichai Denies Google Stifles Competition at Epic Antitrust Trial," *The Guardian*, November 14, 2023, https://www.theguardian.com/technology/2023/nov/14/google-ceo-sundar-pichai-fortnite-epic-games-monopoly-antitrust-trial.
22. Dominic Rushe and Kari Paul, "U.S. Justice Department Sues Google over Accusation of Illegal Monopoly," *The Guardian*, October 20, 2020, https://www.theguardian.com/technology/2020/oct/20/us-justice-department-antitrust-lawsuit-against-google.
23. John Naughton, "Google's Monopoly on the Search Engine Market Finally Timed Out?" *The Guardian*, September 30, 2023, https://www.theguardian.com/commentisfree/2023/sep/30/google-antitrust-us-department-of-justice-court-case-monopoly.
24. Ibid.
25. *The Guardian* staff and agencies, "Justice Department Alleges Google Tried to 'Eliminate' Ad Markets Rivals in Lawsuit," *The Guardian*, January 24, 2023, https://www.theguardian.com/

technology/2023/jan/24/justice-department-lawsuit-google-antitrust-law-advertising.

26. Michael Liedtke, "Google Will Pay $700 Million to Settle State Claims Its Android App Store Stifled Competition," *Fortune*, December 19, 2023, https://fortune.com/2023/12/19/google-settle-700-million-antitrust-state-android-app-store/.

27. Dara Kerr, "U.S. v. Google: As Landmark 'Monopoly Power' Trial Closes, Here's What to Look For," *NPR*, May 2, 2024, https://www.npr.org/2024/05/02/1248152695/google-doj-monopoly-trial-antitrust-closing-arguments.

28. Nick Robins-Early, "A US Judge Ruled that Google Built an Illegal Monopoly. What Happens Next?" *The Guardian*, August 6, 2024, https://www.theguardian.com/technology/article/2024/aug/06/google-antitrust-monopoly-ruling.

29. Nick Robins-Early, "Google's Second Antitrust Suit Brought by US Begins, Over Online Ads." *The Guardian*, September 9, 2024, https://www.theguardian.com/technology/article/2024/sep/09/google-antitrust-lawsuit-online-ads; David McCabe, "U.S. Prepares to Challenge Google's Online Ad Dominance," *The New York Times*, September 9, 2024, https://www.nytimes.com/2024/09/09/technology/google-ads-antitrust-trial.html.

30. Bobby Allyn, "Judge Allows Federal Trade Commission's Latest Suit against Facebook to Move Forward," *NPR*, January 11, 2022, https://www.npr.org/2022/01/11/1072169787/judge-allows-federal-trade-commissions-latest-suit-against-facebook-to-move-forw.

31. Bryce Covert, "Anti-Monopoly Power," *The Nation*, December 25, 2023/January 1, 2024, 36–41.

32. *The Guardian* staff and agencies, "'Different set of rules.'"

33. Allyn, "Judge Allows Federal Trade Commission's Latest Suit."

34. Federal Trade Commission, "FTC Says Ring Employees Illegally Surveilled Customers, Failed to Stop Hackers from Taking Control of Users' Cameras," *ftc.gov*, May 31, 2023, https://www.ftc.gov/news-events/news/press-releases/2023/05/ftc-says-ring-employees-illegally-surveilled-customers-failed-stop-hackers-taking-ntrol-users.

35. Alexander Fabino, "Ring Ordered to Refund 55,000 Customers in Mid-2024: Do You Qualify?" *Newsweek*, November 15, 2023, https://www.newsweek.com/ftc-ring-refund-privacy-breach-settlement-2024-1844178.

36. Lauren Feiner and Annie Palmer, "Amazon to Pay over $30 Million in FTC Settlements over Ring, Alexa Privacy Violations," *CNBC*, May 31, 2023, https://www.cnbc.com/2023/05/31/ftc-sues-amazon-over-ring-doorbell-privacy-violations.html.

37. Nick Robins-Early, "A Threat to Amazon and a Test of the FTC: Is This Big Tech's Antitrust Reckoning?" *The Guardian*, September 26, 2023, https://www.theguardian.com/technology/2023/sep/26/amazon-antitrust-lawsuit-analysis-big-tech.

38. Bartz, "Amazon Faces Landmark Monopoly Lawsuit."

39. Michael Liedtke et al., "Justice Department Sues Apple, Alleging It Illegally Monopolized the Smartphone Market," *APnews*, March 21, 2024, https://apnews.com/article/apple-antitrust-monopoly-app-store-justice-department-822d7e8f5cf53a2636795fcc33ee1fc3.

40. "Regulation (EU) 2022/1925 of the European Parliament and of the Council of 14 September 2022 on Contestable and Fair Markets in the Digital Sector and Amending Directives (EU) 2019/1937 and (EU) 2020/1828 (Digital Markets Act)." December 10, 2022. https://eur-lex.europa.eu/eli/reg/2022/1925/oj.

41. "Regulation (EU) 2022/2065 of the European Parliament and of the Council of 19 October 2022 on a Single Market For Digital Services and Amending Directive 2000/31/EC (Digital Services Act)," October 19, 2022, https://eur-lex.europa.eu/legal-content/EN/TXT/?uri=CELEX:32022R2065.

42. European Commission, "About the Digital Markets Act,"accessed July 8, 2024, https://digital-markets-act.ec.europa.eu/about-dma_en.

43. European Commission, "Digital Services Act (DSA) overview," accessed July 8, 2024, https://commission.europa.eu/strategy-and-policy/priorities-2019-2024/europe-fit-digital-age/digital-services-act_en.

44. Lisa O'Carroll, "How the EU Digital Services Act Affects Facebook, Google and Others," *The Guardian*, August 25, 2023, https://www.theguardian.com/world/2023/aug/25/how-the-eu-digital-services-act-affects-facebook-google-and-others.

45. While the changes have been marketed as a win for users, they are also benefiting Apple's advertising product, Apple Search Ads, which marketers are turning to for mobile ads that drive app installations. Kif Leswing, "Apple's Ad Privacy Change Impact Shows the Power It Wields over Other Industries," *CNCB*, November 13, 2021, https://www.cnbc.com/2021/11/13/apples-privacy-changes-show-the-power-it-holds-over-other-industries.html.

46. John Leonard, "Advertising Boss Calls Apple, Privacy Advocates, Politicians 'Extremists,'" *Computing*, January 30, 2023, https://www.computing.co.uk/news/4066252/advertising-boss-calls-apple-privacy-advocates-politicians-extremists.

47. Dan Milmo, "Meta Dealt Blow by EU Ruling That Could Result in Data Use 'Opt-In,'" *The Guardian*, January 4, 2023, https://www.theguardian.com/technology/2023/jan/04/meta-dealt-blow-eu-ruling-data-opt-in-facebook-instagram-ads.

48. Dan Milmo, "Facebook Owner Meta Fined €1.2bn for Mishandling User Information," *The Guardian*, May 22, 2023, https://www.theguardian.com/technology/2023/may/22/facebook-fined-mishandling-user-information-ireland-eu-meta.

49. Natasha Lomas, "CJEU Ruling on Meta Referral Could Close the Chapter on Surveillance

Capitalism," *TechCrunch*, July 4, 2023, https://techcrunch.com/2023/07/04/cjeu-meta-superprofiling-decision/.

50. Dan Milmo, "Facebook and Instagram Users in Europe Can Pay for Ad-Free Versions," *The Guardian*, October 30, 2023, https://www.theguardian.com/technology/2023/oct/30/meta-facebook-instagram-europe-pay-ad-free.

51. Lotje Beek, "Een Betaald Account Is Geen Oplossing Voor Meta's Privacypobleem," *NRC*, November 10, 2023, https://www.nrc.nl/nieuws/2023/11/10/een-betaald-account-is-geen-oplossing-voor-metas-privacyprobleem-a4180473.

52. Foo Yun Chee and Bart H. Meijer, "Apple, Google, Meta Targeted in EU's First Digital Market Act Probes," *Reuters*, March 25, 2024, https://www.reuters.com/business/media-telecom/eu-investigate-apple-google-meta-potential-digital-markets-act-breaches-2024-03-25/.

53. This meant that the standard commission for app developers using IAP had ultimately been reduced from 30% to 17% + 3%, which is by 10%. For smaller app developers and subscriptions after their first year, the reduction was just 2%. Adam Satariano and Tripp Mickle, "Apple Overhauls App Store in Europe, in Response to New Digital Law," *The New York Times*, January 25, 2024, https://www.nytimes.com/2024/01/25/technology/apple-app-store-europe.html; Javier Espinoza and Michael Acton, "Apple Set to Be the First Big Tech Group to Face Charges under EU Digital Law,"*The Financial Times*, June 14, 2024, https://www.ft.com/content/31a996d5-b472-4357-953e-ace078494604.

54. One such instance occurred in the Netherlands, where an investigation by the Dutch Competition Authority found, in 2022, that Apple had committed an abuse of dominance. Authority for Consumers & Markets, "ACM to Assess Adjusted Proposal of Apple Regarding Its Conditions for Dating Apps," March 28, 2022, https://www.acm.nl/en/publications/acm-assess-adjusted-proposal-apple-regarding-its-conditions-dating-apps.

55. Satariano and Mickle, "Apple Overhauls App Store in Europe."

56. Damien Geradin, "When Apple Takes the European Commission for Fools: An Initial Overview of Apple's New Terms and Conditions for iOS App Distribution in the EU," *The Platform Law Blog*, January 26, 2024, https://theplatformlaw.blog/2024/01/26/when-apple-takes-the-european-commission-for-fools-an-initial-overview-of-apples-new-terms-and-conditions-for-ios-app-distribution-in-the-eu/.

57. *Ibid.*

58. *Ibid.*

59. Jay Peters, "Apple and Meta Could Face Charges for Violating Tech Rules," *The Verge*, June 14, 2024, https://www.theverge.com/2024/6/14/24178724/apple-meta-eu-charges-dma; Espinoza and Acton, "Apple Set to Be the First Big Tech Group to Face Charges under EU Digital Law."

60. Supantha Mukherjee, "Apple Changes EU App Store Policy After Commission Probe," *Reuters*, August 8, 2024, https://www.reuters.com/technology/apple-changes-eu-app-store-policy-after-commission-probe-2024-08-08/.

61. Nick Robins-Early, "U.S. to Argue Google Abused Power to Monopolize Internet Search as Antitrust Trial Begins," *The Guardian*, September 12, 2023, https://www.theguardian.com/us-news/2023/sep/12/google-antitrust-trial-day-1?ref=upstract.com.

62. *The Guardian* staff and agencies, "Google Sued for $2.3bn by European Media Groups over Digital Add Losses," *The Guardian*, February 28, 2024, https://www.theguardian.com/technology/2024/feb/28/news-media-europe-google-lawsuit-ad-revenue.

63. Lisa O'Carroll, "X to Be Investigated for Allegedly Breaking EU Law on Hate Speech and Fake News," *The Guardian*, December 18, 2023, https://www.theguardian.com/technology/2023/dec/18/x-to-be-investigated-for-allegedly-breaking-eu-laws-on-hate-speech-and-fake-news.

64. In April 2023, the Irish watchdog had already fined TikTok €345 million for breaches of EU data law in the platform's handling of children's accounts. In the same month, the UK information commissioner fined the company £12.7 million for illegally processing the data of children aged under 13. Tom Gerken and Liv McMahon, "TikTok Fined $345m over Children's Data Privacy," *BBC*, September 15, 2023, https://www.bbc.com/news/technology-66819174.

65. Clara van de Wiel, "Brussel Treft Apple Met Een Van Hoogste Boetes Ooit: 1,8 Miljard Euro, Voor Machtsmisbruik," *NRC*, March 4, 2024, https://www.nrc.nl/nieuws/2024/03/04/brussel-treft-apple-met-een-van-hoogste-boetes-ooit-18-miljard-voor-machtsmisbruik-a4191983.

66. Qtd. in Lisa O'Carroll and Dan Milmo, "EU Fines Apple €1.8bn over App Store Restrictions on Music Streaming," *The Guardian*, March 4, 2024, https://www.theguardian.com/business/2024/mar/04/eu-fines-apple-18bn-over-app-store-restrictions-on-music-streaming.

67. Maarten Schinkel, "Waarom Amerikaanse Techbedrijven Gevaarlijk Diepe Zakken Hebben," *NRC*, March 7, 2024, https://www.nrc.nl/nieuws/2024/03/07/waarom-amerikaanse-techbedrijven-gevaarlijk-diepe-zakken-hebben-a4192181.

Chapter 10

1. "Artificial Intelligence Act," Briefing EU Legislation in Progress, European Parliamentary Research Service, PE 698.792, November 18, 2021, https://www.europarl.europa.eu/RegData/etudes/BRIE/2021/698792/EPRS_BRI(2021)698792_EN.pdf.

2. Fabio Cozzi et al., "The EU Artificial Intelligence Act: What's the Impact?" *McDermott, Will & Emery*, July 31, 2023, https://www.mwe.com/insights/the-eu-artificial-intelligence-act-whats-the-impact/.

3. Bernd Carsten Stahl, "EU Is Cracking Down on AI, but Leaves a Loophole for Mass Surveillance," *The Conversation*, April 21, 2021, https://theconversation.com/eu-is-cracking-down-on-ai-but-leaves-a-loophole-for-mass-surveillance-159421.

4. Levine, *Surveillance Valley*.

5. Stijn Bronzwaer, "Techlobby Wil Europese AI-Wetgeving Afzwakken," *NRC*, https://www.nrc.nl/nieuws/2023/12/05/techlobby-wil-europese-ai-wetgeving-afzwakken-a4183321.

6. Seth Lazar, "The U.S. Is Racing Ahead in Its Bid to Control Artificial Intelligence—Why Is the EU So Far Behind?" *The Guardian*, November 28, 2023, https://www.theguardian.com/commentisfree/2023/nov/28/united-states-artificial-intelligence-eu-ai-washington.

7. Andreas Rinke, "Exclusive: Germany, France and Italy Reach Agreement on AI Regulation," *Reuters*, November 19, 2023, https://www.reuters.com/technology/germany-france-italy-reach-agreement-future-ai-regulation-2023-11-18/.

8. Thorsten Ammann, and Constantin Orth, "EU AI Act Update—Latest Developments and Potential Roadblocks Ahead," *DLA Piper*, November 29, 2023, 2023, https://www.technologyslegaledge.com/2023/11/eu-ai-act-update-latest-developments-and-potential-roadblocks-ahead/.

9. Ibid.

10. Dan Milmo and Johana Bhuiyan, "Biden Hails 'Bold Action' of U.S. Government with Order on Safe Use of AI," *The Guardian*, October 30, 2023, https://www.theguardian.com/technology/2023/oct/30/biden-orders-tech-firms-to-share-ai-safety-test-results-with-us-government.

11. Ibid.

12. Paul Sawers, "Biden Issues Executive Order to Set Standards for AI Safety and Security," *TechCrunch*, October 30, 3023, https://techcrunch.com/2023/10/30/president-biden-issues-executive-order-to-set-standards-for-ai-safety-and-security/.

13. Lauren Leffer, "Biden's Executive Order on AI Is a Good Start, Experts Say, but Not Enough," *Scientific American*, October 31, 2023, https://www.scientificamerican.com/article/bidens-executive-order-on-ai-is-a-good-start-experts-say-but-not-enough/.

14. Qtd. in Milmo and Bhuiyan, "Biden Hails 'Bold Action.'"

15. At an earlier stage, OpenAI's CEO Sam Altman had on several occasions proposed amendments to water down the text of the AI Act. Nevertheless, the adopted final version does contain wording that requires strict regulation of "high-impact" foundation models with "systemic risk." Billy Perrigo, "Exclusive: OpenAI Lobbied the E.U. To Water Down AI Regulation," *TIME*, June 30, 2023, https://time.com/6288245/openai-eu-lobbying-ai-act/.

16. European Parliament, "Artificial Intelligence Act: MEPs Adopt Landmark Law," European Parliament news release, March 13, 2024, https://www.europarl.europa.eu/news/en/press-room/20240308IPR19015/artificial-intelligence-act-meps-adopt-landmark-law. The AI regulation is fully applicable two years after its entry into force, except for: bans on prohibited practices, which apply six months after the enactment date; codes of practices (nine months); general-purpose AI rules including governance (one year); and obligations for high-risk systems (three years).

17. Natasha Lomas, "EU Lawmakers Bag Late Night Deal of 'Global First' AI Rules," *TechCrunch*, December 8, 2023, https://techcrunch.com/2023/12/08/eu-ai-act-political-deal/.

18. Clara van de Wiel, "Het Is Nu Duidelijk Hoe Europa AI Aan Banden Wil Leggen (En Daarmee Wereldwijd Een Norm Stelt," *NRC*, December 9, 2023, https://www.nrc.nl/nieuws/2023/12/09/het-is-nu-duidelijk-hoe-europa-ai-aan-banden-wil-leggen-en-daarmee-wereldwijd-een-norm-stelt-a4183806.

19. David Evan Harris, "Europe Has Made a Great Leap Forward in Regulating AI. Now the Rest of the World Must Step Up." *The Guardian*, December 13, 2023, https://www.theguardian.com/commentisfree/2023/dec/13/europe-regulating-ai-artificial-intelligence-threat.

20. European Parliament, "Artificial Intelligence Act: Deal on Comprehensive Rules for Trustworthy AI," press release, December 9, 2023, https://www.europarl.europa.eu/news/en/press-room/20231206IPR15699/artificial-intelligence-act-deal-on-comprehensive-rules-for-trustworthy-ai.

21. Lisa O'Carroll, "EU Agrees Tough Limits on Police Use of AI Biometric Surveillance," *The Guardian*, December 11, 2023, https://www.theguardian.com/world/2023/dec/11/eu-agrees-tough-rules-on-police-use-of-ai-biometric-surveillance-ban.

22. Lomas. "EU Lawmakers Bag Late Night Deal."

23. Qtd. in van de Wiel, "Het Is Nu Duidelijk."

24. Qtd. in Lomas, "EU Lawmakers Bag Late Night Deal."

25. Kelvin Chan, "Europe Reaches a Deal on the World's First Comprehensive AI Rules," *The Washington Post*, December 8, 2023, https://www.washingtonpost.com/business/2023/12/08/ai-act-europe-regulation/2a4c44f2-961c-11ee-9d5c-d462c9032daa_story.html.

26. Lomas, "EU Lawmakers Bag Late Night Deal."

27. "High impact" was defined as the cumulative amount of computer transactions used for their training measured in "floating point operations per second" (Flops) when it is greater than 10^{25} (ten to the power of 25).

28. In finance, a unicorn is a privately held startup company with a current valuation of USD 1 billion or more.

29. Courtney Radsch, "The Real Story of the OpenAI Debacle Is the Tyranny of Big Tech," *The Guardian*, November 27, 2023, https://www.theguardian.com/commentisfree/2023/nov/27/openai-microsoft-big-tech-monopoly.

30. In contrast to a user interface, which connects a computer to a person, an application programming interface connects computers or pieces of software to each other. It is not intended to be used directly by a person (the end user) other than a computer programmer who is incorporating such programming into the software.

31. Ibid.

32. Dan Milmo, "Google Says New AI Model Gemini Outperforms ChatGPT in Most Tests," *The Guardian*, December 6, 2023, https://www.theguardian.com/technology/2023/dec/06/google-new-ai-model-gemini-bard-upgrade.

33. Ibid.

34. Tom Lamont, "Humanity's Remaining Timeline? It Looks More Like Five Years Than 50': Meet the Neo-Luddites Warning of an AI Apocalypse," *The Guardian*, February 17, 2024, https://www.theguardian.com/technology/2024/feb/17/humanitys-remaining-timeline-it-looks-more-like-five-years-than-50-meet-the-neo-luddites-warning-of-an-ai-apocalypse.

35. Blake Montgomery, "Sora: OpenAI Lanches Tool That Instantly Creates Video from Text," *The Guardian*, February 15, 2024, https://www.theguardian.com/technology/2024/feb/15/openai-sora-ai-model-video.

36. Alexi Duggins, "TV Soaps Could Be Made by AI Writers within Three Years, Director Warns," *The Guardian*, February 22, 2024, https://www.theguardian.com/tv-and-radio/2024/feb/22/james-hawes-select-committee-tv-soaps-made-using-ai.

37. John Naughton, "OpenAI's New Video Generation Tool Could Learn a Lot from Babies," *The Guardian*, February 24, 2024, https://www.theguardian.com/commentisfree/2024/feb/24/openai-video-generation-tool-sora-babies-ai-artificial-intelligence.

38. Qtd. in Alex Hern, "AI Race Heats up as OpenAI, Google and Mistral Release New Models," *The Guardian*, April 10, 2024 2024, https://www.theguardian.com/technology/2024/apr/10/ai-race-heats-up-as-openai-google-and-mistral-release-new-models.

39. In the fiscal first quarter of 2024, Nvidia's revenue more than tripled and its data center business grew by more than 400% from a year earlier. Nvidia gained more than $1.1 trillion in value in the first half year of 2024 alone. At the end of 2022, Nvidia was worth $359 billion. Halfway through 2024, it is worth $2.33 trillion. This is only $500 billion less than Apple and $900 billion less than Microsoft, the two most valuable companies based in the U.S. Edward Helmore, "Nvidia Reports Strategic Growth as AI Boom Shows No Signs of Stopping," *The Guardian*, May 22, 2024, https://www.theguardian.com/technology/article/2024/may/22/nvidia-quarterly-earnings.

40. John Naughton, "From Boom to Burst, the AI Bubble Is Only Heading in One Direction," *The Guardian*, April 13, 2024, https://www.theguardian.com/commentisfree/2024/apr/13/from-boom-to-burst-the-ai-bubble-is-only-heading-in-one-direction.

41. Dan Milmo and Nick Robins-Early, "Google Rolls out AI-Generated, Summarized Search Results in U.S.," *The Guardian*, May 14, 2024, https://www.theguardian.com/technology/article/2024/may/14/google-ai-search-results. Experts have pointed out that Google is in fact catching up with what a startup such as Perplexity (founded in 2022) has been doing quite well for some time. Perplexity is an AI chatbot and search assistant focused on providing in-depth, reliable research and information. It emphasizes using trustworthy sources like academic studies and allows users to fine-tune their searches to exclude less reliable sources. Perplexity provides detailed, context-aware answers with precise source attributions, enabling users to verify the credibility of the information. Perplexity also offers more control and customization options for users to refine their searches and sources. A key distinction is that Perplexity aims to provide a complete, standalone answer with citations, while Google AI Overviews supplements traditional search results with an AI-generated summary. Steven Vaughan-Nichols, "5 Reasons Why I Prefer Perplexity over Every Other Chatbot," *ZDNET*, May 30, 2024, https://www.zdnet.com/article/5-reasons-why-i-prefer-perplexity-over-every-other-ai-chatbot/.

42. André Spicer, "Beware the 'Botshit': Why Generative AI Is Such a Real and Imminent Threat to the Way We Live," *The Guardian*, January 3, 2024, https://www.theguardian.com/commentisfree/2024/jan/03/botshit-generative-ai-imminent-threat-democracy.

43. Chris Stokel-Walker, "ChatGPT on Your Phone? The Four Reasons Why This Is Happening Far Too Early," *The Guardian*, June 13, 2024. For an excellent introduction into how AI is disrupting and reshaping different sectors (healthcare, manufacturing, agriculture, finance, consulting) and more generally people's everyday lives, as well as the ethical concerns all of this raises, see Chris Stokel-Walker, *How AI Ate the World* (Kingston upon Thames: Canbury, 2023).

44. Rhiannon Williams, "Why Google's AI Overviews Get Things Wrong," *MIT Technology Review*, May 31, 2024, https://www.technologyreview.com/2024/05/31/1093019/why-are-googles-ai-overviews-results-so-bad/; Phillip Adcock, "6 Limitations of AI & Why It Won't Quite Take over in 2023," *AdCockSolutions*, March 14, 2024, https://www.adcocksolutions.com/post/6-limitations-of-ai-why-it-wont-quite-take-over-in-2023.

45. The new tool had used satirical sources like the Onion or joke Reddit posts to generate answers. Nick Robins-Early, "Google to Refine AI-Generated Search Summaries in Response to Bizarre Results," *The Guardian*, May 31, 2024, https://www.theguardian.com/technology/article/2024/may/31/google-ai-summaries-sge-changes.

46. Will Oremus, "Google's Weird AI Answers Hint at a Fundamental Problem," *The Washington Post*, May 29, 2024, https://www.washingtonpost.

com/politics/2024/05/29/google-ai-overview-wrong-answers-unfixable/.

47. Edward Helmore and Kari Paul, "New York Times Sues OpenAI and Microsoft for Copyright Infringement," *The Guardian*, December 27, 2023, https://www.theguardian.com/media/2023/dec/27/new-york-times-openai-microsoft-lawsuit.

48. Lamont, "Humanity's Remaining Timeline?"

49. Nick Robins-Early, "The Intercept, Raw Story and Alternet Sue OpenAI for Copyright Infringement," *The Guardian*, February 28, 2024, https://www.theguardian.com/technology/2024/feb/28/media-outlets-sue-openai-copyright-infringement.

50. Margaret Simons, "If Meta's Intransigence Isn't Enough, AI Poses an Even Greater Threat to Journalism," *The Guardian*, March 1, 2024, https://www.theguardian.com/commentisfree/2024/mar/02/if-metas-intransigence-isnt-enough-ai-poses-an-even-greater-threat-to-journalism.

51. Dan Milmo, "'Impossible' to Create AI Tools Like ChatGPT without Copyrighted Materials, OpenAI Says," *The Guardian*, January 8, 2024, https://www.theguardian.com/technology/2024/jan/08/ai-tools-chatgpt-copyrighted-material-openai.

52. These include Eleuther AI, a grassroots nonprofit research group that began as a loose-knit Discord Collective, now working with the University of Toronto; FineWeb (released by the AI startup Hugging Face), a filtered version of the Common Crawl—the dataset maintained by the nonprofit with the same name—composed of billions upon billions of web pages. A few efforts to release open training data sets, like the group LAION's image sets, have run up against copyright, data privacy and other serious ethical and legal challenges. But some of the more dedicated data curators have promised to do better. The Pile v2, for instance, removes problematic copyrighted material found in its predecessor's dataset, The Pile. Kyle Wiggers, "AI Training Data Has a Price Tag That Only Big Tech Can Afford," *TechCrunch*, June 1, 2024, https://techcrunch.com/2024/06/01/ai-training-data-has-a-price-tag-that-only-big-tech-can-afford/.

53. Gartner, a market-research consultancy, predicts that search engine volume will drop 25% by 2026 due to AI chatbots and other virtual agents, which will replace user queries in traditional search engines. Raptive, which provides digital media, audience and advertising services to about 5,000 websites, estimates that publishers could see a decline of up to 60% in organic search traffic, which could result in about $2 billion in losses to creators—with some websites losing up to two-thirds of their traffic. Casey Newton, "Google's Broken Link to the Web," *Platformer*, May 14, 2024, https://www.platformer.news/google-io-ai-search-sundar-pichai/.

54. *Ibid.*

55. Simons, "If Meta's Intransigence Isn't Enough."

56. Nick Robins-Early, "Google Remains Focused on Its Long Quest for Your Eyeballs," *The Guardian*, May 19, 2024, https://www.theguardian.com/technology/article/2024/may/19/google-ai-overview-attention.

57. Federal Trade Commission, "FTC Launches Inquiry into Generative AI Investments and Partnerships," *ftc.gov*, January 25, 2024, https://www.ftc.gov/news-events/news/press-releases/2024/01/ftc-launches-inquiry-generative-ai-investments-partnerships; Blake Montgomery, "Microsoft's Activision Acquisition and Bets on AI Yield High Quarterly Revenue," *The Guardian*, January 30, 2024, https://www.theguardian.com/technology/2024/jan/30/microsoft-revenue-activision-ai-surpassed-earnings.

58. In March 2024, Microsoft hired Mustafa Suleyman, Inflection AI's CEO (and co-founder of both DeepMind and Inflection AI), as the boss of a new AI division and agreed to pay his company $650 million to license its AI software. Microsoft then also hired a large number of staffers from Inflection AI. Thomas Barrabi, "Feds to Launch Antitrust Probe into Microsoft, OpenAI, Nvidia over 'Monopoly Choke Points,'" *New York Post*, May 13, 2024, https://www.msn.com/en-us/money/other/doj-to-scrutinize-big-tech-s-ai-dominance-with-urgency-antitrust-chief-says/ar-BB1nKIFG.

59. As qtd. in *Ibid.*

60. Cecilia Rikap, "Antitrust Policy and Artifical Intelligence: Some Neglected Issues," *ScheerPost*, June 12, 2024, https://scheerpost.com/2024/06/12/antitrust-policy-and-artificial-intelligence-some-neglected-issues/.

61. Cecilia Rikap, "Intellectual Monopolies as a New Pattern of Innovation and Technological Regime," *Industrial and Social Change* (December 7, 2023), https://doi.org/10.1093/icc/dtad077.

62. Rikap, "Antitrust Policy and Artificial Intelligence."

63. A complementor is a company that sells a product or service that complements the products or services of another company.

64. Cédric Durand and William Milberg, "Intellectual Monopoly in Global Value Chains," *Review of International Political Economy* 27, no. 5 (2019): 1–26, https://doi.org/10.1080/09692290.2019.1660703.

65. Rikap, "Antitrust Policy and Artificial Intelligence."

66. *Ibid.*

67. A *Guardian* editorial of late December 2023 even suggested that "the nearest historical parallel is humankind's acquisition of nuclear power," and that the challenge posed by AI was "arguably greater." "To get from a theoretical understanding of how to split the atom to the assembly of a reactor of bomb is hard and expensive. Malicious applications of code online can be transmitted and replicated with virtual efficiency." *Guardian* editorial, "The Guardian View on the AI Conundrum: What It Means to Be Human Is Elusive," *The Guardian*, December 29, 2023, https://www.theguardian.

com/commentisfree/2023/dec/29/the-guardian-view-on-the-ai-conundrum-what-it-means-to-be-human-is-elusive.

68. Qtd. in Lisa O'Carroll, "EU Agrees 'Historic' Deal with World's First Laws to Regulate AI," *The Guardian*, December 8, 2023, https://www.theguardian.com/world/2023/dec/08/eu-agrees-historic-deal-with-worlds-first-laws-to-regulate-ai.

69. Van de Wiel, "Brussel Treft Apple."

70. Faiola et al. "E.U. reaches Deal."

71. Harris, "Europe Made a Great Leap Forward."

72. John Naughton, "For All the Hype in 2023, We Still Don't Know What AI's Long-Term Impact Will Be," *The Guardian*, December 30, 2023, https://www.theguardian.com/commentisfree/2023/dec/30/ai-artifical-intelligence-2023-long-term-impact-nvidia-h100-microsoft.

73. U.S. Department of Energy, "Data Centers and Servers," https://www.energy.gov/eere/buildings/data-centers-and-servers. (accessed June 3, 2024).

74. John Naughton, "Why AI Is a Disaster for the Climate," *The Guardian*, December 23, 2023, https://www.theguardian.com/commentisfree/2023/dec/23/ai-chat-gpt-environmental-impact-energy-carbon-intensive-technology?ref=upstract.com.

Chapter 11

1. Nick Srnicek, "The Only Way to Rein in Big Tech Is to Treat Them as a Public Service," *The Guardian*, April 23, 2019, https://www.theguardian.com/commentisfree/2019/apr/23/big-tech-google-facebook-unions-public-ownership.

2. Tarnoff, *Internet for People*, 152.

3. Muldoon, *Platform Socialism*, 66.

4. Tarnoff, *Internet for People*, 153.

5. Muldoon, *Platform Socialism*, 67.

6. *Ibid.*, 2.

7. Tarnoff, *Internet for People*, 153. The term "inequality machine" has been used more specifically to characterize the workings of higher education in the U.S. today. See Paul Tough, *The Inequality Machine: How College Divides Us* (Boston: Mariner Books, 2021).

8. It is true that left-wing movements have also benefited from social media. In the U.S., Occupy Wall Street, Black Lives Matter, and the campaigns of presidential candidate Bernie Sanders most likely would not have reached the scale they did without Twitter and Facebook. The shift from traditional media to the more distributed informational worlds of social media has created more room for various left-wing ideas to circulate. But the Right disposes of many more resources to exploit this shift: deep-pocketed donors, a sophisticated media operation—a right-wing media ecosystem, from talk radio to cable TV (Fox News) to news sites in which different modes overlap and interpenetrate—and strong control of the Republican Party. Tarnoff, *Internet for People*, 141–42.

9. *Ibid.*, 153–54.

10. Spicer, "Beware the 'Botshit.'"

11. John Naughton, "Silicon Valley's Business Model Is Incompatible Wth the Moderation of Online Horror and Hatred," *The Guardian*, April 27, 2024, https://www.theguardian.com/commentisfree/2024/apr/27/silicon-valleys-business-model-is-incompatible-with-the-moderation-of-online-horror-and-hatred.

12. Kari Paul, "Reversal of Content Policies at Alphabet, Meta and X Threaten Democracy, Warn Experts," *The Guardian*, December 7, 2023, https://www.theguardian.com/media/2023/dec/07/2024-elections-social-media-content-safety-policies-moderation?ref=upstract.com.

13. Qtd. in *Ibid.*

14. Tarnoff, *Internet for People*, 154.

15. Fuchs and Unterberger, "The Public Service Media."

16. Bongers, "Amerikanen Zijn Hypocriet over TikTok."

17. The latest wave of new municipal networks ranges from conduit-only networks like the one in West Des Moines, Iowa that offers residents a choice of broadband providers; institutional networks such as the I-net of the city of Alexandria, VA, built to serve local government operation, and the city partnering with Ting in providing fiber-to-the-home service citywide; to open-access networks like Yellowstone Fiber in Bozeman, Montana; as well as the massive municipal fiber-to-the-home (FTTH) network under construction in Knoxville, Tenn. It will take the Knoxville Utilities Board (KUB) seven to ten years to complete. Once completed, KUB Fiber will be one of the largest municipal broadband networks in the nation, rivaling its Chattanooga Neighbor EPB Fiber and the multi-state footprint of UTOPIA Fiber. Sean Gonsalves, "New Municipal Broadband Networks Skyrocket in Post-Pandemic America as Alternative to Private Monopoly Model," *Community Networks*, January 18, 2024, https://communitynets.org/content/new-municipal-broadband-networks-skyrocket-post-pandemic-america-alternative-private.

18. By 2024, ISP lobbyists had convinced 17 states to pass protectionist laws impeding local efforts to build such municipal networks. Some of these laws even bar communities from striking public/private network deals.

19. Gonsalves, "New Municipal Broadband Networks."

20. There has been a substantial amount of intertribal collaboration around the current broadband build-outs, with a few notable forerunners to emulate. One of those is the Tohono O'odham Nation in southern Arizona (with internet access through its local utility authority since 1998). Aside from this one, a few others, including the Gila River Indian Community in Arizona and the Mohawk Nation in Akwesasne in upstate New York, have been operating independent telecom

services that predate the recent federal funding wave. Evan Malgren, "Internet for the People: One Native Tribe's Experiment in Self-Sufficient, Tribally Own Internet," *The Nation*, February 24, 2024, 48–53.

21. *Ibid.*, 50.
22. Tarnoff, *Internet for People*, 155–56.
23. *Ibid.*, 157.
24. Muldoon, *Platform Socialism*, 148.
25. Ethan Zuckerman, "What Is Digital Public Infrastructure?" *Center for Journalism & Liberty*, November 17, 2020, https://www.journalismliberty.org/publications/what-is-digital-public-infrastructure.
26. *Ibid.*
27. A decentralized and federated social media space can be explored using a social media browser, which aggregates feeds from each of the social networks, allows the user to control how they appear, and lets him/her post to any networks where s/he has appropriate permissions as a member. Gobo.social is a good example of such a social media browser. *Ibid.*
28. Tarnoff, *Internet for People*, 158–59.
29. *Ibid.*, 160.
30. McChesney, *Digital Disconnect*, 219–32.
31. Victor Pickard, *Democracy without Journalism? Confronting the Misinformation Society* (Oxford: Oxford University Press, 2019), 156–57.
32. Paul Romer, "A Tax That Could Fix Big Tech," *The New York Times*, May 6, 2018, https://www.nytimes.com/2019/05/06/opinion/tax-facebook-google.html; Christian Fuchs, *The Online Advertising Tax as the Foundation of a Public Service Internet* (London: University of Westminster Press, 2018).
33. Zuckerman, "What Is Digital Public Infrastructure?"
34. 8kun, previously called 8chan, is an American far-right imageboard website composed of user-created message boards. An owner moderates each board, with minimal interaction from site administration. Gab is an American alt-tech microblogging and social networking service known for its far-right user base. Julia Carrie Wong, "8chan: The Far-Right Website Linked to the Rise in Hate Crimes," *The Guardian*, August 4, 2019, https://www.theguardian.com/technology/2019/aug/04/mass-shootings-el-paso-texas-dayton-ohio-8chan-far-right-website. Greta Jasser et al., "'Welcome to #Gabfam': Far-Right Virtual Community on Gab," *New Media & Society* 25, no. 7 (2023): 1728–45, https://doi.org/10.1177/14614448211024546.
35. Tarnoff, *Internet for People*, 163.
36. Zuckerman, "What Is Digital Public Infrastructure?"
37. Srnicek, "The Only Way to Rein in Big Tech."
38. Nathan Schneider, "Denver Taxi Drivers Are Turning Uber's Disruption on Its Head," *The Nation*, September 7, 2016, https://www.thenation.com/article/archive/denver-taxi-drivers-are-turning-ubers-disruption-on-its-head/.
39. The Platform Cooperativism Consortium is a "think-and-do tank" for the platform cooperativism movement initially based at The New School in New York City. As a global network of researchers, platform co-ops, independent software developers, artists, designers, lawyers, activists, publishing outlets, and funders, it engages in research, advocacy, education, and technology-based projects. It was launched in November 2016 at the occasion of the "Building the Cooperative Internet" conference. Platform Cooperativism Consortium, "Platform Cooperativism Consortium: A Hub That Starts, Grows, and Converts Platform Coops." https://platform.coop/about/pcc/; Nathan Schneider, "The Rise of a Cooperatively Owned Internet," *The Nation*, October 13, 2016, https://www.thenation.com/article/archive/the-rise-of-a-cooperatively-owned-internet/.
40. Damion Bunders et al., "The Feasibility of Platform Cooperatives in the Gig Economy," *Journal of Co-operative Organization and Management* 10, no. 1 (2022): 1–8, https://doi.org/10.1016/j.jcom.2022.100167.
41. Tarnoff, *Internet for People*, 164.
42. *Ibid.*, 165–66.
43. Muldoon, *Platform Socialism*, 2.
44. *Ibid.*, 3.
45. Cole joined the Fabian society during his university years but became disillusioned with the top-down nationalization projects and lack of support for workplace democracy. Muldoon borrowed from Cole's writings to envision the platform economy as a federated network of democratic associations governed by active citizens. As he acknowledges, these writings also offered him valuable ideas about how worker control over the labor process can be balanced with the interests of broader community participation. Neurath became a member of the German Social Democratic Party during the German Revolution of 1918–19 and was in charge of the central economic planning office in Munich during the short-lived Bavarian Soviet Republic. Muldoon acknowledges that Neurath's work offered him important insights into economic planning and how democratic allocation of resources could occur through deliberation using the 'common good' ideal as a basis. Muldoon, *Platform Socialism*, 80.
46. Eden Medina, *Cybernetic Revolutionaries: Technology and Politics in Allende's Chili* (Cambridge: MIT Press, 2014).
47. In a recent nine-part podcast series, *The Santiago Boys*, Morozov presents the Chile project, Cybersyn, as a source of inspiration for European countries in creating their own digital technologies today. He sees parallels between then and now: Allende started a program that today could be defined as aiming at technological sovereignty or technological autonomy. See https://the-santiago-boys.com.
48. Muldoon, *Platform Socialism*, 75–76.
49. Morozov, "Digital Socialism?"
50. Muldoon, *Platform Socialism*, 98.
51. *Ibid.*, 68–69.
52. Josh Simons and Dipayan Ghosh, "Utilities

for Democracy: Why and How the Algorithmic Infrastructure of Facebook and Google Must Be Regulated," *Brookings*, August 2020, 10, https://www.brookings.edu/wp-content/uploads/2020/08/FP_20200908_facebook_google_algorithm_simons_ghosh.pdf.

53. William J. Novak, "Law and the Social Control of American Capitalism," *Emory Law Journal* 60, no. 2 (2010): 377–405, https://scholarlycommons.law.emory.edu/elj/vol60/iss2/4.

54. K. Sabeel Rahman and Zephyr Teachout, "From Private Bads to Public Goods: Adapting Public Utility Regulation for Informational Infrastructure," *The Knight First Amendment Institute*, February 4, 2020, https://knightcolumbia.org/content/from-private-bads-to-public-goods-adapting-public-utility-regulation-for-informational-infrastructure.

55. Muldoon, *Platform Socialism*, 69–70.

56. *Ibid.*, 80–100.

57. *Ibid.*, 71.

58. The first of these gatherings, held in Barcelona, showed the breadth of these municipalist movements that had begun to identify with some broader phenomena—ranging from Cambiamo Messina dal Basso in Messina, Italy; Zagreb je Naš in Croatia; Miasto Jes Nasze in Warsaw, Poland; Ne Davimo Beograd in Belgrade, Serbia; the Autonomous Government of Rojava in northeastern Syria; Beirut Madinati in Lebanon; the Umbrella Movement in Hong Kong; Movimiento Autonomista in Valparaíso, Chile; Ciudad Futura in Rosario, Argentina; Cooperation Richmond in California and Cooperation Jackson in Mississippi, USA; and a number of other Spanish cities such as Marea Atlantica in A Coruña and Zaragoza en Común in Zaragoza. Bertie Russell, "Beyond the Local Trap: New Municipalism and the Rise of the Fearless Cities," *Antipode* 51, no. 3: 989–1010, https://onlinelibrary.wiley.com/doi/full/10.1111/anti.12520.

59. Mark Purcell, "Urban Democracy and the Local Trap," *Urban Studies* 43, no. 11 (2006): 1931, 1941, http://www.jstor.org/stable/43197418.

60. Muldoon, *Platform Socialism*, 101–02.

61. Laura Roth and Bertie Russell, "Translocal Solidarity and the New Municipalism," *Roar Magazine* 8 (2018), https://roarmag.org/magazine/municipalist-movement-internationalism-solidarity/.

62. However, criticisms have arisen about the ideological biases in Wikipedia. Wikipedia's co-founder Larry Sanger has described what he saw as the downfall of the once community-run, neutral, encyclopedia into a mouthpiece for the U.S. establishment. Glen Greenwald, "Wikipedia Co-Founder Calls Wikipedia 'Most Biased Encyclopedia,'" *ScheerPost*, August 3, 2023, https://scheerpost.com/2023/08/03/glenn-greenwald-wikipedia-co-founder-calls-wikipedia-most-biased-encyclopedia/. There have also been various academic studies of this issue, which are summarized in Wikipedia itself. See https://en.wikipedia.org/wiki/Ideological_bias_on_Wikipedia.

63. Muldoon, *Platform Socialism*, 85.

64. *Ibid.*, 87.

65. *Ibid.*, 122–23.

66. *Ibid.*, 126.

67. *Ibid.*, 127.

68. After President Biden had proposed a 25% tax on individuals with assets over $100 million, a growing number of venture capitalists supported Donald Trump in the 2024 presidential election. Evgeny Morozov, "Silicon Valley Wants Unfettered Control of the Tech Market. That's Why It's Cosying up to Trump," *The Guardian*, June 26, 2024, https://www.theguardian.com/commentisfree/article/2024/jun/26/silicon-valley-tech-market-donald-trump-joe-biden-wealth-tax-big-tech-venture-capitalists.

69. Thanks to this model, Americans now pay some of the highest drug prices in the world. Yet when politicians have tried to curb these excessive costs, they have been met with accusations from the VC industry that they are undermining progress.Venture capitalists have been eager to emphasize the role they play in delivering "progress." Through podcasts, conferences and publications, they have successfully recast their interests as those of humanity at large. *Ibid.* For a clear expression of this worldview, see *The Techno-Optimist Manifesto*, a treatise by Marc Andreessen, co-founder of the VC firm Andreessen Horowitz, in which he ultimately considers free markets "the most effective way to organize a technological economy." Marc Andreessen, "The Techno-Optmist Manifesto." *a16z*, October 16, 2023, https://a16z.com/the-techno-optimist-manifesto/.

70. Muldoon, *Platform Socialism*, 132–33.

71. *Ibid.*, 135.

72. *Ibid.*, 133. There has been a move to commercialize the Reddit platform further, with an initial public offering (IPO) launched on March 21, 2024, which will likely lead to Reddit's deterioration as a democratic virtual community. Nick Robins-Early and agencies, "Reddit Shares Priced at $34 in Largest IPO by Social Media Company in Years," *The Guardian*, March 20, 2024, https://www.theguardian.com/technology/2024/mar/20/reddit-stock-market-debut-new-york-exchange; Hussein Kesvani, "First It Was Facebook, Then Twitter. Is Reddit About to Become Rubbish Too?" *The Guardian*, March 20, 2024, https://www.theguardian.com/commentisfree/2024/mar/20/facebook-twitter-reddit-rubbish-ipo.

73. Tim Christiaens, "Review of James Muldoon, *Platform Socialism: How to Reclaim Our Digital Futures*," *Marx & Philosophy Review of Books*, January 17, 2022, https://marxandphilosophy.org.uk/reviews/19828_platform-socialism-how-to-reclaim-our-digital-futures-by-james-muldoon-reviewed-by-tim-christiaens/.

74. Mueller, *Breaking Things at Work*, 105.

75. Anonymous is a decentralized international activist and hacktivist collective and movement primarily known for its various cyber attacks against several governments, government institutions and government agencies, corporations and

the Church of Scientology. Anonymous originated in 2003 on the imageboard 4chan representing the concept of many online and offline community users simultaneously existing as an "anarchic," digitized "global brain" or "hive mind." Anonymous members can sometimes be distinguished in public by the wearing of Guy Fawkes masks in the style portrayed in the graphic novel and film *V for Vendetta*. Gabriella Coleman, *Hacker, Hoaxer, Whistleblower, Spy: The Many Faces of Anonymous* (London: Verso Books, 2014); E. Gabriella Coleman, "From Busting Cults to Breeding Cults: Anonymous H/Acktivism vs. The (a)Nomynous Far Right and QAnon," *HAU: Journal of Ethnographic Theory* 13, no. 2 (2023): 248–63, https://doi.org/10.1086/727758.

76. Couldry and Mejias, *Costs of Connection*, 195.

77. Lamont, "Humanity's Remaining Timeline?"

78. Ibid.

79. Qtd. in Brian Merchant, *Blood in the Machine: The Origins of the Rebellion against Big Tech* (New York: Little, Brown, 2023), 389.

80. Qtd. in Lamont, "Humanity's Remaining Timeline?"

81. Zachary Loeb, "Specters of Ludd (Review of Gavin Mueller, *Breaking Things at Work*)," *boundary 2*, September 28, 2021, https://www.boundary2.org/2021/09/zachary-loeb-specters-of-ludd-review-of-gavin-mueller-breaking-things-at-work/.

82. Christiaens, "Review of James Muldoon, *Platform Socialism*."

83. Loeb, "Specters of Ludd."

84. Lawrence Robinson, Melinda Smith, and Jeanne Segal, "Smartphone and Internet Addiction," *Helpguide.org*, updated April 29, 2024, https://www.helpguide.org/articles/addictions/smartphone-addiction.htm.

85. Mueller, *Breaking Things at Work*, 93–136; Merchant, *Blood in the Machine*, 388–412.

Chapter 12

1. Muldoon, *Platform Socialism*, 119–20.

2. Anu Bradford, *Digital Empires: The Global Battle to Regulate Technology* (New York: Oxford University Press, 2023), 79–80, 94–96.

3. Ibid., 77–91.

4. Ibid., 153–72.

5. Michael Kwet, "Digital Colonialism: U.S. Empire and the New Imperialism in the Global South," *Race & Class* 60, no. 4 (2020): 3–26, https://doi.org/10.1177/0306396818823172.

6. "Project Loon" was started by Google and, until 2021, still operated in some countries to test the capacity of using large balloons to spread digital connectivity. Jay Peters, "Alphabet is Shutting Down Loon, Its Internet Balloon Company," *The Verge*, January 21, 2021, https://www.theverge.com/2021/1/21/22243484/alphabet-google-shutting-down-loon-internet-balloon-company-x.

7. Free Basics was introduced in India in 2015, but thanks to a successful campaign called Save the Internet launched by local activists, coders, and policy wonks, India's regulatory authority decided to uphold internet neutrality, forbid zero-rating (the practice of providing "free" online services access), and effectively banned Free Basics in February 2016. Olivia Solon, "'It's Digital Colonialism': How Facebook's Free Internet Service Has Failed Its Users," *The Guardian*, July 27, 2017, https://www.theguardian.com/technology/2017/jul/27/facebook-free-basics-developing-markets; Toussaint Nothias, "Access Granted: Facebook's Free Basics in Africa," *Media, Culture & Society* 42, no. 3 (2020): 329–48, https://doi.org/10.1177/0163443719890530.

8. Renata Avila, "Against Digital Colonialism," in *Platforming Equality: Policy Changes for the Digital Economy*, ed. James Muldoon and Will Stronge (London: Autonomy, 2020), 47–57.

9. Marcus Willett, "China's Investment in Digital Technologies and the Digital Great Game," in *The Digital Silk Road: China's Technological Rise and the Geopolitics of Cyberspace*, ed. David Gordon and Meia Nouwens (London: Routledge, 2022), 23–36; Robert Koepp, "Locating the Digital Silk Road in the Belt and Road Initiatives," in *The Digital Silk Road*, ed. Gordon and Nouwens, 37–49.

10. Joshua Kurlantzick, "Assessing China's Silk Road: A Transformative Approach to Technology or a Danger to Freedoms?" *Council on Foreign Relations*, December 18, 2020, https://www.cfr.org/blog/assessing-chinas-digital-silk-road-transformative-approach-technology-financing-or-danger.

11. Bradford, *Digital Empires*, 28.

12. Avila, "Against Digital Colonialism."

13. Yashmin Ismail and Rashmi Jose, "E-Commerce Takes Centre Stage at World Trade Organization in Run-up to MC13," *The International Institute for Sustainable Development (IISD)*, January 11, 2024, https://www.iisd.org/articles/policy-analysis/e-commerce-developments-wto-mc13.

14. Ricaurte, "Data Epistemologies, Coloniality of Power, and Resistance."

15. Charles Taylor, *Modern Social Imaginaries* (Durham: Duke University Press, 2004); Couldry and Mejias, "Decolonial Turn," 12.

16. Couldry and Mejias, *Costs of Connection*, 198–99.

17. Couldry and Mejias, "Decolonial Turn," 12–13.

18. Toussaint Nothias, "How to Fight Digital Colonialism," *Boston Review*, November 14, 2022, https://www.bostonreview.net/articles/how-to-fight-digital-colonialism/.

19. Larry Diamond, "Democracy's Arc: From Resurgent to Imperiled," *Journal of Democracy* 33, no. 1 (2022): 163–79, https://www.journalofdemocracy.org/articles/democracys-arc-from-resurgent-to-imperiled/.

20. James Muldoon, "Data-Owning Democracy or Digital Socialism?" *Critical Review of International Social and Political Philosophy* (September

2022): 18–19, https://doi.org/10.1080/13698230.2022.2120737.

21. Bradford, *Digital Empires*, 26–27, 133–34, 242–46; Charlene Barshefsky, "EU Digital Protectionism Risks Damaging Ties with the U.S.," *Financial Times*, August 2, 2020, https://www.ft.com/content/9edea4f5-5f34-4e17-89cd-f9b9ba698103.

22. Bradford, *Digital Empires*, 24.

23. *Ibid.*, 364.

24. *Ibid.*, 25, 94–99.

25. Adrian Shahbaz, "The Digital Silk Road and Normative Values," in *The Digital Silk Road*, ed. Gordon and Nouwens, 118–30.

26. Bradford, *Digital Empires*, 290–302.

27. For further details, see Bruce Schneier, *Data and Goliath: The Hidden Battles to Collect Your Data and Control Your World* (New York: W.W. Norton, 2015), 62–87; Levine, *Surveillance Valley*.

28. Bradford, *Digital Empires*, 365–66.

29. This includes heavy U.S. government pressure on some of its allies to stop exporting advanced chipmaking machinery to China. The Netherlands has been told that its semiconductor giant ASML, the largest tech firm in Europe by valuation, can no longer sell China certain types of its advanced immersion lithography machines for producing high-quality computer chips. The restrictions that the Dutch government agreed to put on sales of ASML's deep ultraviolet (DUV) lithography machines were officially put into place on January 1, 2024. Japan, too, is subject to similar restrictions regarding the same types of machines manufactured by its domestic semiconductor industry. Marc Hijink, "VS Leggen Strengere Exportrestricties Naar China Op Aan ASML," *NRC*, October 17, 2023, https://www.nrc.nl/nieuws/2023/10/17/vs-leggen-strengere-exportrestricties-naar-china-op-aan-asml-a4177595; Jack Simpson, "ASML Halts Hi-Tech Chip-Making Exports to China Repeatedly after U.S. Request," *The Guardian*, January 2, 2024, https://www.theguardian.com/technology/2024/jan/02/asml-halts-hi-tech-chip-making-exports-to-china-reportedly-after-us-request?ref=biztoc.com. However, in August 2024, the United States granted to its allies, the Netherlands, Japan and South Korea, exemptions to the stricter rules for the export of advanced chips technology to China it then introduced. The stricter rules could hit the exports of chip companies to China from, among others, Taiwan, Singapore, Israel and Malaysia. Naïm Derbali, "ASML uit het vizier van verscherpte Amerikaanse exportregels naar China," NRC, July 31, 2024, https://www.nrc.nl/nieuws/2024/07/31/asml-uit-het-vizier-van-verscherpte-amerikaanse-exportregels-naar-china-a4861348. On the other hand, however, at least eleven state-linked Chinese entities were then using (or had shown interest in using) cloud services provided by Amazon Web Services or its rival Microsoft's Azure to access advanced U.S-made chips (Nvidia A100 and H100 chips) and advanced AI models of OpenAI and Anthropic. Providing access to such technologies through the cloud was not a violation of U.S. regulations then, but the U.S. government was working hard on new export restrictions to block this detour via the cloud. But compliance with such regulations is difficult, because it is not always clear to the tech companies who their ultimate customer is. Reuters, "List of Chinese Entities Who Have Turned to the Cloud for Access to Restricted US Tech," *Reuters*, August 23, 2024, https://www.reuters.com/technology/list-chinese-entities-who-have-turned-cloud-access-restricted-us-tech-2024-08-23/.

30. Bradford, *Digital Empires*, 26, 380.

31. *Ibid.*: 367–68.

32. *Ibid.*, 324–26.

33. *Ibid.*, 28.

34. *Ibid.*, 139–41, 376.

35. *Ibid.*, 377–78.

36. *Ibid.*, 391.

37. *Ibid.*, 378.

38. *Ibid.*, 126.

39. *Ibid.*, 381.

40. Jared Cohen and Richard Fontaine, "Uniting the Techno-Democracies: How to Build Digital Cooperation," *Foreign Affairs*, November/December 2020, https://www.foreignaffairs.com/articles/united-states/2020-10-13/uniting-techno-democracies.

41. Anja Manuel, "The Tech 10: A Flexible Approach for International Technology Governance," in *Domestic and International (Dis) Order: A Strategic Response*, ed. Leah Bitounis and Niamh King (Aspen: Aspen Institute, 2020), 71–74, https://www.aspeninstitute.org/wp-content/uploads/2020/11/Chapter-10_Manuel_The-Tech-10.pdf.

42. Adrian Shahbaz, "The Rise of Digital Authoritarianism," *Freedom House*, 2018, https://freedomhouse.org/report/freedom-net/2018/rise-digital-authoritarianism.

43. Louis Menand, "American Democracy Was Never Designed to Be Democratic," *The New Yorker*, August 15, 2022, https://www.newyorker.com/magazine/2022/08/22/american-democracy-was-never-designed-to-be-democratic-eric-holder-our-unfinished-march-nick-seabrook-one-person-one-vote-jacob-grumbach-laboratories-against-democracy.

44. See "Summit for Democracy," U.S. Department of State, February 2021, https://www.state.gov/summit-for-democracy-2021/.

45. See "A Declaration for the Future of the Internet," White House, April 28, 2022, https://www.whitehouse.gov/briefing-room/statements-releases/2022/04/28/fact-sheet-united-states-and-60-global-partners-launch-declaration-for-the-future-of-the-internet/.

46. Bradford, *Digital Empires*, 387.

47. *Ibid.*, 387–88; David Gordon and Meia Nouwens, "Conclusion," in *The Digital Silk Road*, ed. Gordon and Nouwens, 171.

48. On May 14, 2024 the White House announced this packet of measures designed to protect U.S. manufacturers from cheap imports

from China. "Fact Sheet: President Biden Takes Action to Protect American Workers and Businesses from China's Unfair Trade Practices," White House, May 14, 2024, https://www.whitehouse.gov/briefing-room/statements-releases/2024/05/14/fact-sheet-president-biden-takes-action-to-protect-american-workers-and-businesses-from-chinas-unfair-trade-practices/; Larry Elliott, "Biden Announces 100% Tariff on Chinese-Made Electric Vehicles," *The Guardian*, May 14, 2022, https://www.theguardian.com/business/article/2024/may/14/joe-biden-tariff-chinese-made-electric-vehicles.

49. Reuters, "Chinese Industry to Seek Probe into EU Pork Imports, Global Times Reports," *Reuters*, May 27, 2024, https://www.reuters.com/markets/commodities/chinese-industry-seek-probe-into-eu-pork-imports-global-times-reports-2024-05-27/.

50. Lisa O'Carroll, "EU to Put Tariffs of up to 38% on Chinese Electric Vehicles as Trade War Looms," *The Guardian*, June 12, 2024, https://www.theguardian.com/business/article/2024/jun/12/eu-import-tariffs-chinese-evs-electric-vehicles-trade-war.

51. In Germany, the U.S. chipmaker Intel and the Taiwanese chipmaker TSMC are building factories aided by EU subsidies. Marike Stellinga, "China Wil Niet Afhankelijk Zijn Van De VS En Steekt Daarom Opnieuw Tientallen Miljarden in Zijn Chipsector," *NRC*, May 27, 2024, https://www.nrc.nl/nieuws/2024/05/27/china-wil-niet-afhankelijk-zijn-van-de-vs-en-steekt-daarom-opnieuw-tientallen-miljarden-in-zijn-chipsector-a4200097.

52. Milo van Bokkum, "China Dreigt Met Wraakmaatregelen Door Europees Staatssteunonderzoek," *NRC*, May 27, 2024, https://www.nrc.nl/nieuws/2024/05/27/china-dreigt-met-wraakmaatregelen-door-europees-staatssteunonderzoek-a4200086.

53. The new investments in the Chinese chip industry amount to €44 billion, after China previously had invested some $140 billion. The U.S. has handed out and set up, roughly $110 billion in subsidies, favorable loans and tax credits in this sector, and the EU likewise invested $46 billion. Stellinga, "China Wil Niet Afhankelijk Zijn."

54. Timothy McLaughlin, "The Most Dangerous Conflict No One Is Talking About," *The Atlantic*, December 2, 2023, https://www.theatlantic.com/international/archive/2023/12/south-china-sea-philippines-dispute-explained/676218/; Marvin Ott, "The South China Sea in Strategic Terms," *Asia Dispatches*, May 14, 2019, https://www.wilsoncenter.org/blog-post/the-south-china-sea-strategic-terms.

55. Larry Elliott, "Most Difficult Global Outlook since 1930s Heralds End of U.S.-Led World Order," *The Guardian*, April 21, 2024, https://www.theguardian.com/business/2024/apr/21/most-difficult-global-outlook-for-a-century-heralds-end-of-us-led-world-order-imf.

Bibliography

Abbate, Janet. *Inventing the Internet*. Boston: MIT Press, 2000.

ACLU. "In Big Win, Settlement Ensures Clearview AI Complies with Groundbreaking Illinois Biometric Privacy Law." ACLU news release. May 9, 2022. https://www.aclu.org/press-releases/big-win-settlement-ensures-clearview-ai-complies-with-groundbreaking-illinois.

———. *War Comes Home: The Excessive Militarization of American Police*. New York: American Civil Liberties Union, 2015. https://www.aclu.org/report/war-comes-home-excessive-militarization-american-police.

Adcock, Phillip, "6 Limitations of AI & Why It Won't Quite Take over in 2023." *AdCockSolutions*. March 14, 2024. https://www.adcocksolutions.com/post/6-limitations-of-ai-why-it-wont-quite-take-over-in-2023.

Agre, Philip E. "Surveillance and Capture: Two Models of Privacy." *The Information Society* 10, no. 2 (1994): 101–27. https://www.tandfonline.com/doi/abs/10.1080/01972243.1994.9960162.

"Airbnb Q4 2022 and Full-Year Financial Results." Airbnb.com. February 14, 2023. https://news.airbnb.com/airbnb-q4-2022-and-full-year-financial-results/.

Ajuntament de Barcelona. "Decidim Barcelona." Accessed June 9, 2024. https://ajuntament.barcelona.cat/digital/en/technology-accessible-everyone/accessible-and-participatory/accessible-and-participatory-5.

Allen, Marshall. "You Snooze, You Lose: Insurers Make the Old Adage Literally True." *ProPublica*. November 21, 2018. https://www.propublica.org/article/you-snooze-you-lose-insurers-make-the-old-adage-literally-true.

Allyn, Bobby. "Judge Allows Federal Trade Commission's Latest Suit against Facebook to Move Forward." *NPR*. January 11, 2022. https://www.npr.org/2022/01/11/1072169787/judge-allows-federal-trade-commissions-latest-suit-against-facebook-to-move-forw.

Altenried, Moritz. "The Platform as Factory: Crowdwork and the Hidden Labour Behind Artificial Intelligence." *Capital and Class* 44, no. 2 (2020): 145–58. https://doi.org/10.1177/0309816819899410.

Ammann, Thorsten, and Constantin Orth, "EU AI Act Update—Latest Developments and Potential Roadblocks Ahead." *DLA Piper*. November 29, 2023. https://www.technologyslegaledge.com/2023/11/eu-ai-act-update-latest-developments-and-potential-roadblocks-ahead/.

Amnesty International. "USA: NYPD Ordered to Hand over Documents Detailing Surveillance of Black Lives Matter Protests Following Lawsuit." *Amnesty International*. August 1, 2022. https://www.amnesty.org/en/latest/news/2022/08/usa-nypd-black-lives-matter-protests-surveillance/.

Amoore, Louise. "Algorithmic War: Everyday Geographies of the War on Terror." *Antipode* 41, no. 1 (2009): 49–69. https://doi.org/10.1111/j.1467-8330.2008.00655.x.

Andreessen, Marc. "The Techno-Optimist Manifesto." a16z. October 16, 2023. https://a16z.com/the-techno-optimist-manifesto/.

Andrejevic, Mark. "The Big Data Divide." *International Journal of Communication* 8 (2014): 1673–89. https://ijoc.org/index.php/ijoc/article/viewFile/2161/1163.

———. *Ispy: Surveillance and Power in the Interactive Era*. Lawrence: University Press of Kansas, 2007.

"Apple's Independent Repair Provider Program Expands Globally." Apple news release. March 29, 2021. https://www.apple.com/newsroom/2021/03/apples-independent-repair-provider-program-expands-globally/.

Arthur, Charles. "Your Face Belongs to Us by Kashmir Hill Review—Nowhere to Hide." *The Guardian*. September 21, 2023. https://www.theguardian.com/books/2023/sep/21/your-face-belongs-to-us-by-kashmir-hill-review-nowhere-to-hide.

"Artificial Intelligence Act," Briefing EU Legislation in Progress, European Parliamentary Research Service, PE 698.792. November 2021. https://www.europarl.europa.eu/RegData/etudes/BRIE/2021/698792/EPRS_BRI(2021)698792_EN.pdf.

Arvidson, Adam, and Eleanor Colleoni. "Value in Informational Capitalism and on the Internet." *The Information Society* 28, no. 3 (2012): 135–50. https://doi.org/10.1080/01972243.2012.669449.

Atkin, Douglas. *The Culting of Brands: When*

Customers Become True Believers. New York: Portfolio, 2004.

Auletta, Ken. *Googled: The End of the World as We Know It.* New York—London: Penguin, 2010.

Authority for Consumers & Markets, "ACM to Assess Adjusted Proposal of Apple Regarding Its Conditions for Dating Apps." March 28, 2022. https://www.acm.nl/en/publications/acm-assess-adjusted-proposal-apple-regarding-its-conditions-dating-apps.

Avila, Renata. "Against Digital Colonialism." In *Platforming Equality: Policy Changes for the Digital Economy,* edited by James Muldoon and Will Stronge, 47–57. London: Autonomy, 2020.

AWS. "Healthcare.Gov Case Study—Amazon Web Services (AWS)." Accessed January 20, 2024. https://aws.amazon.com/solutions/case-studies/healthcare-gov/.

Aydin, Rebecca. "How 3 Guys Turned Renting Air Mattresses in Their Apartment into a $31 Billion Company, Airbnb." *Business Insider.* September 19, 2019. https://www.businessinsider.in/How-3-guys-turned-renting-an-air-mattress-in-their-apartment-into-a-25-billion-company/articleshow/51114238.cms.

Ayres, Ian, and Alan Schwartz. "The No-Reading Problem in Consumer Contract Law." *Stanford Law Review* 66, no. 3 (2014): 545–610. https://www.stanfordlawreview.org/print/article/the-no-reading-problem-in-consumer-contract-law/.

Baker, Tom, and Jonathan Simon. "Embracing Risk." In *Embracing Risk: The Changing Culture of Insurance and Responsibility,* edited by Tom Baker and Jonathan Simon, 1–26. Chicago: University of Chicago Press, 2002.

Bar-Gill, Oren. *Seduction by Contract: Law, Economics, and Psychology in Consumer Markets.* Oxford: Oxford University Press, 2012.

Baraniuk, Chris. "How Algorithms Run Amazon's Warehouses." *BBC.* August 1, 2015. https://www.bbc.com/future/article/20150818-how-algorithms-run-amazons-warehouses.

Barcelona Urban Platform. "Commitment to Boosting Innovation in Barcelona." June 21, 2021. https://ajuntament.barcelona.cat/digital/sites/default/files/commitment_to_boosting_innovation_in_barcelona_bithabitat.pdf.

Barrabi, Thomas. "Feds to Launch Antitrust Probe into Microsoft, OpenAI, Nvidia over 'Monopoly Choke Points.'" *New York Post.* May 13, 2024. https://www.msn.com/en-us/money/other/doj-to-scrutinize-big-tech-s-ai-dominance-with-urgency-antitrust-chief-says/ar-BB1nKIFG.

Barshefsky, Charlene. "EU Digital Protectionism Risks Damaging Ties with the US." *Financial Times.* August 2, 2020. https://www.ft.com/content/9edea4f5-5f34-4e17-89cd-f9b9ba698103.

Bartz, Diane. "Amazon Faces Landmark Monopoly Lawsuit by FTC." *MSN.* September 26, 2023. https://www.msn.com/en-us/news/us/amazon-faces-landmark-monopoly-lawsuit-by-ftc/ar-AA1hmZF7.

Bateson, Gregory. *Steps to an Ecology of Mind.* San Francisco: Chandler Pub Co., 1972.

Beek, Lotje. "Een Betaald Account Is Geen Oplossing Voor Meta's Privacyprobleem." *NRC.* November 10, 2023. https://www.nrc.nl/nieuws/2023/11/10/een-betaald-account-is-geen-oplossing-voor-metas-privacyprobleem-a4180473.

Beniger, James R. *The Control Revolution: Technological and Economic Origins of the Information Society.* Cambridge: Harvard University Press, 1986.

Benjamin, Ruha. *Race after Technology: Abolitionist Tools for the New Jim Code.* London: Polity Books, 2019.

Bennett, Colin J. *The Privacy Advocates: Resisting the Spread of Surveillance.* Cambridge: MIT Press, 2008.

Berg, Janine, Marianne Furrer, Ellie Harmon, Rani Uma, and M. Six Silberman. *Digital Labour Platforms and the Future of Work: Toward Decent Work in the Online World.* International Labour Organization, 2018. https://www.ilo.org/global/publications/books/WCMS_645337/lang—en/index.htm.

Bhuiyan, Johana. "Amazon Ring Says US Police Will Now Need Warrant to Access User Footage." *The Guardian.* January 24, 2024. https://www.theguardian.com/technology/2024/jan/24/police-warrant-amazon-ring-footage.

———. "How Chinese Firm Linked to Repression of Uyghurs Aids Israeli Surveillance in West Bank." *The Guardian.* November 11, 2023. https://www.theguardian.com/technology/2023/nov/11/west-bank-palestinians-surveillance-cameras-hikvision.

———. "'Ready for Some Help?': How a Controversial Technology Firm Courted Portland Police." *The Guardian.* May 3, 2023. https://www.theguardian.com/us-news/2023/may/03/oregon-police-gunshot-detection-shotspotter.

Bivens, Josh. *The Economic Costs and Benefits of Airbnb. No Reason for Local Policymakers to Let Airbnb Bypass Tax for Regulatory Obligations.* Economic Policy Institute, 2019. https://files.epi.org/pdf/157766.pdf.

Blystone, Dan. "The History of Uber: How the Controversial Ride-Sharing Company Came to Dominate Its Market Worldwide." *Investopedia.* April 18, 2023. https://www.investopedia.com/articles/personal-finance/111015/story-uber.asp.

Boewe, Jorn, and Johannes Schulten. *The Long Struggle of the Amazon Employees: Laboratories of Resistance.* New York: Rosa Luxemburg Stiftung, 2017.

Borsook, Paulina. *Cyberselfish: A Critical Romp through the Terribly Libertarian Culture of High Tech.* New York: Public Affairs, 2000.

Bosker, Bianca. "Facebook's Mark Zuckerberg 2005 Interview Reveals Ceo's Doubts." *Huffington Post.* August 11, 2011. https://www.huffpost.com/entry/facebook-mark-zuckerberg-2005-interview_n_924628.

Botsman, Rachel. "Big Data Meets Big Brother as China Moves to Rate Its Citizens." *Wired.* October 12, 2017. https://www.wired.co.uk/

article/chinese-government-social-credit-score-privacy-invasion.

Bradford, Anu. *Digital Empires: The Global Battle to Regulate Technology*. New York: Oxford University Press, 2023.

Bratton, Benjamin H. *The Stack: On Software and Sovereignty*. Cambridge: MIT Press, 2016.

Brayne, Sarah. "Big Data Surveillance: The Case of Policing." *American Sociological Review* 82, no. 5 (2017): 977–1008. https://doi.org/10.1177/0003122417725865.

———. *Predict and Surveil: Data, Discretion, and the Future of Policing*. New York: Oxford University Press, 2021.

———. "Surveillance and System Avoidance: Criminal Justice Contact and Institutional Attachment." *American Sociological Review* 79, no. 3 (2014): 367–91. https://doi.org/10.1177/0003122414530398.

Brenner, Robert. *The Boom and the Bubble*. London: Verso, 2001.

Bridges, Lauren. "Ring Is the Largest Civilian Surveillance Network the US Has Ever Seen." *The Guardian*. May 18, 2021. https://www.theguardian.com/commentisfree/2021/may/18/amazon-ring-largest-civilian-surveillance-network-us?ICID=ref_fark.

Brin, Sergey, and Lawrence Page. "The Anatomy of a Large-Scale Hypertextual Web Search Engine." *Computer Networks and ISDN Systems* 30, no. 1–7: 1–20. https://doi.org/10.1016/S0169-7552(98)00110-X.

Bronowicka, Joanna, and Mirela Ivanova. "The Algorithmic Boss: Guessing, Gaming, Reframing and Contesting Rules in App-Based Management." In *Augmented Exploitation: Artificial Intelligence, Automation and Work*, edited by Phoebe V. Moore and Jamie Woodcock, 149–62. London: Pluto Press, 2021.

Bronzwaer, Stijn. "Clearview AI: Het Bedrijf Dat Met Gezichtsherkenning Verder Gaat Waar Zelfs Google En Facebook Stoppen." *NRC*. October 5, 2023. https://www.nrc.nl/nieuws/2023/10/05/clearview-ai-het-bedrijf-dat-met-gezichtsherkenning-verder-gaat-waar-zelfs-google-en-facebook-stoppen-a4176272.

———. "Techlobby Wil Europese AI-Wetgeving Afzwakken." *NRC*. December 5, 2023. https://www.nrc.nl/nieuws/2023/12/05/techlobby-wil-europese-ai-wetgeving-afzwakken-a4183321.

Brooke, Heather. *The Revolution Will Be Digitised: Dispatches from the Information War*. London: Heinemann, 2011.

Brown, Alleen. "Federal Agencies Pushed Extreme View of Cop City Protesters, Records Show." *The Guardian*. December 6, 2023. https://www.theguardian.com/us-news/2023/dec/06/cop-city-atlanta-georgia-environment-protesters-terrorism.

Bruder, Jessica. "We're Watching You Work. Labor Is Fighting Employer's Techno-Utopian Dream of a Perfectly Efficient—and Totally Surveilled—Workforce." *The Nation*. June 15, 2015, 28–30.

Brussee, Vincent. *Social Credit: The Warring States of China's Emerging Data Empire*. Singapore: Palgrave Macmillan, 2023.

Bullington, Jonathan, and Emily Lane. "How a Tech Firm Brought Data and Worry to New Orleans Crime Fighting." *NOLA.com | The Times Picayune*. March 1, 2018. https://www.nola.com/news/crime_police/how-a-tech-firm-brought-data-and-worry-to-new-orleans-crime-fighting/article_33b8bf05-722f-5163-9a0c-774aa69b6645.html.

Bunders, Damion, Martijn Arets, Koen Frenken, and Tine de Moor. "The Feasibility of Platform Cooperatives in the Gig Economy." *Journal of Co-operative Organization and Management* 10, no. 1 (2022): 1–8. https://doi.org/10.1016/j.jcom.2022.100167.

Burawoy, Michael. *Manufacturing Consent: Changes in the Labor Process under Monopoly Capital*. Chicago: University of Chicago Press, 1979.

Burt, Ronald S. "The Network Entrepreneur." In *Entrepreneurship: The Social Science View*, edited by Richard Swedberg, 281–307. Oxford: Oxford University Press, 2000.

Cadwalladr, Carole. "'Capitalism Is Dead. Now We Have Something Much Worse': Yanis Varoufakis on Extremism, Starmer and the Tyranny of Big Tech." *The Guardian*. September 24, 2023. https://www.theguardian.com/world/2023/sep/24/yanis-varoufakis-technofeudalism-capitalism-ukraine-interview.

Cantor, Matthew. "Idle No More: How Mouse Jigglers Are Taking on Nosy Bosses." *The Guardian*. March 6, 2023. https://www.theguardian.com/technology/2023/mar/05/idle-no-more-how-automatic-mouse-jigglers-are-taking-on-nosy-bosses.

Caprotti, Federico, and Robert Cowley. "Varieties of Smart Urbanism in the UK: Discursive Logics, the State and Local Urban Context." *Transactions of the Institute of British Geographers* 44, no. 3 (2019): 587–601. http://dx.doi.org/10.1111/tran.12284.

Caro, Robert. *The Power Broker: Robert Moses and the Fall of New York*. New York: Vintage Books, 1975.

Carr, Nicholas. *The Glass Cage: How Our Computers Are Changing Us*. New York: W.W. Norton, 2015.

Cavada, Marianna, Chris Rogers, and Dexter Hunt. "Smart Cities: Contradicting Definitions and Unclear Measures." *The 4th World Sustainability Forum 2014—Conference Proceedings Paper*, 1–12. https://sciforum.net/manuscripts/2454/manuscript.pdf.

Cecchinato, Maria E., Sandy J.J. Gould, and Frederick Harry Pitts. "Self-Tracking and Surveillance at Work: Insights from Human-Computer Interaction and Social Science." In *Augmented Exploitation: Artificial Intelligence, Automation and Work*, edited by Phoebe V. Moore and Jamie Woodcock, 127–37. London: Pluto Press, 2021.

Cerf, Vint, and Robert Kahn. "A Protocol for Packet Network Intercommunication." *IEEE*

Transactions on Communications 22, no. 5 (1974): 637–48.

Chan, Kelvin. "Europe Reaches a Deal on the World's First Comprehensive AI Rules." *The Washington Post*. December 8, 2023. https://www.washingtonpost.com/business/2023/12/08/ai-act-europe-regulation/2a4c44f2-961c-11ee-9d5c-d462c9032daa_story.html.

Chapell, Bill. "What's at Stake in Atlanta's 'Cop City' Protests." *NPR*. March 7, 2023. https://www.npr.org/2023/03/07/1161343394/atlanta-cop-city-protests-explained.

Chien, Shiuh-Shen, and Max D. Woodworth. "China's Urban Speed Machine: The Politics of Speed and Time in a Period of Rapid Urban Growth." *International Journal of Urban and Regional Research* 42, no. 4 (2018): 723–37. https://onlinelibrary.wiley.com/doi/10.1111/1468-2427.12610.

"China: Police 'Big Data' Systems Violate Privacy, Target Dissent." *Human Rights Watch*. November 19, 2017. https://www.hrw.org/news/2017/11/19/china-police-big-data-systems-violate-pr.ivacy-target-dissent.

Chiu, Allyson. "She Installed a Ring Camera in Her Children's Room for 'Peace of Mind.' A Hacker Accessed It and Harassed Her 8-Year-Old Daughter." *The Washington Post*. December 12, 2019. https://www.washingtonpost.com/nation/2019/12/12/she-installed-ring-camera-her-childrens-room-peace-mind-hacker-accessed-it-harassed-her-year-old-daughter/.

Christiaens, Tim. "Review of James Muldoon, *Platform Socialism: How to Reclaim Our Digital Futures*," *Marx & Philosophy Review of Books*. January 17, 2022. https://marxandphilosophy.org.uk/reviews/19828_platform-socialism-how-to-reclaim-our-digital-futures-by-james-muldoon-reviewed-by-tim-christiaens/.

Christophers, Brett. "Capitalism Has Always Been 'Rogue.'" *Jacobin* (March 2020). https://jacobin.com/2020/03/surveillance-capitalism-shoshana-zuboff-review/.

———. *Rentier Capitalism*. London: Verso, 2020.

Clark, Jennifer. "What Cities Need Now: Smart Cities Haven't Brought the Tangible Improvements That Many Hoped They Would. What Comes Next?" *MIT Technology Review*. April 28, 2021. https://www.technologyreview.com/2021/04/28/1023104/smart-cities-urban-technology-pandemic-covid/.

Cohen, Jared, and Richard Fontaine. "Uniting the Techno-Democracies: How to Build Digital Cooperation." *Foreign Affairs*. November/December 2020. https://www.foreignaffairs.com/articles/united-states/2020-10-13/uniting-techno-democracies.

Cohen, Julie E. "The Biopolitical Public Domain: The Legal Construction of the Surveillance Economy." *Philosophy & Technology* 31, no. 2 (2018): 213–33. http://dx.doi.org/10.1007/s13347-017-0258-2.

Coleman, E. Gabriella. "From Busting Cults to Breeding Cults: Anonymous H/Acktivism vs. The (a)Nomynous Far Right and QAnon." *HAU: Journal of Ethnographic Theory* 13, no. 2 (2023): 248–63. https://doi.org/10.1086/727758.

Coleman, Gabriella. *Hacker, Hoaxer, Whistleblower, Spy: The Many Faces of Anonymous*. London: Verso Books, 2014.

Cord, Florian, and Simon Schleusener. "Looking Backward at the Present, 2020–1990: Deleuze's 'Postscript on Control Societies.'" *Coils of the Serpent: Journal for the Study of Contemporary Power* 6 (2020): 1–6. https://ul.qucosa.de/api/qucosa%3A72917/attachment/ATT-0/.

Corkery, Michael, and Jessica Silver-Greenberg. "Miss a Payment? Good Luck Moving That Car." *The New York Times*. September 24, 2014. https://archive.nytimes.com/dealbook.nytimes.com/2014/09/24/miss-a-payment-good-luck-moving-that-car/.

Couldry, Nick, and Ulises A. Mejias. *The Costs of Connectivity: How Data Is Colonizing Human Life and Appropriating It for Capitalism*. Stanford: Stanford University Press, 2019.

———. "The Decolonial Turn in Data and Technology Research, What Is at Stake and Where Is It Heading?" *Information, Communication & Society* 26, no. 4 (2023): 716–802. https://www.tandfonline.com/doi/full/10.1080/1369118X.2021.1986102.

Covert, Bryce. "Anti-Monopoly Power." *The Nation*. December 25, 2023/January 1, 2024, 38–41.

Cozzi, Fabio, Stefano Mechelli, Lorraine Maisner-Boché, Massimiliano Moruzzi, and Pilar Arzuago. "The EU Artificial Intelligence Act: What's the Impact?" *McDermott, Will & Emery*. July 31, 2023. https://www.mwe.com/insights/the-eu-artificial-intelligence-act-whats-the-impact/.

Crain, Matthew. *Profit over Privacy: How Surveillance Advertising Conquered the Internet*. Minneapolis: University of Minnesota Press, 2021.

Crawford, Kate, Jessica Lingel, and Tero Karppi. "Our Metrics, Ourselves: A Hundred Years of Self-Tracking from the Weight Scale to the Worst Device." *European Journal of Cultural Studies* 18, no. 4–5 (2015): 479–96. https://doi.org/10.1177/1367549415584857.

Cypher, James M. "The Political Economy of Systemic U.S. Militarism." *Monthly Review* 73, no. 11 (2022). https://monthlyreview.org/2022/04/01/the-political-economy-of-systemic-u-s-militarism-2/.

Datta, Ayona. "Three Big Challenges for Smart Cities and How to Solve Them." *The Conversation*. June 9, 2016. https://theconversation.com/three-big-challenges-for-smart-cities-and-how-to-solve-them-59191.

Davenport, Thomas H. "How Big Data Is Helping the NYPD Solve Crimes Faster." *Fortune*. July 17, 2016. https://fortune.com/2016/07/17/big-data-nypd-situational-awareness/.

Davies, Dave. "Surveillance and Local Police: How Technology Is Evolving Faster Than Regulation." *NPR*. January 27, 2021. https://www.npr.org/2021/01/27/961103187/surveillance-and-local-police-how-technology-is-evolving-faster-than-regulation.

Davis, Mike. *Planet of Slums*. New York: Verso, 2007.

"A Declaration for the Future of the Internet." White House. April 28, 2022. https://www.whitehouse.gov/briefing-room/statements-releases/2022/04/28/fact-sheet-united-states-and-60-global-partners-launch-declaration-for-the-future-of-the-internet/.

Derbali, Naïm. "ASML uit het vizier van verscherpte Amerikaanse exportregels naar China." *NRC*. July 31, 2024. https://www.nrc.nl/nieuws/2024/07/31/asml-uit-het-vizier-van-verscherpte-amerikaanse-exportregels-naar-china-a4861348.

Dhingra, Mani. "What's So Smart About Smart Cities?" *RTE Brainstorm*. March 22, 2023. https://www.rte.ie/brainstorm/2023/0321/1364457-whats-so-smart-about-smart-cities/.

Diamond, Larry. "Democracy's Arc: From Resurgent to Imperiled." *Journal of Democracy* 33, no. 1 (2022): 163–79. https://www.journalofdemocracy.org/articles/democracys-arc-from-resurgent-to-imperiled/.

Didier, Emmanuel. "Globalization of Quantitative Policing: Between Management and Statactivism." *Annual Review of Sociology* 44 (2018): 515–35. https://doi.org/10.1146/annurev-soc-060116-053308.

DigiChina. "State Council Notice Concerning Issuance of the Planning Outline for the Construction of a Social Credit System (2014–2020)." *DigiChina*. June 14, 2014. https://digichina.stanford.edu/work/planning-outline-for-the-construction-of-a-social-credit-system-2014-2020/.

Douglas Heaven, Will. "Predictive Policing Algorithms Are Racist. They Need to Be Dismantled." *MIT Technology Review*. July 17, 2020. https://www.technologyreview.com/2020/07/17/1005396/predictive-policing-algorithms-racist-dismantled-machine-learning-bias-criminal-justice/.

Dubal, Veena. "A Brief History of the Gig." *Logic(s)*. May 4, 2020. https://logicmag.io/security/a-brief-history-of-the-gig/.

Duggins, Alexi. "TV Soaps Could Be Made by AI Writers within Three Years, Director Warns." *The Guardian*. February 22, 2024. https://www.theguardian.com/tv-and-radio/2024/feb/22/james-hawes-select-committee-tv-soaps-made-using-ai.

Durand, Cédric, and William Milberg. "Intellectual Monopoly in Global Value Chains." *Review of International Political Economy* 27, no. 5 (2019): 1–26. https://doi.org/10.1080/09692290.2019.1660703.

Dyson, Esther. *Release 2.0: A Design for Living in the Digital Age*. New York: Broadway Books, 1997.

Elliott, Larry. "Biden Announces 100% Tariff on Chinese-Made Electric Vehicles." *The Guardian*. May 14, 2022. https://www.theguardian.com/business/article/2024/may/14/joe-biden-tariff-chinese-made-electric-vehicles.

———. "Most Difficult Global Outlook since 1930s Heralds End of US-Led World Order." *The Guardian*. April 21, 2024. https://www.theguardian.com/business/2024/apr/21/most-difficult-global-outlook-for-a-century-heralds-end-of-us-led-world-order-imf.

Ericson, Richard V., and Kevin D. Haggerty. *Policing the Risk Society*. Oxford: Oxford University Press, 1997.

Espinoza, Javier, and Michael Acton. "Apple Set to Be the First Big Tech Group to Face Charges under EU Digital Law." *The Financial Times*. June 14, 2024. https://www.ft.com/content/31a996d5-b472-4357-953e-ace078494604.

Eubanks, Virginia. *Automating Inequality: How High-Tech Tools Profile, Police, and Punish the Poor*. New York: St. Martin's Press, 2018.

European Commission. "About the Digital Markets Act." Accessed July 8, 2024. https://digital-markets-act.ec.europa.eu/about-dma_en.

———. "Digital Services Act (DSA) overview." Accessed July 8, 2024. https://commission.europa.eu/strategy-and-policy/priorities-2019-2024/europe-fit-digital-age/digital-services-act_en.

European Parliament. "Artificial Intelligence Act: Deal on Comprehensive Rules for Trustworthy AI." News release. December 9, 2023. https://www.europarl.europa.eu/news/en/press-room/20231206IPR15699/artificial-intelligence-act-deal-on-comprehensive-rules-for-trustworthy-ai.

———. "Artificial Intelligence Act: MEPs Adopt Landmark Law." Press release. March 13, 2024. https://www.europarl.europa.eu/news/en/press-room/20240308IPR19015/artificial-intelligence-act-meps-adopt-landmark-law.

Evans, Scarlett. "Illinois Bans Drones from Using Facial Recognition." *IOT World Today*. June 29, 2023. https://www.iotworldtoday.com/security/illinois-bans-drones-from-using-facial-recognition-#close-modal.

Ewen, Stuart. *Captains of Consciousness: Advertising and the Social Roots of the Consumer Culture*. New York: McGraw-Hill, 1976.

Ewen, Stuart, and Elizabeth Ewen. *Channels of Desire*. New York: McGraw-Hill, 1982.

Fabino, Alexander. "Ring Ordered to Refund 55,000 Customers in Mid-2024: Do You Qualify?" *Newsweek*. November 15, 2023. https://www.newsweek.com/ftc-ring-refund-privacy-breach-settlement-2024-1844178.

"Fact Sheet: President Biden Takes Action to Protect American Workers and Businesses from China's Unfair Trade Practices." White House. May 14, 2024. https://www.whitehouse.gov/briefing-room/statements-releases/2024/05/14/fact-sheet-president-biden-takes-action-to-protect-american-workers-and-businesses-from-chinas-unfair-trade-practices/.

Fasman, John. *We See It All: Liberty and Justice in an Age of Perpetual Surveillance*. New York: Public Affairs, 2021.

Federal Trade Commission. "FTC Launches Inquiry into Generative AI Investments and Partnerships." *ftc.gov*. January 25, 2024. https://www.ftc.gov/news-events/news/press-releases/2024/01/ftc-launches-inquiry-generative-ai-investments-partnerships.

———. "FTC Says Ring Employees Illegally Surveilled Customers, Failed to Stop Hackers from Taking Control of Users' Cameras." *ftc.gov*. May 31, 2023. https://www.ftc.gov/news-events/news/press-releases/2023/05/ftc-says-ring-employees-illegally-surveilled-customers-failed-stop-hackers-taking-ntrol-users.

Feiner, Lauren, and Annie Palmer. "Amazon to Pay over $30 Million in FTC Settlements over Ring, Alexa Privacy Violations." *CNBC*. May 31, 2023. https://www.cnbc.com/2023/05/31/ftc-sues-amazon-over-ring-doorbell-privacy-violations.html.

Ferguson, Andrew Guthrie. "High-Tech Surveillance Amplifies Police Bias and Overreach." *The Conversation*. June 12, 2020. https://theconversation.com/high-tech-surveillance-amplifies-police-bias-and-overreach-140225.

———. *The Rise of Big Data Policing: Surveillance, Race, and the Future of Law Enforcement*. New York: New York University Press, 2017.

Fischels, Josie. "A Look Back at the Very First Website Ever Launched, 30 Years Later." *NPR*. August 6, 2021. https://www.npr.org/2021/08/06/1025554426/a-look-back-at-the-very-first-website-ever-launched-30-years-later.

FLIA. "Notice of the State Council Issuing the New Generation of Artificial Intelligence Development Plan." State Council Document No. 35. July 2017. https://flia.org/notice-state-council-issuing-new-generation-artificial-intelligence-development-plan/.

Ford, Martin. *Rise of the Robots: Technology and the Threat of a Jobless Future*. New York: Basic Books, 2015.

Foster, John B., and Brett Clark. "Notes from the Editors." *Monthly Review* 70, no. 2 (2018). https://monthlyreview.org/category/2018/volume-70-issue-02-june/.

Foster, John B., and Robert W. McChesney. "Surveillance Capitalism. Monopoly-Finance Capital, the Military-Industrial Complex, and the Digital Age." *Monthly Review* 66, no. 3 (2014). https://monthlyreview.org/2014/07/01/surveillance-capitalism/.

Foucault, Michel. *The Birth of Biopolitics: Lectures at the College of France, 1978–1979*. New York: Picador, 2008.

———. *Discipline and Punish: The Birth of the Prison*. New York: Vintage Books, 1995.

Fourcade, Marion, and Kieran Healy. "Seeing Like a Market." *Socio-Economic Review* 15, no. 1 (2017): 9–29. https://doi.org/10.1093/ser/mww033.

Frischmann, Brett, and Evan Selinger. *Re-Engineering Humanity*. Cambridge: Cambridge University Press, 2019.

From, Al. *Democrats and the Return to Power*. New York: Palgrave Macmillan, 2013.

Fu, Diana, and Rui Hou. "Rating Citizens with China's Social Credit System." In *CPC Futures: The New Era of Socialism with Chinese Characteristics*, edited by Frank N. Pieke and Bert Hofman, 78–85. Singapore: East Asian Institute, National University of Singapore, 2022. https://epress.nus.edu.sg/cpcfutures/9789811852060-10.pdf.

Fuchs, Christian. "Labor in Informational Capitalism and on the Internet." *The Information Society* 26, no. 3 (2010): 179–96. https://doi.org/10.1080/01972241003712215.

———. *The Online Advertising Tax as the Foundation of a Public Service Internet*. London: University of Westminster Press, 2018.

Fuchs, Christian, and Klaus Unterberger, eds. *The Public Service Media and Public Service Internet Manifesto*. London: University of Westminster Press, 2021.

Gandy, Oscar H. *The Panoptic Sort: A Political Economy of Personal Information*. Oxford: Oxford University Press, [1993] 2021.

Gangadharan, Seeta P., and Jedszej Niklas. "Decentering Technology in Discourse on Discrimination." *Information, Communication & Society* 22, no. 7 (2021): 882–99. https://doi.org/10.1080/1369118X.2019.1593484.

Geradin, Damien, "When Apple Takes the European Commission for Fools: An Initial Overview of Apple's New Terms and Conditions for iOS App Distribution in the EU," *The Platform Law Blog*. January 26, 2024. https://theplatformlaw.blog/2024/01/26/when-apple-takes-the-european-commission-for-fools-an-initial-overview-of-apples-new-terms-and-conditions-for-ios-app-distribution-in-the-eu/.

Gerken, Tom, and Liv McMahon. "TikTok Fined $345m over Children's Data Privacy." *BBC*. September 15, 2023. https://www.bbc.com/news/technology-66819174.

Gershgorn, Dave. "China's Sharp Eyes' Program Aims to Surveil." *One Zero*. March 2, 2021. https://onezero.medium.com/chinas-sharp-eyes-program-aims-to-surveil-100-of-public-space-ddc22d63e015.

Gilliom, John, and Torin Monahan. *SuperVision: An Introduction to the Surveillance Society*. Chicago: University of Chicago Press, 2012.

Goldsmith, Stephen. "As the Chorus of Dumb City Advocates Increases, How Do We Define the Truly Smart City?" *Data-Smart City Solutions*. September 16, 2021. https://datasmart.hks.harvard.edu/chorus-dumb-city-advocates-increases-how-do-we-define-truly-smart-city.

Gonsalves, Sean. "New Municipal Broadband Networks Skyrocket in Post-Pandemic America as Alternative to Private Monopoly Model." *Community Networks*. January 18, 2024. https://communitynets.org/content/new-municipal-broadband-networks-skyrocket-post-pandemic-america-alternative-private.

Gordon, David, and Meia Nouwens. "Conclusion." In *The Digital Silk Road: China's Technological Rise and the Geopolitics of Cyberspace*, edited by

David Gordon and Meia Nouwens, 157–73. New York: Routledge, 2022.

Graham, Mark, and Jamie Woodcock. *The Gig Economy: A Critical Introduction*. London: Polity Press, 2019.

Graham, Steven. *Cities under Siege: The New Military Urbanism*. London: Verso Books, 2011.

Gray, Mary L., and Suri Siddarth. *Ghost Work: How to Stop Silicon Valley from Building a New Global Underclass*. Boston: Houghton Mifflin, 2019.

Greenbaum, Joan. *Windows on the Workplace: Technology, Jobs, and the Organization of Office Work*. New York: Monthly Review Press, 2004.

Greenfield, Adam. *Against the Smart City: A Pamphlet*. New York: Do projects, 2013.

Greenstein, Shane. "Commercialization of the Internet: The Interaction of Public Policy and Private Choices or Why Introducing the Market Worked So Well." In *Innovation Policy and the Economy*, edited by Adam B. Jaffe, Josh Lerner and Scott Stern, 151–61. Cambridge: MIT Press, 2001.

Greenwald, Glen. "Wikipedia Co-Founder Calls Wikipedia 'Most Biased Encyclopedia.'" *ScheerPost*. August 3, 2023. https://scheerpost.com/2023/08/03/glenn-greenwald-wikipedia-co-founder-calls-wikipedia-most-biased-encyclopedia/.

Griffin, Carl J. "Enclosure as Internal Colonisation: The Subaltern Commons, *Terra Nullius* and the Settling of England's Waste." *Transactions of the RHS* 1 (December 2023): 95–120. https://doi.org/10.1017/S0080440123000014.

Guardian editorial. "The Guardian View on Microworking: Younger, Educated Workers Left Powerless." *The Guardian*. August 21, 2022. https://www.theguardian.com/commentisfree/2022/aug/21/the-guardian-view-on-microworking-younger-educated-workers-left-powerless.

———. "Google Sued for $2.3bn by European Media Groups over Digital Add Losses." *The Guardian*. February 28, 2024. https://www.theguardian.com/technology/2024/feb/28/news-media-europe-google-lawsuit-ad-revenue.

———. "The Guardian View on the AI Conundrum: What It Means to Be Human Is Elusive." *The Guardian*. December 29, 2023. https://www.theguardian.com/commentisfree/2023/dec/29/the-guardian-view-on-the-ai-conundrum-what-it-means-to-be-human-is-elusive.

The Guardian staff and agencies. "'Different Set of Rules': How FTC Head Lina Khan Is Fighting Tech Giants Such as Amazon." *The Guardian*. September 27, 2023. https://www.theguardian.com/technology/2023/sep/27/ftc-head-lina-khans-fight-against-amazon-has-been-years-in-the-making.

———. "Justice Department Alleges Google Tried to 'Eliminate' Ad Markets Rivals in Lawsuit." *The Guardian*. January 24, 2023. https://www.theguardian.com/technology/2023/jan/24/justice-department-lawsuit-google-antitrust-law-advertising.

Gurley, Lauren Kaori. "Internal Documents Show Amazon's Dystopian System for Tracking Workers Every Minute of Their Shifts." *Vice*. June 2, 2022. https://www.vice.com/en/article/5dgn73/internal-documents-show-amazons-dystopian-system-for-tracking-workers-every-minute-of-their-shifts.

Hafner, Katie, and Matthew Lyon. *Where the Wizards Stay up Late: The Origins of the Internet*. New York: Simon & Schuster, [1996] 2006.

Haggart, Blayne. "Evaluating Scholarship, or Why I Won't Be Teaching Shoshana Zuboff's the Age of Surveillance Capitalism." *Blayne Haggart's Orangespace*. February 15, 2019. https://blaynehaggart.com/2019/02/15/evaluating-scholarship-or-why-i-wont-be-teaching-shoshana-zuboffs-the-age-of-surveillance-capitalism/.

Haggerty, Kevin D., and Richard V. Ericson. "The Surveillant Assemblage." *British Journal of Sociology* 51, no. 4 (2000): 605–22. https://doi.org/10.1080/00071310020015280.

Han, Hahrie, and Elizabeth McKenna. *Groundbreakers: How Obama's 2.2 Million Volunteers Transformed Campaigning in America*. Oxford: Oxford University Press, 2015.

Hanna, Alex, Emily Denton, Razvan Amironesei, Andrew Smart, and Hilary Nicole. "Lines of Sight." *Logic(s)*. December 20, 2020. https://logicmag.io/commons/lines-of-sight/.

Hansen, Evan. "Doubleclick under Email Attack for Consumer Profiling Plans." *CNET*. February 2, 2000. https://www.cnet.com/tech/services-and-software/doubleclick-under-email-attack-for-consumer-profiling-plans/.

Harris, David Evan. "Europe Has Made a Great Leap Forward in Regulating AI. Now the Rest of the World Must Step Up." *The Guardian*. December 13, 2023. https://www.theguardian.com/commentisfree/2023/dec/13/europe-regulating-ai-artificial-intelligence-threat.

Hart, David. "On the Origins of Google." *National Science Foundation*, August 17, 2004. http://www.nsf.gov/discoveries/disc_summ.jsp?cntn_id=100660&org=NSF.

Harvey, David. *A Brief History of Neoliberalism*. Oxford: Oxford University Press, 2005.

Hati, Sri Rahaju Hijrah, Tengku Ezni Balqial, Arga Hananto, and Elevita Yuliati. "A Decade of Systematic Literature Review on Airbnb, the Sharing Economy from a Multiple Stakeholder Perspective." *Heliyon* 7 (2021). https://www.semanticscholar.org/paper/A-decade-of-systematic-literature-review-on-Airbnb:-Hati-Balqiah/652ed1dc827b5b48182c98a18cf392b7ad8e8dbb.

Hauburisn, Christopher. "Automation Is Coming for Truckers. But First, They're Being Watched." *Vox*. November 20, 2017. https://www.vox.com/videos/2017/11/20/16670266/trucking-eld-surveillance.

Hawkins, Andrew J. "Sidewalk Labs Shuts Down Toronto Smart City Project." *The Verge*. May 7, 2020. https://www.theverge.com/2020/5/7/21250594/alphabet-sidewalk-labs-toronto-quayside-shutting-down.

Hayes, Adam. "Smart Home: Definition, How They Work, Pros and Cons." *Investopedia*. September 14, 2022. https://www.investopedia.com/terms/s/smart-home.asp.

Head, Simon. *Mindless: Why Smarter Machines Are Making Dumber Humans*. New York: Basic Books, 2014.

Helmond, Anne. "The Platformization of the Web: Making Web Data Platform Ready." *Social Media + Society* 1, no. 2 (2015): 1–11. https://doi.org/10.1177/2056305115603080.

Helmore, Edward. "Nvidia Reports Strategic Growth as AI Boom Shows No Signs of Stopping." *The Guardian*. May 22, 2024. https://www.theguardian.com/technology/article/2024/may/22/nvidia-quarterly-earnings.

Helmore, Edward, and Kari Paul. "New York Times Sues OpenAI and Microsoft for Copyright Infringement." *The Guardian*. December 27, 2023. https://www.theguardian.com/media/2023/dec/27/new-york-times-openai-microsoft-lawsuit.

Hern, Alex. "AI Race Heats up as OpenAI, Google and Mistral Release New Models." *The Guardian*. April 10, 2024 2024. https://www.theguardian.com/technology/2024/apr/10/ai-race-heats-up-as-openai-google-and-mistral-release-new-models.

Hesmondhalgh, David. *The Cultural Industries*. Los Angeles: Sage, 2013.

Higginbotham, Stacey. "Why Insurance Companies Want to Subsidize Your Smart Home." *MIT Technological Review*. October 12, 2016. https://www.technologyreview.com/2016/10/12/157045/why-insurance-companies-want-to-subsidize-your-smart-home/.

Hijink, Marc. "Apple Doet Een Greep Naar De Advertentiedollars Van De iPhone." *NRC*. September 8, 2022. https://www.nrc.nl/nieuws/2022/09/08/apple-doet-een-greep-naar-de-advertentiedollars-van-de-iphone-a4141255.

———. "VS Leggen Strengere Exportrestricties Naar China Op Aan ASML." *NRC*. October 17, 2023. https://www.nrc.nl/nieuws/2023/10/17/vs-leggen-strengere-exportrestricties-naar-china-op-aan-asml-a4177595.

Hill, Kashmir. *Your Face Belongs to Us: The Secret Startup Dismantling Your Privacy*. New York: Simon & Schuster, 2023.

Holt, Jennifer. *Empires of Entertainment: Media Industries and the Politics of Deregulation, 1980–1996*. New Brunswick: Rutgers University Press, 2011.

Hoofnagle, Chris J. "Big Brother's Little Helpers: How Choicepoint and Other Commercial Data Brokers Collect and Package Your Data for Law Enforcement." *North Carolina Journal of International Law* 29, no. 4 (2003): 595–637. https://scholarship.law.unc.edu/ncilj/vol29/iss4/1.

Horowitz, Robert. *Irony of Regulatory Reform: The Deregulations of American Telecommunications*. Oxford: Oxford University Press, 1989.

Hu, Tung-Hui. *A Prehistory of the Cloud*. Cambridge: MIT Press, 2015.

Hughes, John. "Airbnb SWOT Analysis." *Business Chronicler*. Accessed January 31, 2024. https://businesschronicler.com/swot/airbnb-swot-analysis/.

Hussain, Zainab, and Joshua Franklin. "Valuation Surges Past $100 Billion in Biggest U.S. IPO of 2020." *Reuters*. December 10, 2020. https://www.reuters.com/article/airbnb-ipo-idUSKBN28K261.

Husueh, Vicki. "Cultivating and Challenging the Common: Lockean Property, Indigenous Traditionalisms, and the Problem of Exclusion." *Contemporary Political Theory* 5 (2006): 193–214. https://doi.org/10.1057/palgrave.cpt.9300233.

Huws, Ursula. "Logged Labour: A New Paradigm of Work Organisation?" *Work Organisation, Labour and Globalisation* 10, no. 1 (2016): 7–26. https://doi.org/10.13169/workorgalaboglob.10.1.0007.

Hvistendahl, Mara. "Inside China's Vast New Experiment in Social Ranking." *Wired*. December 14, 2017. https://www.wired.com/story/age-of-social-credit/.

Hwang, Tim. *Subprime Attention Crisis: Advertising and the Time Bomb at the Heart of the Internet*. New York: FSG Originals x Logic, 2020.

"Inside Airbnb." Airbnb.com, accessed May 12, 2024. http://insideairbnb.com/.

Irani, Lilly, and M. Six Silberman. "Turkopticon: Interrupting Worker Invisibility in Amazon Mechanical Turk." *Proceedings of the SIGCHI Conference on Human Factors in Computing Systems* (2013): 611–20. http://crowdsourcing-class.org/readings/downloads/ethics/turkopticon.pdf.

Ismail, Yashmin, and Rashmi Jose. "E-Commerce Takes Centre Stage at World Trade Organization In Run-up to MC13." *The International Institute for Sustainable Development (IISD)*. January 11, 2024. https://www.iisd.org/articles/policy-analysis/e-commerce-developments-wto-mc13.

Jasser, Greta, Jordan McSwiney, Ed Pertwee, and Savvas Zannettou. "'Welcome to #Gabfam': Far-Right Virtual Community on Gab." *New Media & Society* 25, no. 7 (2023): 1728–45. https://doi.org/10.1177/14614448211024546.

Jesse, Matthias, and Dietmar Jannach. "Digital Nudging with Recommender Systems: Survey and Future Directions." *Computers in Human Behavior Reports* 3 (2021): 1–14. https://doi.org/10.1016/j.chbr.2020.100052.

Jolly, Jasper, and Graeme Wearden. "Landmark Moment as Uber Unveils Its First Annual Profit as Limited Company." *The Guardian*. February 7, 2024. https://www.theguardian.com/technology/2024/feb/07/landmark-moment-as-uber-unveils-first-annual-profit-as-limited-company.

Jones, Callum. "Amazon Profits Surge on Strong Trading Season and Cloud Computing Growth." *The Guardian*. February 1, 2024. https://www.theguardian.com/technology/2024/feb/01/amazon-earnings-q4.

———. "Fears of Employee Displacement as Amazon Brings Robots into Warehouses."

The Guardian. October 19, 2023. https://www.theguardian.com/technology/2023/oct/18/amazon-robot-warehouses-digit-workers.

Jones, Phil. *Work without the Worker: Labor in the Age of Platform Capitalism*. London: Verso, 2021.

Jones, Phil, and James Muldoon. *Rise and Grind: Microwork and Hustle Culture in the UK*. Thinktank Autonomy. June 2022. https://autonomy.work/wp-content/uploads/2022/06/riseandgrind11.pdf.

Jouvenal, Justin. "The New Way Police Are Surveilling You: Calculating Your Threat Score." *The Washington Post*. January 10, 2016. https://www.washingtonpost.com/local/public-safety/the-new-way-police-are-surveilling-you-calculating-your-threat-score/2016/01/10/e42bccac-8e15-11e5-baf4-bdf37355da0c_story.html.

Kaaz, Kim J., Alex Hoffer, Masha Saeidi, Anita Sarma, and Rakesh B. Bobba. "Understanding User Perceptions of Privacy, and Configuration Challenges in Home Automation." In *2017 IEEE Symposium on Visual Languages and Human-Centric Computing (VL/HCC)*, 297–301. Raleigh, NC: IEEE, 2017.

Kahn, Linda M. "Sources of Tech Platform Power." *Georgetown Law & Technology Review* 2, no. 2 (2018): 325–34. https://georgetownlawtechreview.org/sources-of-tech-platform-power/GLTR-07-2018/.

Kantor, Jodi. "Working Anything but 9 to 5 Scheduling: Technology Leaves Low-Income Parents with Hours of Chaos." *The New York Times*. August 13, 2014. https://www.nytimes.com/interactive/2014/08/13/us/starbucks-workers-scheduling-hours.html.

Kashkool, Keyvan. *The Making of a Modern Market: Ebay.Com*. Doctoral dissertation, University of California, Berkeley, 2010. https://escholarship.org/uc/item/16t905b5.

Kear, Mark. "Playing the Credit Score Game: Algorithms, 'Positive' Data and the Personification of Financial Objects." *Economy and Society* 46, no. 3–4 (2017): 346–68. https://doi.org/10.1080/03085147.2017.1412642.

Kelly, Kevin. *New Rules for the New Economy: 10 Radical Strategies for a Connected World*. New York: Viking, 1998.

Kennedy, Devin. "The People's Utility." *Logic(s)*. August 1, 2018. https://logicmag.io/failure/the-people's-utility/.

Kerr, Dana. "U.S. v. Google: As Landmark 'Monopoly Power' Trial Closes, Here's What to Look For." *NPR*. May 2, 2024, https://www.npr.org/2024/05/02/1248152695/google-doj-monopoly-trial-antitrust-closing-arguments.

Kesvani, Hussein. "First It Was Facebook, Then Twitter. Is Reddit About to Become Rubbish Too?" *The Guardian*. March 20, 2024. https://www.theguardian.com/commentisfree/2024/mar/20/facebook-twitter-reddit-rubbish-ipo.

Kim, Nancy S. *Wrap Contracts: Foundations and Ramifications*. New York: Oxford University Press, 2013.

Kingsley-Hughes, Adrian. "The Android 'Toxic Hellstew' Survival Guide." *ZDnet*. June 9, 2014. https://www.zdnet.com/article/the-android-toxic-hellstew-survival-guide/.

Kirkpatrick, David. *The Facebook Effect: The Inside Story of the Company That Is Connecting the World*. New York: Simon & Schuster, 2010.

Kitchin, Rob. "Conceptualising Smart Cities." *Urban Research & Practice* 5, no. 1 (2022): 155–59. https://doi.org/10.1080/17535069.2022.2031143.

———. "Decentering Smart Cities." In *Equality in the City: Imaginaries of the Smart Future*, edited by Susan Flynn, 260–66. Bristol: Intellect, 2021.

———. *Getting Smarter About Smart Cities: Improving Data Privacy and Data Security*. Dublin, Ireland: Data Protection Unit, Department of the Taoiseach, 2016. https://www.academia.edu/27367984/Getting_smarter_about_smart_cities_Improving_data_privacy_and_data_security.

———. "Making Sense of Smart Cities: Addressing Present Shortcomings." *Cambridge Journal of Regions, Economy and Society* 8, no. 1 (2015): 131–36. https://doi.org/10.1093/cjres/rsu027.

Klein, Naomi. *This Changes Everything: Capitalism Vs. the Climate*. New York: Simon & Schuster, 2015.

Klyce, John. "Safety vs. Privacy Issues Arise as Fedex Express Installs Driver-Facing Cameras in Vehicles." *Memphis Business Journal.*, October 18, 2021. https://www.bizjournals.com/memphis/news/2021/10/18/fedex-express-installing-driver-cameras-delivery.html.

Knight, Will. "China's AI Awakening." *MIT Technology Review*, October 10, 2017. https://www.technologyreview.com/2017/10/10/148284/chinas-ai-awakening/.

Koepp, Robert. "Locating the Digital Silk Road in the Belt and Road Initiatives." In *The Digital Silk Road: China's Technological Rise and the Geopolitics of Cyberspace*, edited by David Gordon and Meia Nouwens, 37–49. New York: Routledge, 2022.

Kramer, Adam D.I., Jamie E. Guillory, and Jeffrey T. Hancock. "Experimental Evidence of Massive-Scale Emotional Contagion through Social Networks." *Proceedings of the National Academy of Sciences of the United States of America* 111, no. 24 (2014): 8788–90. https://doi.org/10.1073/pnas.1320040111.

Kravits, Derek, and Marshall Allen. "Your Medical Devices Are Not Keeping Your Health Data to Yourselves." *ProPublica*, November 21, 2018. https://www.propublica.org/article/your-medical-devices-are-not-keeping-your-health-data-to-themselves.

Kristol, David M. "Cookies: Standards, Privacy, and Politics." *ACM Transactions on Internet Technology* 1, no. 2 (2001): 155–98. https://dl.acm.org/doi/10.1145/502152.502153.

Kurenkow, Audrey. "A Brief History of Neural Nets and Deep Learning." *Skynet Today*.

September 27, 2020. https://www.skynettoday.com/overviews/neural-net-history.

Kurlantzick, Joshua. "Assessing China's Silk Road: A Transformative Approach to Technology or a Danger to Freedoms?" *Council on Foreign Relations*. December 18, 2020. https://www.cfr.org/blog/assessing-chinas-digital-silk-road-transformative-approach-technology-financing-or-danger.

Kwet, Michael. "Digital Colonialism: US Empire and the New Imperialism in the Global South." *Race & Class* 60, no. 4 (2020): 3–26. https://doi.org/10.1177/0306396818823172.

Lakhani, Nina. "Investment Fund Links to Atlanta Police and 'Cop City' Project Revealed." *The Guardian*. March 22, 2023. https://www.theguardian.com/us-news/2023/mar/22/investment-fund-links-atlanta-police-cop-city-project.

Lamont, Tom. "'Humanity's Remaining Timeline? It Looks More Like Five Years Than 50': Meet the Neo-Luddites Warning of an AI Apocalypse." *The Guardian*. February 17, 2024. https://www.theguardian.com/technology/2024/feb/17/humanitys-remaining-timeline-it-looks-more-like-five-years-than-50-meet-the-neo-luddites-warning-of-an-ai-apocalypse.

Lamoureux, Edward Lee. *Privacy Surveillance and the New Media You*. New York: Peter Lang, 2016.

Landau, Susan. *Surveillance or Security? The Risks Posed by New Wiretapping Technologies*. Cambridge: MIT Press, 2010.

Langley, Paul, and Andrew Leyshon. "Platform Capitalism: The Intermediation and Capitalization of Digital Economic Circulation." *Finance and Society* 3, no. 1 (2017): 11–31. https://doi.org/10.2218/finsoc.v3i1.1936.

Lazar, Seth. "The US Is Racing Ahead in Its Bid to Control Artificial Intelligence—Why Is the EU So Far Behind?" *The Guardian*. November 28, 2023. https://www.theguardian.com/commentisfree/2023/nov/28/united-states-artificial-intelligence-eu-ai-washington.

Lecher, Colin. "Amazon Automation Tracks and Fires Warehouse Workers for Productivity." *The Verge*. April 25, 2019. https://www.theverge.com/2019/4/25/18516004/amazon-warehouse-fulfillment-centers-productivity-firing-terminations.

Lee, Ben. "Putting the 'Capitalism' in Surveillance Capitalism." *Current Affairs*. May 2021. https://www.currentaffairs.org/2021/05/putting-the-capitalism-in-surveillance-capitalism.

Lee, Min Kyung, Daniel Kushbit, Eva Metsky, and Laura A. Dabbish. "Working with Machines: The Impact of Algorithmic and Data-Driven Management on Human Workers." *Proceedings of the 33rd Annual ACM Conference on Human Factors in Computing Systems* (April 18, 2015): 1603–12. https://doi.org/10.1145/2702123.2702548.

Leffer, Lauren. "Biden's Executive Order on AI Is a Good Start, Experts Say, but Not Enough." *Scientific American*, October 31, 2023. https://www.scientificamerican.com/article/bidens-executive-order-on-ai-is-a-good-start-experts-say-but-not-enough/.

Lehdonvirta, Vili, "From Millions of Tasks to Thousands of Jobs: Bringing Digital Work to Developing Countries," *blogs.worldbank.org*. January 31, 2012. https://blogs.worldbank.org/psd/from-millions-of-tasks-to-thousands-of-jobs-bringing-digital-work-to-developing-countries.

Leonard, John. "Advertising Boss Calls Apple, Privacy Advocates, Politicians 'Extremists.'" *Computing*. January 30, 2023. https://www.computing.co.uk/news/4066252/advertising-boss-calls-apple-privacy-advocates-politicians-extremists.

Lester, Toby. "The Reinvention of Privacy." *Atlantic Monthly* 287, no. 3 (March 2001): 27–39. https://www.theatlantic.com/magazine/archive/2001/03/the-reinvention-of-privacy/302140/.

Leswing, Kif. "Apple's Ad Privacy Change Impact Shows the Power It Wields over Other Industries." *CNCB*. November 13, 2021. https://www.cnbc.com/2021/11/13/apples-privacy-changes-show-the-power-it-holds-over-other-industries.html.

Levine, E.S., Jessica Tisch, Anthony Tasso, and Joy Michael. "The New York City Police Department's Domain Awareness System." *Interfaces* 47, no. 1 (2017): 70–84. https://dl.acm.org/doi/10.1287/inte.2016.0860.

Levine, Yasha. *Surveillance Valley: The Secret Military History of the Internet*. London: Icon Books Ltd, 2019.

Levy, Karen E.C. "The Contexts of Control: Information, Power, and Truck-Driving Work." *The Information Society* 31, no. 2 (2015): 166–74. https://doi.org/10.1080/01972243.2015.998105.

Levy, Steven. *In the Plex: How Google Thinks, Works, and Shapes Our Lives*. New York: Simon & Schuster, 2011.

Lichtenstein, Nelson. *The Retail Revolution: How Wal-Mart Created a Brave New World of Business*. New York: Metropolitan Books, 2009.

———. "Wal-Mart's Authoritarian Culture." *The New York Times*. June 21, 2011. https://www.nytimes.com/2011/06/22/opinion/22Lichtenstein.html?_r=1.

Liedtke, Michael. "Google Will Pay $700 Million to Settle State Claims Its Android App Store Stifled Competition." *Fortune*. December 19, 2023. https://fortune.com/2023/12/19/google-settle-700-million-antitrust-state-android-app-store/.

Liedtke, Michael, Lindsay Whitehurst, Mike Balsamo, and Frank Bajak. "Justice Department Sues Apple, Alleging It Illegally Monopolized the Smartphone Market." *APnews*. March 21, 2024. https://apnews.com/article/apple-antitrust-monopoly-app-store-justice-department-822d7e8f5cf53a2636795fcc33ee1fc3.

Lipton, Beryl. "Neigborhood Watch Out: Cops Are Incorporating Private Cameras into Their Real-Time Surveillance Works." *Electronic*

Frontier Foundation. May 11, 2023. https://www.eff.org/deeplinks/2023/05/neighborhood-watch-out-cops-are-incorporating-private-cameras-their-real-time.

Loeb, Zachary. "Specters of Ludd (Review of Gavin Mueller, *Breaking Things at Work*)." *boundary 2*. September 28, 2021. https://www.boundary2.org/2021/09/zachary-loeb-specters-of-ludd-review-of-gavin-mueller-breaking-things-at-work/.

Lomas, Natasha. "CJEU Ruling on Meta Referral Could Close the Chapter on Surveillance Capitalism." *TechCrunch*, July 4, 2023. https://techcrunch.com/2023/07/04/cjeu-meta-superprofiling-decision/.

———. "EU Lawmakers Bag Late Night Deal of 'Global First' AI Rules." *TechCrunch*. December 8, 2023. https://techcrunch.com/2023/12/08/eu-ai-act-political-deal/.

Lovejoy, Ben. "GDPR-Style Privacy Law in California Acting as a 'Blueprint' for Other States." *9to5mac.com*. December 27, 2019. https://9to5mac.com/2019/12/27/gdpr-style-privacy-law/.

Lupton, Deborah. *The Quantified Self: A Sociology of Self-Tracking*. Cambridge: Cambridge University Press, 2016.

———, ed. *Self-Tracking, Health and Medicine: Sociological Perspectives*. London: Routledge, 2018.

Lyon, David. *Surveillance and Social Sorting: Privacy, Risk and Digital Discrimination*. New York: Routledge, 2002.

MacArthur Justice Center. "Shotspotter Generated over 40,000 Dead-End Police Deployments in Chicago in 21 Months, According to New Study." May 3, 2021. https://www.macarthurjustice.org/shotspotter-generated-over-40000-dead-end-police-deployments-in-chicago-in-21-months-according-to-new-study/.

Maheshwari, Sapna. "Smart TV in Millions of U.S. Homes Track More Than What's on Tonight." *The New York Times*. July 5, 2018. https://www.nytimes.com/2018/07/05/business/media/tv-viewer-tracking.html.

Malcomson, Scott. *Splinternet: How Geopolitics and Commerce Are Fragmenting the World Wide Web*. New York—London: OR Books, 2016.

Malgren, Evan. "Internet for the People: One Native Tribe's Experiment in Self-Sufficient, Tribally Own Internet." *The Nation*. February 24, 2024, 48–53.

Mann, Steve, Jason Nolan, and Barry Wellman. "Sousveillance: Inventing and Using Wearable Computing Devices for Data Collection in Surveillance Environments." *Surveillance and Society* 1, no. 3 (2003): 331–55. https://ojs.library.queensu.ca/index.php/surveillance-and-society/article/view/3344.

Manuel, Anja. "The Tech 10: A Flexible Approach for International Technology Governance." In *Domestic and International (Dis)Order: A Strategic Response*, edited by Leah Bitounis and Niamh King, 71–76. Aspen: Aspen Institute, 2020. https://www.aspeninstitute.org/wp-content/uploads/2020/11/Chapter-10_Manuel_The-Tech-10.pdf.

Marchand, Roland. *Advertising the American Dream: Making Way for Modernity, 1920–1940*. Berkeley: University of California Press, 1985.

Markoff, John. "Building the Superhighway." *The New York Times*. January 24, 1993. https://www.nytimes.com/1993/01/24/business/building-the-electronic-superhighway.html.

———. *Machines of Loving Grace: The Quest for Common Ground between Humans and Robots*. New York: Ecco, 2015.

———. *What the Doormouse Said … How the Sixties Counterculture Shaped the Personal Computer*. New York: Viking Penguin, 2005.

Mason, Sarah. "Chasing the Pink." *Logic(s)*. January 1, 2019. https://logicmag.io/play/chasing-the-pink/.

Mathis, Sommer, and Alexandra Kanik. "Why You'll Be Hearing a Lot Less About 'Smart Cities.'" *City Monitor*. February 18, 2021. https://citymonitor.ai/government/why-youll-be-hearing-a-lot-less-about-smart-cities.

McCabe, David. "U.S. Prepares to Challenge Google's Online Ad Dominance." *The New York Times*. September 9, 2024. https://www.nytimes.com/2024/09/09/technology/google-ads-antitrust-trial.html.

McChesney, Robert W. *Digital Disconnect: How Capitalism Is Turning the Internet against Democracy*. New York: The New Press, 2013.

———. *The Problem of the Media: U.S. Communication Politics in the 21st Century*. New York: Monthly Review Press, 2004.

McClelland, Mac. "I Was a Warehouse Wage Slave." *Mother Jones*. March/April, 2012. https://www.motherjones.com/politics/2012/02/mac-mcclelland-free-online-shipping-warehouses-labor/.

McCue, Colleen. *Data Mining and Predictive Analysis*. Oxford: Butterworth-Heinemann, 2007.

McGuirk, Justin. "Honeywell, I'm Home! The Internet of Things and the New Domestic Landscape." *e-flux*. https://www.e-flux.com/journal/64/60855/honeywell-i-m-home-the-internet-of-things-and-the-new-domestic-landscape/.

McLaren, Duncan, and Julian Agyeman. *Sharing Cities: A Case for Truly Smart and Sustainable Cities*. Cambridge: MIT Press, 2015.

McLaughlin, Timothy. "The Most Dangerous Conflict No One Is Talking About." *The Atlantic*. December 2, 2023. https://www.theatlantic.com/international/archive/2023/12/south-china-sea-philippines-dispute-explained/676218/.

Medina, Eden. *Cybernetic Revolutionaries: Technology and Politics in Allende's Chili*. Cambridge: MIT Press, 2014.

Melman, Seymour. *Pentagon Capitalism*. New York: McGraw-Hill, 1971.

Menand, Louis. "American Democracy Was Never Designed to Be Democratic." *The New Yorker*. August 15, 2022. https://www.newyorker.com/magazine/2022/08/22/american-democracy-was-never-designed-to-be-democratic-eric-holder-our-unfinished-march-nick-seabrook-one-person-one-vote-jacob-grumbach-laboratories-against-democracy.

Merchant, Brian. *Blood in the Machine: The Origins of the Rebellion against Big Tech*. New York: Little, Brown and Company, 2023.

Miller, Mike. "Alinsky for the Left: The Politics of Community Organizing," *Dissent*. Winter 2010. https://www.dissentmagazine.org/article/alinsky-for-the-left-the-politics-of-community-organizing/.

Milmo, Dan. "Facebook and Instagram Users in Europe Can Pay for Ad-Free Versions." *The Guardian*. October 30, 2023. https://www.theguardian.com/technology/2023/oct/30/meta-facebook-instagram-europe-pay-ad-free.

———. "Facebook Owner Meta Fined €1.2bn for Mishandling User Information." *The Guardian*. May 22, 2023. https://www.theguardian.com/technology/2023/may/22/facebook-fined-mishandling-user-information-ireland-eu-meta.

———. "Google Faces $25bn Lawsuit in UK and EU over Digital Advertising." *The Guardian*. September 13, 2022. https://www.theguardian.com/technology/2022/sep/13/google-lawsuit-uk-eu-digital-advertising.

———. "Google Fails to Overturn €4bn Fine over Android Bundling." *The Guardian*. September 14, 2022. https://www.theguardian.com/technology/2022/sep/14/google-fails-to-overturn-4bn-fine-over-android-bundling.

———. "Google Says New AI Model Gemini Outperforms ChatGPT in Most Tests." *The Guardian*. December 6, 2023. https://www.theguardian.com/technology/2023/dec/06/google-new-ai-model-gemini-bard-upgrade.

———. "'Impossible' to Create AI Tools Like ChatGPT without Copyrighted Materials, OpenAI Says." *The Guardian*. January 8, 2024. https://www.theguardian.com/technology/2024/jan/08/ai-tools-chatgpt-copyrighted-material-openai.

———. "Meta Dealt Blow by EU Ruling That Could Result in Data Use 'Opt-In.'" *The Guardian*. January 4, 2023. https://www.theguardian.com/technology/2023/jan/04/meta-dealt-blow-eu-ruling-data-opt-in-facebook-instagram-ads.

Milmo, Dan, and Johana Bhuiyan. "Biden Hails 'Bold Action' of US Government with Order on Safe Use of AI." *The Guardian*. October 30, 2023. https://www.theguardian.com/technology/2023/oct/30/biden-orders-tech-firms-to-share-ai-safety-test-results-with-us-government.

Milmo, Dan, and Nick Robins-Early. "Google Rolls out AI-Generated, Summarized Search Results in US." *The Guardian*. May 14, 2024. https://www.theguardian.com/technology/article/2024/may/14/google-ai-search-results.

Minocher, Xerxes, and Caelyn Randall. "Predictable Policing: New Technology, Old Bias, and Future Resistance in Big Data Surveillance." *Convergence* 26, no. 5–6 (2020): 1108–24. https://doi.org/10.1177/1354856520933838.

Minton, Anna. "New York Is Breaking Free of Airbnb's Clutches. This Is How the Rest of the World Can Follow Suit." *The Guardian*. September 23, 2023. https://www.theguardian.com/commentisfree/2023/sep/27/new-york-airbnb-renters-cities-law-ban-properties.

Monahan, Torin, and Priscilla M. Regan. "Zones of Opacity: Data Fusion in Post-9/11 Security Organizations." *Canadian Journal of Law and Society* 27, no. 3 (2012): 301–17. http://dx.doi.org/10.1017/S0829320100010528.

Montgomery, Blake. "Microsoft's Activision Acquisition and Bets on AI Yield High Quarterly Revenue." *The Guardian*. January 30, 2024. https://www.theguardian.com/technology/2024/jan/30/microsoft-revenue-activision-ai-surpassed-earnings.

———. "Sora: OpenAI Lanches Tool That Instantly Creates Video from Text." *The Guardian*. February 15, 2024. https://www.theguardian.com/technology/2024/feb/15/openai-sora-ai-model-video.

Montgomery, Kathryn C. *Generation Digital: Politics, Commerce, and Childhood in the Age of the Internet*. Cambridge: MIT Press, 2007.

Morozov, Evgeny. "Capitalism's New Clothes." *The Baffler*. February 4, 2019. https://thebaffler.com/latest/capitalisms-new-clothes-morozov.

———. "Digital Socialism? The Calculation Debate in the Age of Big Data." *New Left Review*. March-June 2019. https://newleftreview.org/issues/ii116/articles/evgeny-morozov-digital-socialism.

———. "Silicon Valley Wants Unfettered Control of the Tech Market. That's Why It's Cosying up to Trump," *The Guardian*. June 26, 2024. https://www.theguardian.com/commentisfree/article/2024/jun/26/silicon-valley-tech-market-donald-trump-joe-biden-wealth-tax-big-tech-venture-capitalists.

———. *To Save Everything, Click Here: The Folly of Technological Solutionism*. New York: Public Affairs, 2013.

Morozov, Evgeny, and Francesca Bria. *Rethinking the Smart City: Democratizing the Smart City*. New York: Rosa Luxemburg Stiftung, New York Office, 2018.

Mosco, Vincent. *The Political Economy of Communication*. London: Sage, 1996.

Mozur, Paul. "Inside China's Dystopian Dreams: A.I., Shame and Lots of Cameras." *The New York Times*. July 8, 2018. https://www.nytimes.com/2018/07/08/business/china-surveillance-technology.html.

Mueller, Gavin. *Breaking Things at Work: The Luddites Are Right About Why You Hate Your Job*. London: Verso, 2021.

Mueller, Milton. "A Critique of the 'Surveillance Capitalism' Thesis: Toward a Digital Political Economy." *SSRN*. August 2, 2022. https://dx.doi.org/10.2139/ssrn.4178467.

Mukherjee, Supantha. "Apple Changes EU App Store Policy After Commission Probe." *Reuters*. August 8, 2024. https://www.reuters.com/

technology/apple-changes-eu-app-store-policy-after-commission-probe-2024-08-08/.

Muldoon, James. "Data-Owning Democracy or Digital Socialism?." *Critical Review of International Social and Political Philosophy* (September 2022): 1–22. https://doi.org/10.1080/13698230.2022.2120737.

———. *Platform Socialism: How to Reclaim Our Digital Future from Big Tech*. London: Pluto Press, 2022.

Musa, Sam. "Smart Cities—a Road Map for Development." *IEEE Potentials* 37, no. 2 (2018): 19–23. https://doi.org/10.1109/MPOT.2016.2566099.

Myslewski, Rik. "The Internet of Things Helps Insurance Firms Reward, Punish." *The Register*. May 23, 2014. https://www.theregister.com/2014/05/23/the_internet_of_things_helps_insurance_firms_reward_punish/.

Naughton, John. "For All the Hype in 2023, We Still Don't Know What AI's Long-Term Impact Will Be." *The Guardian*. December 30, 2023. https://www.theguardian.com/commentisfree/2023/dec/30/ai-artifical-intelligence-2023-long-term-impact-nvidia-h100-microsoft.

———. "From Boom to Burst, the AI Bubble Is Only Heading in One Direction." *The Guardian*. April 13, 2024. https://www.theguardian.com/commentisfree/2024/apr/13/from-boom-to-burst-the-ai-bubble-is-only-heading-in-one-direction.

———. "Google's Monopoly on the Search Engine Market Finally Timed Out?" *The Guardian*, September 30, 2023. https://www.theguardian.com/commentisfree/2023/sep/30/google-antitrust-us-department-of-justice-court-case-monopoly.

———. "OpenAI's New Video Generation Tool Could Learn a Lot from Babies." *The Guardian*. February 24, 2024. https://www.theguardian.com/commentisfree/2024/feb/24/openai-video-generation-tool-sora-babies-ai-artificial-intelligence.

———. "Silicon Valley's Business Model Is Incompatible Wth the Moderation of Online Horror and Hatred." *The Guardian*. April 27, 2024. https://www.theguardian.com/commentisfree/2024/apr/27/silicon-valleys-business-model-is-incompatible-with-the-moderation of onlinc-horror-and-hatred.

———. *What You Really Need to Know About the Internet. From Gutenberg to Zuckerberg*. London: Quercus, 2012.

———. "Why AI Is a Disaster for the Climate." *The Guardian*. December 23, 2023. https://www.theguardian.com/commentisfree/2023/dec/23/ai-chat-gpt-environmental-impact-energy-carbon-intensive-technology?ref=upstract.com.

Neff, Gina, and Dawn Nafus. *Self-Tracking*. Cambridge: MIT Press, 2016.

Newton, Casey. "Google's Broken Link to the Web." *Platformer*. May 14, 2024. https://www.platformer.news/google-io-ai-search-sundar-pichai/.

Ng, Alfred. "Amazon's Helping Police Build a Surveillance Network with Ring Doorbells." *CNET*. June 5, 2019. https://www.cnet.com/home/smart-home/features/amazons-helping-police-build-a-surveillance-network-with-ring-doorbells/.

Nijssen, Timo. "Zonder Blauwe Hesjes, Maar Met Instagram Wil De FNV Fietskoeriers Verenigen." *NRC*. February 8, 2024. https://www.nrc.nl/nieuws/2024/02/08/zonder-blauwe-hesjes-maar-met-instagram-wil-de-fnv-fietskoeriers-verenigen-a4189603.

Noble, Safiya Umoja. *Algorithms of Oppression*. New York: New York University Press, 2018.

Nothias, Toussaint. "Access Granted: Facebook's Free Basics in Africa." *Media, Culture & Society* 42, no. 3 (2020): 329–48. https://doi.org/10.1177/0163443719890530.

———. "How to Fight Digital Colonialism." *Boston Review*. November 14, 2022. https://www.bostonreview.net/articles/how-to-fight-digital-colonialism/.

Novak, William J. "Law and the Social Control of American Capitalism." *Emory Law Journal* 60, no. 2 (2010). https://scholarlycommons.law.emory.edu/elj/vol60/iss2/4.

NYCLU. "Stop and Frisk Data in the De Blasio Era (2019)." *New York Civil Liberties Union*. March 14, 2019. https://www.nyclu.org/en/publications/stop-and-frisk-de-blasio-era-2019.

O'Carroll, Lisa. "EU Agrees 'Historic' Deal with World's First Laws to Regulate AI." *The Guardian*. December 8, 2023. https://www.theguardian.com/world/2023/dec/08/eu-agrees-historic-deal-with-worlds-first-laws-to-regulate-ai.

———. "EU Agrees Tough Limits on Police Use of AI Biometric Surveillance." *The Guardian*. December 11, 2023. https://www.theguardian.com/world/2023/dec/11/eu-agrees-tough-rules-on-police-use-of-ai-biometric-surveillance-ban.

———. "EU to Put Tariffs of up to 38% on Chinese Electric Vehicles as Trade War Looms." *The Guardian*. June 12, 2024. https://www.theguardian.com/business/article/2024/jun/12/eu-import-tariffs-chinese-evs-electric-vehicles-trade-war.

———. "How the EU Digital Services Act Affects Facebook, Google and Others." *The Guardian*. August 25, 2023. https://www.theguardian.com/world/2023/aug/25/how-the-eu-digital-services-act-affects-facebook-google-and-others.

———. "X to Be Investigated for Allegedly Breaking EU Law on Hate Speech and Fake News." *The Guardian*. December 18, 2023. https://www.theguardian.com/technology/2023/dec/18/x-to-be-investigated-for-allegedly-breaking-eu-laws-on-hate-speech-and-fake-news.

O'Carroll, Lisa, and Dan Milmo. "EU Fines Apple €1.8bn over App Store Restrictions on Music Streaming." *The Guardian*. March 4, 2024. https://www.theguardian.com/business/2024/mar/04/eu-fines-apple-18bn-over-app-store-restrictions-on-music-streaming.

Oladipo, Gloria. "Robot Dogs Deployed in New York Building Collapse Revive Surveillance

Fears." *The Guardian.* April 26, 2023. https://www.theguardian.com/technology/2023/apr/26/robot-dogs-new-york-building-collapse-surveillance.

Olla, Akin. "Why Is Georgia Prosecuting Leftwing Activists with the Same Law as Trump?" *The Guardian.* September 20, 2023. https://www.theguardian.com/commentisfree/2023/sep/20/georgia-prosecuting-cop-city-activists-rico-trump.

O'Neil, Cathy. *Weapons of Math Destruction: How Big Data Increases Inequality and Threatens Democracy.* New York: Crown Publishers, 2016.

Ong, Thuy. "Amazon Patents Wristband That Track Warehouse Employees' Hands in Real Time." *The Verge.* February 1, 2018. https://www.theverge.com/2018/2/1/16958918/amazon-patents-trackable-wristband-warehouse-employees.

O'Reilly, Tim. "What Is Web 2.0: Design Patterns and Business Models for the Next Generation of Software." *Communications and Strategies,* no. 1 (First Quarter 2007): 17–37. https://papers.ssrn.com/sol3/papers.cfm?abstract_id=1008839.

Oremus, Will. "Google's Weird AI Answers Hint at a Fundamental Problem." *The Washington Post.* May 28, 2024. https://www.washingtonpost.com/politics/2024/05/29/google-ai-overview-wrong-answers-unfixable/.

Ott, Marvin. "The South China See in Strategic Terms." *Asia Dispatches.* May 14, 2019. https://www.wilsoncenter.org/blog-post/the-south-china-sea-strategic-terms.

Pasquale, Frank. *The Black Box Society: The Secret Algorithms That Control Money and Information.* Cambridge: Harvard University Press, 2016.

Paul, Kari."Dozens Sue Amazon's Ring after Camera Hack Leads to Threats and Racial Slurs." *The Guardian.* December 23, 2020. https://www.theguardian.com/technology/2020/dec/23/amazon-ring-camera-hack-lawsuit-threats.

———. "Reversal of Content Policies at Alphabet, Meta and X Threaten Democracy, Warn Experts." *The Guardian.* December 23, 2023. https://www.theguardian.com/media/2023/dec/07/2024-elections-social-media-content-safety-policies-moderation?ref=upstract.com.

Pearlstein, Steven. "Here's the Inside Story of How Congress Failed to Rein in Big Tech." *The Washington Post.* July 6, 2023. https://www.washingtonpost.com/opinions/2023/07/06/congress-facebook-google-amazon-apple-regulation-failure/.

Peppet, Scott R. "Unraveling Privacy: The Personal Prospectus and the Threat of a Full-Disclosure Future." *Northwestern University Law Review* 105, no. 3 (2011): 1153–204. https://scholarlycommons.law.northwestern.edu/cgi/viewcontent.cgi?article=1157&context=nulr.

Perrigo, Billy. "Exclusive: OpenAI Lobbied the E.U. To Water Down AI Regulation." *TIME.* June 30, 2023. https://time.com/6288245/openai-eu-lobbying-ai-act/.

Peters, Jay. "Alphabet is Shutting Down Loon, Its Internet Balloon Company."*The Verge.* January 21, 2021. https://www.theverge.com/2021/1/21/22243484/alphabet-google-shutting-down-loon-internet-balloon-company-x.

———. "Apple and Meta Could Face Charges for Violating Tech Rules." *The Verge.* June 14, 2024. https://www.theverge.com/2024/6/14/24178724/apple-meta-eu-charges-dma.

Pickard, Victor. *Democracy without Journalism? Confronting the Misinformation Society.* Oxford: Oxford University Press, 2019.

Picon, Antoine. *Smart Cities: A Spatialised Intelligence.* West Sussex, UK: Wiley, 2015.

Platform Cooperativism Consortium. "Platform Cooperativism Consortium: A Hub That Starts, Grows, and Converts Platform Coops." Accessed May 21, 2024. https://platform.coop/about/pcc/.

Polanyi, Karl. *The Great Transformation.* Boston: Beacon Press, [1944] 2001.

Prassl, Jeremias. *Humans as a Service: The Promise and Perils of Work in the Gig Economy.* Oxford: Oxford University Press, 2018.

Pratt, Timothy. "Atlanta Police Surveil People Opposing 'Cop City': 'There's This Constant Stalking Feeling.'" *The Guardian.* May 29, 2024. https://www.theguardian.com/us-news/article/2024/may/29/atlanta-police-cop-city-surveillance.

———. "Georgia Police and FBI Conduct Swat-Style Raids on 'Cop City' Activists' Homes." *The Guardian.* February 10, 2024. https://www.theguardian.com/us-news/2024/feb/10/georgia-police-fbi-raids-cop-city-activists-atlanta.

———. "'Our Struggles Are Connected': Atlanta Protesters Link Cop City to Gaza War." *The Guardian.* May 13, 2024. https://www.theguardian.com/us-news/article/2024/may/13/cop-city-emory-atlanta-israel.

———. "Outcry as Atlanta Refuses to Handle Petitions over 'Cop City' Police Campus." *The Guardian.* September 11, 2023. https://www.theguardian.com/us-news/2023/sep/11/atlanta-petition-police-cop-city-georgia.

Price, Catherine. "You Have One Life: Do You Really Want to Spend It Looking at Your Phone?" *The Guardian.* January 2, 2024. https://www.theguardian.com/lifeandstyle/2024/jan/02/smartphones-attention-economy-reclaim-free-time.

Progressive Corporation. *Linking Driving Behavior to Automobile Accidents and Insurance Rates: An Analysis of Five Billion Miles Driven.* Mayfield, OH: Progressive Corporation, 2012. https://goodtimesweb.org/documentation/2012/snapshot_report_final_070812.pdf.

Purcell, Mark. "Urban Democracy and the Local Trap." *Urban Studies* 43, no. 11 (2006): 1921–41. http://www.jstor.org/stable/43197418.

Pureswaran, Veena, and Paul Brody. *Device Democracy: Saving the Internet of Things.* IBM Executive Report Electronics Industry,

2014. https://www.ibm.com/downloads/cas/Y5ONA8EV.

Putnam, Robert. *Bowling Alone:The Collapse and Revival of American Community*. New York: Simon & Schuster, 2000.

Quijano, Anibal. "Coloniality and Modernity/Rationality." *Cultural Studies* 2, no. 2–3 (2007): 168–78. https://www.tandfonline.com/doi/abs/10.1080/09502380601164353.

Quinn, Alexander J., Benjamin B. Bederson, Tom Yeh, and Jimmy Lin. "Crowdflow: Integrating Machine Learning with Mechanical Turk for Speed-Cost-Quality Flexibility" (2020). https://www.semanticscholar.org/paper/CrowdFlow-:-Integrating-Machine-Learning-with-Turk-Quinn-Bederson/53db354fdcfa8fe8d22c1d5ae84b9e86e6f57abe.

Radin, Margaret Jane. *Boilerplate: The Fine Print, Vanishing Rights, and the Rule of Law*. Princeton: Princeton University Press, 2013.

Radsch, Courtney. "The Real Story of the OpenAI Debacle Is the Tyranny of Big Tech." *The Guardian*. November 27, 2023. https://www.theguardian.com/commentisfree/2023/nov/27/openai-microsoft-big-tech-monopoly.

Rahman, K. Sabeel, and Zephyr Teachout. "From Private Bads to Public Goods: Adapting Public Utility Regulation for Informational Infrastructure." *The Knight First Amendment Institute*. February 4, 2020. https://knightcolumbia.org/content/from-private-bads-to-public-goods-adapting-public-utility-regulation-for-informational-infrastructure.

Rappoport, John. "How Private Insurers Regulate Public Police." *Harvard Law Review* 130, no. 6 (2017): 1539–614. https://harvardlawreview.org/print/vol-130/how-private-insurers-regulate-public-police/.

"Regulation (EU) 2022/1925 of the European Parliament and of the Council of 14 September 2022 on Contestable and Fair Markets in the Digital Sector and Amending Directives (EU) 2019/1937 and (EU) 2020/1828 (Digital Markets Act)." December 10, 2022. https://eur-lex.europa.eu/eli/reg/2022/1925/oj.

"Regulation (EU) 2022/2065 of the European Parliament and of the Council of 19 October 2022 on a Single Market For Digital Services and Amending Directive 2000/31/EC (Digital Services Act)." October 19, 2022. "https://eur-lex.europa.eu/legal-content/EN/TXT/?uri=CELEX:32022R2065.

Reuters. "Chinese Industry to Seek Probe into EU Pork Imports, Global Times Reports." *Reuters*. May 27, 2024. https://www.reuters.com/markets/commodities/chinese-industry-seek-probe-into-eu-pork-imports-global-times-reports-2024-05-27/.

———. "List of Chinese Entities Who Have Turned to the Cloud for Access to Restricted US Tech." *Reuters*. August 23, 2024. https://www.reuters.com/technology/list-chinese-entities-who-have-turned-cloud-access-restricted-us-tech-2024-08-23/.

Rheingold, Howard. *The Virtual Community:Homesteading on the Electronic Frontier*. Reading, MA: Addison Wesley, 1993.

Ricaurte, Paola. "Data Epistemologies, the Coloniality of Power, and Resistance." *Television & New Media* 20, no. 4 (2019): 350–65. https://doi.org/10.1177/1527476419831640.

Rikap, Cecilia. "Antitrust Policy and Artifical Intelligence: Some Neglected Issues." *ScheerPost*, June 12, 2024. https://scheerpost.com/2024/06/12/antitrust-policy-and-artificial-intelligence-some-neglected-issues/.

———. "Intellectual Monopolies as a New Pattern of Innovation and Technological Regime." *Industrial and Social Change*. December 7, 2023. https://doi.org/10.1093/icc/dtad077.

Rinke, Andreas. "Exclusive: Germany, France and Italy Reach Agreement on AI Regulation." *Reuters*. November 20, 2023. https://www.reuters.com/technology/germany-france-italy-reach-agreement-future-ai-regulation-2023-11-18/.

Robins, Kevin, and Frank Webster. "Capitalism: Information, Technology, and Everyday Life." In *The Political Economy of Information*, edited by Vincent Mosco and Janet Wasko, 44–75. Madison: University of Wisconsin Press, 1998.

———. *Times of the Technoculture: From the Information Society to the Virtual Life*. London: Routledge, 1999.

Robins-Early, Nick. "Google Remains Focused on Its Long Quest for Your Eyeballs." *The Guardian*. May 19, 2024. https://www.theguardian.com/technology/article/2024/may/19/google-ai-overview-attention.

———. "Google to Refine AI-Generated Search Summaries in Response to Bizarre Results." *The Guardian*. May 31, 2024. https://www.theguardian.com/technology/article/2024/may/31/google-ai-summaries-sge-changes.

———. "Google's Second Antitrust Suit Brought by US Begins, Over Online Ads." *The Guardian*.September 9, 2024. https://www.theguardian.com/technology/article/2024/sep/09/google-antitrust-lawsuit-online-ads.

———. "The Intercept, Raw Story and Alternet Sue OpenAI for Copyright Infringement." *The Guardian*. February 28, 2024. https://www.theguardian.com/technology/2024/feb/28/media-outlets-sue-openai-copyright-infringement.

———. "Sundar Pichai Denies Google Stifles Competition at Epic Antitrust Trial." *The Guardian*. November 14, 2023. https://www.theguardian.com/technology/2023/nov/14/google-ceo-sundar-pichai-fortnite-epic-games-monopoly-antitrust-trial.

———. "A Threat to Amazon and a Test of the FTC: Is This Big Tech's Antitrust Reckoning?" *The Guardian*. September 26, 2023. https://www.theguardian.com/technology/2023/sep/26/amazon-antitrust-lawsuit-analysis-big-tech.

———. "A US Judge Ruled that Google Built an Illegal Monopoly. What Happens Next? *The Guardian*. August 6, 2024. https://www.

theguardian.com/technology/article/2024/aug/06/google-antitrust-monopoly-ruling.

———. "US to Argue Google Abused Power to Monopolize Internet Search as Antitrust Trial Begins." *The Guardian*. September 12, 2023. https://www.theguardian.com/us-news/2023/sep/12/google-antitrust-trial-day-1?ref=upstract.com.

Robins-Early, Nick and agencies. "Reddit Shares Priced at $34 in Largest IPO by Social Media Company in Years." *The Guardian*. March 20, 2024. https://www.theguardian.com/technology/2024/mar/20/reddit-stock-market-debut-new-york-exchange.

Robinson, Lawrence, Melinda Smith, and Jeanne Segal. "Smartphone and Internet Addiction." *Helpguide.org*. Updated April 29, 2024. https://www.helpguide.org/articles/addictions/smartphone-addiction.htm.

Romano, Angela. "Asia." In *Public Sentinel: News Media and Governance Reform*, edited by Pippa Norris, 353–75. New York: World Bank Publications, 2020.

Romer, Paul. "A Tax That Could Fix Big Tech." *The New York Times*. May 6, 2018. https://www.nytimes.com/2019/05/06/opinion/tax-facebook-google.html.

Rosenblat, Alex. *Uberland: How Algorithms Are Rewriting the Rules of Work*. Berkeley: University of California Press, 2018.

Rosenblat, Alex, and Luke Stark. "Algorithmic Labor and Information Asymmetries: A Case Study of Uber's Drivers." *International Journal of Communication* 10 (2016): 3758–84. https://papers.ssrn.com/sol3/papers.cfm?abstract_id=2686227.

Roth, Laura, and Bertie Russell. "Translocal Solidarity and the New Municipalism." *Roar Magazine* 8 (2018): 81–93. https://roarmag.org/magazine/municipalist-movement-internationalism-solidarity/.

Ruhl, Charlotte. "Broken Windows Theory of Criminiology." *Simply Psychology*. February 13, 2024. https://www.simplypsychology.org/broken-windows-theory.html.

Rushe, Dominic, and Kari Paul. "US Justice Department Sues Google over Accusation of Illegal Monopoly." *The Guardian*. October 20, 2020. https://www.theguardian.com/technology/2020/oct/20/us-justice-department-antitrust-lawsuit-against-google.

Rushkoff, Douglas. *Throwing Rocks at the Google Bus: How Google Became the Enemy of Prosperity*. New York: Portfolio/Penguin, 2016.

Russell, Bertie. "Beyond the Local Trap: New Municipalism and the Rise of the Fearless Cities." *Antipode* 51, no. 3: 989–1010. https://onlinelibrary.wiley.com/doi/full/10.1111/anti.12520.

Russell, Patrick. "Smart Way Forward: A Review of *Smart Cities: A Spatialised Intelligence*." *Planning Forum* 17 (2016): 109–12. https://sites.utexas.edu/planningforum/files/2021/02/PlanningForumVol17-1.pdf.

Ryan, Johnny. *A History of the Internet and the Digital Future*. London: Reaktion Books, 2010.

Sadowski, Jathan. "'Anyway, the Dashboard Is Dead': On Trying to Build Urban Informatics." *New Media & Society* 26, no. 1 (2021): 313–28. https://doi.org/10.1177/14614448211058455.

———. "The Internet of Landlords: Digital Platforms and New Mechanisms of Rentier Capitalism." *Antipode* 52, no. 2 (2020): 562–80. https://doi.org/10.1111/anti.12595.

———. *Too Smart: How Digital Capitalism Is Extracting Data, Controlling Our Lives, and Taking over the World*. Cambridge: MIT Press, 2020.

———. "When Data Is Capital: Datafication, Accumulation, Extraction." *Big Data & Society* 6, no. 1 (2019): 1–12. https://doi.org/10.1177/2053951718820549.

Sai'di, Ahmad H. "Colonialism and Surveillance." In *Routledge Handbook of Surveillance Studies*, edited by Kirstie Ball, Kevin Haggerty and David Lyon, 151–58. New York: Routledge, 2012.

Sainato, Michael. "Ford Seeks to Remotely Repossess Cars after Missed Payments in US Patent." *The Guardian*. March 3, 2023. https://www.theguardian.com/business/2023/mar/03/ford-reposses-patent-remote-lock.

Salam, Erum. "Uber and Lyft to Pay out $328m to New York Ride-Share Drivers." *The Guardian*. November 2, 2023. https://www.theguardian.com/technology/2023/nov/02/uber-lyft-settlement-new-york-driver-lawsuit.

Salehi, Niloufar, Lily Irani, Michael Bernstein, Ali Alkhatib, Eva Ogbe, Kristy Milland, and Clickhappier. "We Are Dynamo: Overcoming Stalling and Friction in Collective Action for Crowd Workers." *CH '15: Proceedings of the 13rd Annual ACM Conference on Human Factors in Computing Systems* (2015): 1621–30. https://dl.acm.org/doi/10.1145/2702123.2702508.

Satariano, Adam, and Tripp Mickle. "Apple Overhauls App Store in Europe, in Response to New Digital Law." *The New York Times*. January 25, 2024. https://www.nytimes.com/2024/01/25/technology/apple-app-store-europe.html.

Sawers, Paul. "Biden Issues Executive Order to Set Standards for AI Safety and Security." *TechCrunch*, October 30, 3023. https://techcrunch.com/2023/10/30/president-biden-issues-executive-order-to-set-standards-for-ai-safety-and-security/.

Scannell, R. Joshua. "Both a Cyborg and a Goddess: Deep Managerial Time and Informatic Governance." In *Object-Oriented Feminism*, edited by Katherine Behar, 247–74. Minneapolis: University of Minnesota Press, 2016.

Schiller, Dan. *How to Think About Information*. Urbana: University of Illinois Press, 2007.

Schinkel, Maarten. "Waarom Amerikaanse Techbedrijven Gevaarlijk Diepe Zakken Hebben." *NRC*. March 7, 2024. https://www.nrc.nl/nieuws/2024/03/07/waarom-amerikaanse-techbedrijven-gevaarlijk-diepe-zakken-hebben.

Schneider, Nathan. "Denver Taxi Drivers Are Turning Uber's Disruption on Its Head."

The Nation. September 7, 2016. https://www.thenation.com/article/archive/denver-taxi-drivers-are-turning-ubers-disruption-on-its-head/.

———. "The Rise of a Cooperatively Owned Internet." *The Nation*. October 13, 2016. https://www.thenation.com/article/archive/the-rise-of-a-cooperatively-owned-internet/.

Schneier, Bruce. *Data and Goliath: The Hidden Battles to Collect Your Data and Control Your World*. New York: W.W. Norton, 2015.

———. "Mission Creep: When Everything Is Terrorism." *The Atlantic*. July 16, 2013. https://www.theatlantic.com/politics/archive/2013/07/mission-creep-when-everything-is-terrorism/277844/.

Scholz, Trebor, ed. *Digital Labor*. New York: Routledge, 2013.

Schor, Juliet B. *After the Big Gig: How the Sharing Economy Got Hijacked and How to Win It Back*. Oakland: University of California Press, 2020.

Schremmer, Mark. "UPS Receives Five-Year Renewal on ELD Exemption." *Landline*. October 26, 2022. https://landline.media/ups-receives-five-year-renewal-on-eld-exemption/.

Schulman, Stacey Lynn. "Hyperlinks and Marketing Design." In *The Hyperlinked Society*, edited by Joseph Turow and Lokman Tsui, 145–58. Ann Arbor: University of Michigan Press, 2011.

Schwartz, Peter, and Stewart Brand. *The 1989 GBN Scenario Book: Decades of Restructuring*. Emeryville: Global Business Network, 1989.

Sennett, Richard. "No One Likes a City That's Too Smart." *The Guardian*. December 14, 2012. https://www.theguardian.com/commentisfree/2012/dec/04/smart-city-rio-songdo-masdar.

Shah, Rajiv C., and Kesan Jay P. "The Privatization of the Internet's Backbone Network." *Journal of Broadcasting & Electronic Media* 51, no. 1 (2007): 93–107. https://www.tandfonline.com/doi/abs/10.1080/0883815070130807.

Shahbaz, Adrian. "The Digital Silk Road and Normative Values." In *The Digital Silk Road: China's Technological Rise and the Geopolitics of Cyberspace*, edited by David Gordon and Meia Nouwens, 107–30. New York: Routledge, 2022.

———. "The Rise of Digital Authoritarianism." *Freedom House*. 2018. https://freedomhouse.org/report/freedom-net/2018/rise-digital-authoritarianism.

Shanahan, Ed. "Robot Dog's Work in Garage Collapse Should Quiet Critics, Adams Suggests." *The New York Times*. April 25, 2023 (updated April 28, 2023). https://www.nytimes.com/2023/04/25/nyregion/digidog-robot-nypd.html.

Sharon, Tamar, and Dorien Zandbergen. "From Data Fetishism to Quantifying Selves: Self-Tracking Practices and the Other Values of Data." *New Media & Society* 19, no. 11 (2016): 1695–709. https://doi.org/10.1177/1461444816636090.

Simons, Josh, and Dipayan Ghosh. "Utilities for Democracy: Why and How the Algorithmic Infrastructure of Facebook and Google Must Be Regulated." *Brookings*. August 2020, 1–28. https://www.brookings.edu/wp-content/uploads/2020/08/FP_20200908_facebook_google_algorithm_simons_ghosh.pdf.

Simons, Margaret. "If Meta's Intransigence Isn't Enough, AI Poses an Even Greater Threat to Journalism." *The Guardian*, March 1, 2024. https://www.theguardian.com/commentisfree/2024/mar/02/if-metas-intransigence-isnt-enough-ai-poses-an-even-greater-threat-to-journalism.

Simpson, Jack. "ASML Halts Hi-Tech Chip-Making Exports to China Repeatedly after US Request." *The Guardian*. January 2, 2024. https://www.theguardian.com/technology/2024/jan/02/asml-halts-hi-tech-chip-making-exports-to-china-reportedly-after-us-request?ref=biztoc.com.

Skou, Maria, and Nicklas Echsner-Rasmussen. "Smart Cities around the World." *Geoforum Perpektiv* 14, no. 25 (2016): 61–67. https://doi.org/10.5278/ojs.perspektiv.v14i25.1289.

Soderberg, Johan. *Hacking Capitalism: The Free and Open Software Movement*. New York: Routledge, 2008.

Solon, Olivia. "'It's Digital Colonialism': How Facebook's Free Internet Service Has Failed Its Users." *The Guardian*. July 27, 2017. https://www.theguardian.com/technology/2017/jul/27/facebook-free-basics-developing-markets.

Spicer, André. "Beware the 'Botshit': Why Generative AI Is Such a Real and Imminent Threat to the Way We Live." *The Guardian*. January 3, 2024. https://www.theguardian.com/commentisfree/2024/jan/03/botshit-generative-ai-imminent-threat-democracy.

Srnicek, Nick. "The Only Way to Rein in Big Tech Is to Treat Them as a Public Service." *The Guardian*. April 23, 2019. https://www.theguardian.com/commentisfree/2019/apr/23/big-tech-google-facebook-unions-public-ownership.

———. *Platform Capitalism*. Malden, MA: Polity Press, 2017.

Stahl, Bernd Carsten. "EU Is Cracking Down on AI, but Leaves a Loophole for Mass Surveillance." *The Conversation*. April 21, 2021. https://theconversation.com/eu-is-cracking-down-on-ai-but-leaves-a-loophole-for-mass-surveillance-159421.

Stanley, Jay. "New Orleans Program Offers Lessons in Pitfalls of Predictive Policing." *ACLU*. March 15, 2018. https://www.aclu.org/news/privacy-technology/new-orleans-program-offers-lessons-pitfalls-predictive-policing.

Stein, Michael Isaac. "New Orleans City Council Bans Facial Recognition, Predictive Policing and Other Surveillance Tech." *The Lens*. December 18, 2020. https://thelensnola.org/2020/12/18/new-orleans-city-council-approves-ban-on-facial-recognition-predictive-policing-and-other-surveillance-tech/.

Stellinga, Marike. "China Wil Niet Afhankelijk Zijn Van De VS En Steekt Daarom Opnieuw Tientallen Miljarden in Zijn Chipsector." *NRC*. May 27, 2024. https://www.nrc.nl/nieuws/2024/05/27/china-wil-niet-afhankelijk-zijn-van-de-vs-en-

steekt-daarom-opnieuw-tientallen-miljarden-in-zijn-chipsector-a4200097.
Stepanek, Marcia. "Weblining." *Bloomberg*. April 3, 2000. https://www.bloomberg.com/news/articles/2000-04-02/weblining.
Stokel-Walker, Chris. "ChatGPT on Your Phone? The Four Reasons Why This Is Happening Far Too Early." *The Guardian*. June 13, 2024.
———. *How AI Ate the World*. Kingston upon Thames: Canbury, 2023.
Stone, Brad. *The Everything Store: Jeff Bezos and the Age of Amazon*. New York: Little, Brown, 2013.
Streeter, Thomas. "'That Deep Romantic Chasm': Libertarianism, Neoliberalism and the Computer Culture." In *Communication, Citizenship and Social Policy: Rethinking the Limits of the Welfare State*, edited by Andrew Calabrese and Jean-Claude Burgelman, 49–64. Boulder: Rowman & Littlefield, 1999.
Streitfeld, David, and Edward Wyatt. "U.S. Escalates Google's Case by Hiring Noted Outside Lawyer." *The New York Times*. April 27, 2012. https://www.nytimes.com/2012/04/27/technology/google-antitrust-inquiry-advances.html.
"Summit for Democracy," U.S. Department of State, February 2021. https://www.state.gov/summit-for-democracy-2021/.
Sunstein, Cass. *Republic.Com*. Princeton: Princeton University Press, 2002.
Tan, K.W. "As China Shuts out World, Internet Access from Abroad Gets Harder Too." *Freedom House*. June 23, 2022. https://freedomhouse.org/article/china-shuts-out-world-internet-access-abroad-gets-harder-too.
Tarnoff, Ben. *Internet for the People: The Fight for Our Digital Future*. London: Verso, 2022.
Taylor, Astra. "The Automation Charade." *Logic(s)*, August 1, 2018. https://logicmag.io/failure/the-automation-charade/.
Taylor, Charles. *Modern Social Imaginaries*. Durham: Duke University Press, 2004.
Tepper, Taylor, and Michael Adams. "Federal Funds Rate History 1990 to 2024." *Forbes Advisor*. May 10, 2024. https://www.forbes.com/advisor/investing/fed-funds-rate-history/.
Thaler, Richard T., and Cass R. Sunstein. *Nudge: Improving Decisions About Health, Wealth, and Happiness*. New York: Penguin, 2008.
Thelen, Kathleen. "Regulating Uber: The Politics of the Platform Economy in Europe and the United States." *Perspectives on Politics* 16, no. 4 (2018): 938–53.
Thomas, Daniel R., Alistair R. Beresford, and Andrew Rice. "Securing Metrics for the Android Ecosystem." In *Proceedings of the 5th Annual ACM CCS Workshop on Security and Privacy in Smartphones and Mobile Devices—SPSM '15*, 87–98. Cambridge: Computer Laboratory, University of Cambridge, 2015. https://www.cl.cam.ac.uk/~drt24/papers/spsm-scoring.pdf.
Thompson, E.P. *Customs in Common: Studies in Traditional Popular Culture*. New York: Free Press, 1991.
———. *The Making of the English Working Class*. London: Penguin, [1968] 1979.
Tough, Paul. *The Inequality Machine: How College Divides Us*. Boston: Mariner Books, 2021.
Townsend, Anthony M. *Smart Cities: Big Data, Civic Hackers, and the Quest for a New Utopia*. New York: W.W. Norton, 2013.
Turner, Fred. *From Counterculture to Cyberculture: Stewart Brand, the Whole Earth Network, and the Rise of Digital Utopianism*. Chicago: University of Chicago Press, 2008.
Turow, Joseph. *The Daily You: How the New Advertising Industry Is Defining Your Identity and Your Worth*. New Haven: Yale University Press, 2011.
———. *Niche Envy: Marketing Discrimination in the Digital Age*. Cambridge: MIT Press, 2006.
Uchida, Craig D., Swatt, Marc, David Gamero, Jeanine Lopez, Erika Salazar, Elliott King, et al. "The Los Angeles Smart Policing Initiative: Reducing Gun-Related Violence through Operation Laser." BJA Bureau of Justice Assistance. October 1, 2012. https://bja.ojp.gov/sites/g/files/xyckuh186/files/media/document/LosAngelesSPI.pdf.
UK Parliament, "Enclosing the Land." Accessed July 8, 2024. https://www.parliament.uk/about/living-heritage/transformingsociety/towncountry/landscape/overview/enclosingland/.
U.S. Department of Energy. "Data Centers and Servers." Accessed June 3, 2024. https://www.energy.gov/eere/buildings/data-centers-and-servers.
Vallas, Steve, and Juliet Schor. "What Do Platforms Do? Understanding the Gig Economy." *Annual Review of Sociology* 46 (2020): 273–94. https://www.annualreviews.org/doi/abs/10.1146/annurev-soc-121919-054857.
van Bokkum, Milo. "China Dreigt Met Wraakmaatregelen Door Europees Staatssteunonderzoek." *NRC*. May 27, 2024. https://www.nrc.nl/nieuws/2024/05/27/china-dreigt-met-wraakmaatregelen-door-europees-staatssteunonderzoek-a4200086.
———. "Voor Het Eerst Sinds De Oprichting in 2009 Maakt Uber Winst." *NRC*. August 1, 2023. https://www.nrc.nl/nieuws/2023/08/01/voor-het-eerst-sinds-de-oprichting-in-2009-maakt-uber-winst-a4171026.
van de Wiel, Clara. "Brussel Treft Apple Met Een Van Hoogste Boetes Ooit: 1,8 Miljard Euro, Voor Machtsmisbruik." *NRC*. March 4, 2024. https://www.nrc.nl/nieuws/2024/03/04/brussel-treft-apple-met-een-van-hoogste-boetes-ooit-18-miljard-voor-machtsmisbruik-a4191983.
———. "Het Is Nu Duidelijk Hoe Europa AI Aan Banden Wil Leggen (En Daarmee Wereldwijd Een Norm Stelt)." *NRC*. December 9, 2023. https://www.nrc.nl/nieuws/2023/12/09/het-is-nu-duidelijk-hoe-europa-ai-aan-banden-wil-leggen-en-daarmee-wereldwijd-een-norm-stelt-a4183806.
van Dijck, José. *The Culture of Connectivity: A Critical History of Social Media*. Oxford: Oxford University Press, 2013.

van Doorn, Niels. "A New Institution on the Block: On Platform Urbanism and Airbnb Citizenship." *New Media & Society* 22, no. 10 (2020): 1808–26. https://doi.org/10.1177/1461444819884377.

———. "Platform Labor: On the Gendered and Racialized Exploitation of Low-Income Service Work in the 'On-Demand' Economy." *Information, Communication, Society* 20, no. 6 (2017): 898–914. https://doi.org/10.1080/1369118X.2017.1294194.

van Doorn, Niels, and Adam Badger. "Platform Capitalism's Hidden Abode: Producing Data Assets." *Antipode* 52, no. 5 (2020): 1475–95. https://doi.org/10.1111/anti.12641.

van Elteren, Mel. *Managerial Control of American Workers: Methods and Technology from the 1880s to Today.* Jefferson, NC: McFarland, 2017.

Vasudevan, Krishnan, and Ngai Keung Chan. "Gamification and Work Games: Examining Consent and Resistance among Uber Drivers." *New Media & Society* 24, no. 4 (2022): 866–86. https://doi.org/10.1177/14614448221079028.

Vaughan-Nichols, Steven. "5 Reasons Why I Prefer Perplexity over Every Other Chatbot." *ZDNET.* May 30, 2024. https://www.zdnet.com/article/5-reasons-why-i-prefer-perplexity-over-every-other-ai-chatbot/.

Vigo, Julian. "The World Google Controls and Surveillance Capitalism." *Counterpunch.* December 17, 2018. https://www.counterpunch.org/2018/12/17/the-world-google-controls-and-surveillance-capitalism/.

Vincent, James. "Artificial Intelligence Is Going to Supercharge Surveillance." *The Verge.* January 23, 2018. https://www.theverge.com/2018/1/23/16907238/artificial-intelligence-surveillance-cameras-security.

Waldman, Peter, Lizette Chapman, and Jordan Roberson. "Palantir Knows Everything About You." *Bloomberg.* April 19, 2018. https://www.bloomberg.com/features/2018-palantir-peter-thiel/?leadSource=uverify%20wall.

Walker, Edward T. *Grassroots for Hire.* Cambridge: Cambridge University Press, 2014.

Webster, Frank, and Kevin Robins. "'I'll Be Watching You': Comment on Sewell and Wilkinson." *Sociology* 27, no. 2 (1993): 243–52. https://doi.org/10.1177/0038038593027002004.

———. *Information Technology: A Luddite Analysis.* Norwood, NJ: Ablex, 1986.

Weil, David. *The Fissured Workplace: Why Work Became So Bad for So Many and What Can Be Done to Improve It.* Cambridge: Harvard University Press, 2014.

Weinberg, Gabriel. "What If We All Just Sold Non-Creepy Advertising?" *The New York Times.* June 19, 2019. https://www.nytimes.com/2019/06/19/opinion/facebook-google-privacy.html.

Weisbaum, Herb. "The Downside of Connected Tech: Are the Smart Devices in Your Home Spying on You?" *NBC.* December 16, 2019. https://www.nbcnews.com/better/lifestyle/downside-connected-tech-are-smart-devices-your-home-spying-you-ncna1101906.

Weizenbaum, Joseph. *Computer Power and Human Reason: From Judgment to Calculation* New York: W.H. Freeman & Co., 1976.

White House, *The Framework for Global Electronic Commerce.* Washington, D.C.: 1997. https://clintonwhitehouse4.archives.gov/WH/New/Commerce/.

Whitehead, Mark. "Review of Shoshana Zuboff, *The Age of Surveillance Capitalism: The Fight for a Human Future at the New Frontier of Power.* New York: Public Affairs, 2019." *Antipode online.* October 2, 2019. https://antipodeonline.org/wp-content/uploads/2019/10/Book-review_Whitehead-on-Zuboff.pdf.

Whitehead, Mark, Rhys Jones, Rachel Lilley, Jessica Pykett, and Rachel Howell. *Neuroliberalism: Behavioural Government in the 21st Century.* New York: Routledge, 2018.

Wiggers, Kyle. "AI Training Data Has a Price Tag That Only Big Tech Can Afford." *TechCrunch.* June 1, 2024. https://techcrunch.com/2024/06/01/ai-training-data-has-a-price-tag-that-only-big-tech-can-afford/.

Wilkinson-Ryan, Tim. "The Perverse Consequences of Disclosing Standard Terms." *Cornell Law Review* 103, no. 1 (2020): 117–75. https://www.cornelllawreview.org/2020/07/28/the-perverse-consequences-of-disclosing-standard-terms/.

Willett, Marcus. "China's Investment in Digital Technologies and the Digital Great Game." In *The Digital Silk Road: China's Technological Rise and the Geopolitics of Cyberspace*, edited by David Gordon and Meia Nouwens, 23–36. London: Routledge, 2022.

Williams, Rhiannon. "Why Google's AI Overviews Get Things Wrong." *MIT Technology Review.* May 31, 2024. https://www.technologyreview.com/2024/05/31/1093019/why-are-googles-ai-overviews-results-so-bad/.

Wilson, James Q., and George L. Kelling. "Broken Windows." *Atlantic Monthly* 249, no. 3 (1982): 29–38.

Winder, Davey. "How to Stop Your Smart Home Spying on You." *The Guardian.* March 8, 2020. https://www.theguardian.com/technology/2020/mar/08/how-to-stop-your-smart-home-spying-on-you-lightbulbs-doorbell-ring-google-assistant-alexa-privacy.

Winner, Langdon. *The Whale and the Reactor: A Search for Limits in an Age of High Technology.* Chicago: University of Chicago Press, 1986.

Winston, Ali. "Palantir Has Secretly Been Using New Orleans to Test Its Predictive Policing Technology." *The Verge.* February 27, 2018. https://www.theverge.com/2018/2/27/17054740/palantir-predictive-policing-tool-new-orleans-nopd.

Winston, Ali, and Ingrid Burrington. "A Pioneer in Predictive Policing Is Starting a Troubling New Project." *The Verge.* April 2, 2015. https://www.theverge.com/2018/4/26/17285058/predictive-policing-predpol-pentagon-ai-racial-bias.

Wong, Julia Carrie "8chan: The Far-Right Website Linked to the Rise in Hate Crimes." *The

Guardian. August 4, 2019. https://www.theguardian.com/technology/2019/aug/04/mass-shootings-el-paso-texas-dayton-ohio-8chan-far-right-website.

Wood, Alex J., Mark Graham, Vili Lehdonvirta, and Isis Hjorth. "Good Gig, Bad Gig: Autonomy and Algorithmic Control in the Global Gig Economy." *Work, Employment and Society* 33, no. 1 (2019): 56–75. https://journals.sagepub.com/doi/pdf/10.1177/0950017018785616.

Woodcock, Jamie. *The Fight against Platform Capitalism: An Inquiry into the Global Sruggles of the Gig Economy.* London: University of Westminster Press, 2021.

"World Bank Promotes Microwork Opportunities for Jobless Palestinians." World Bank news release. March 26, 2013. https://www.worldbank.org/en/news/press-release/2013/03/26/world-bank-promotes-microwork-opportunities-for-palestinians.

World Economic Forum. *Personal Data: The Emergence of a New Data Asset Class.* Geneva, CH: WE Forum, February 17, 2011. https://www.weforum.org/publications/personal-data-emergence-new-asset-class/.

Wu, John C. "Anatomy of a Dot-Com." *Supply Chain Management Review* 5, no. 6 (2001): 42–51. https://kupdf.net/download/operations-management_5afdf9f5e2b6f57070550a90_pdf.

Wu, Tim. *The Master Switch: The Rise and Fall of Information Empires.* New York: Vintage, 2010.

Yates, Luke. "Understanding the Airbnb 'Movement': How Platform-Sponsored Grassroots Lobbying Is Changing Politics." *Rosa Luxemburg Stiftung News.* October 22, 2021. https://www.rosalux.de/en/news/id/45224/understanding-the-airbnb-movement.

Yun Chee, Foo, and Bart H. Meijer. "Apple, Google, Meta Targeted in EU's First Digital Market Act Probes." *Reuters.* March 25, 2024. https://www.reuters.com/business/media-telecom/eu-investigate-apple-google-meta-potential-digital-markets-act-breaches-2024-03-25/.

Zhang, Phoebe. "Cities in China Most Monitored in the World, Report Finds." *South China Morning Post.* August 19, 2019. https://www.scmp.com/news/china/society/article/3023455/report-finds-cities-china-most-monitored-world.

Zuboff, Shoshana. *The Age of Surveillance Capitalism: The Fight for a Human Future at the New Frontier of Power.* New York: Public Affairs, 2019.

Zuckerberg, Mark, "Building Global Community," Facebook.com. February 16, 2017. https://www.facebook.com/zuck/posts/10154544292806634.

———. "Harvard Commencement Address." *The Harvard Gazette.* May 25, 2017. https://news.harvard.edu/gazette/story/2017/05/mark-zuckerbergs-speech-as-written-for-harvards-class-of-2017/.

Zuckerman, Ethan. "What Is Digital Public Infrastructure?" *Center for Journalism & Liberty.* November 17, 2020. https://www.journalismliberty.org/publications/what-is-digital-public-infrastructure.

Index

Aadhaar identity-card system 22
Abacus Direct 51–53
ABF Freight systems 99; outward- and driver-facing cameras 99–100
accumulation by dispossession 17, 45, 191
Accurint 91
ACLU (American Civil Liberties Union) 147–48
Acxiom 23, 91
ad agencies/networks 3, 40–43, 51, 55, 57–58, 74, 153
AdMeld 157
AdSense 57
ADT 113
advanced semiconductors 216, 221
Advertising Age 37
advocacy-oriented capitalism 10–13
AdWords 56–57
aerial surveillance hardware 148; aerial cameras 138, 144
Afghanistan 129
The Age of Surveillance Capitalism (Zuboff) 3, 7, 11, 13; criticisms of Zuboff's approach 10–14
AI Overviews 179–80, 182; *see also* Perplexity
AI regulations in U.S. 172–73
AI systems 5, 160, 170–73, 176–80, 183, 186–87, 205
Airbnb 3, 23, 60, 63, 73–78, 118, 199, 201, 204; AirBed & Breakfast 73; *Airbnb Magazine* 75; #BelongAnywhere 74; Home Sharing Clubs 75–76; hosts 60, 74–75, 77; public policy teams 76; toolbox of community organizing 75
Aleph Alpha 171, 175, 177
Alexa 80, 113, 160
algorithmic management 2–4, 68–70, 96, 98, 102–4, 205–6, 230n50
Algorithms of Oppression (Noble) 22
Alibaba 23, 59, 92–93, 208
Alipay 93
Allende, Salvador 197; socialist rule (1970–73) 197, 250n47
Alphabet 92, 118, 125, 165–167, 169, 177, 190
ALPR (automatic license plate readers) 133–35, 140–41
alterglobalization movement 125
alternative asset management funds 122
AlterNet 180
Amazon 2–3, 5, 8–9, 11, 23, 45, 59–60, 65–69, 71, 80, 92, 97–98, 100, 105–6, 109, 113–16, 118, 138–39, 150, 153–54, 159–60, 162, 164, 166, 177, 183–85, 189, 195, 208–9, 214, 218, 224n27, 234n3, 235n10, 243n18, 253n29; Amazon Marketplace 65; Amazon Prime 66, 160; Amazon Robotics 98; Amazon Translate 106; brick-and-mortar stores 66; copycat products 66, 160; Digit, humanoid robot 98; Mechanical Turk 105; Amazon Rekognition 139; vehicle cameras facing road and drivers 100
Amazon Ring 113–15, 138–39; Ring camera 113
Amazon Web Services (AWS) 45, 59, 66–67, 105–6, 114, 195, 253n29; Elastic Compute Cloud (EC2) 67; online marketplace for goods 65; Simple Storage Service (S3) 67; universal logistics system 65–66
Amnesty International 144–45, 176; Amnesty International USA 144–45
Amrute, Sareeta 212
Amsterdam 74, 126
"The Anatomy of a Large-Scale Hypertextual Web Search Engine" (Brin and Page) 37
Andrejevic, Mark 47, 54–55
Android: app store 156; devices 156; operating system 155–56, 162–63, 165
Anonymous 204, 251n75
Anthropic 183, 253n29
anti-digital colonialism activists 212
Anti-Monopoly Law (China) 207
anti-monopoly measures 153–54, 189, 192
"Anti-Ring" stickers 204
antitrust investigations of Big Tech's grip on AI development 183–85
antitrust laws 216–17; U.S. antitrust legislation reinvigorated 158–61
antitrust measures 4–5, 152–55, 157–61, 168–69, 188, 192; renewed antitrust movement 214
AOL (America Online) 15, 55, 57–58
app makers mimicking techniques used by slot machines 63
App Tracking Transparency (ATT) 13
Appen 70, 106, 109
Apple 9, 12–13, 15, 22–23, 28, 89, 113, 115, 118, 153–57, 160–61, 163, 165–69, 177, 214, 218, 224n27, 243n18, 243n21, 244n45, 247n39; iMessage app 161; Independent Repair Provider Program 12; iPhone store 156; iPhones 12–13, 22, 161, 163, 165–67, 224n26; iPhones 14 and 14Pro 13; iPods/iTunes 9; MacOS system 13; operating system

iOS 17.4 166; Music 169; Privacy Policy (December 2017) 13; Safari browser 157, 166; Watch 89, 115, 161
application programming interface (API) 105–6, 177
Arkansas 147
armored vehicles 127
Army Files (CONUS) scandal 15
ARPANET 15, 25–27, 34, 67
artificial intelligence 1, 2, 5, 22, 67–68, 79, 82, 87, 92, 98, 118–19, 139, 148, 159, 170–71, 174–75, 177, 185–86, 210; use of AI to exploit specific people's vulnerabilities 219
artificial neural networks (ANNs) 82
assault rifles 127
Associated Press 181, 186
AT&T 26, 121, 226n32
Atkin, Douglas 74
Atlanta Police Foundation (APF) 145
attention economy 63
audit society 123
Australia 17, 219
Austria 162, 168
automated digital technologies forcing people to adapt to operative systems 203
automated recommender systems 8, 163
automatic license plate reader (ALPR) 133–35, 140–41
automatized repossession 88
Avila, Renata 212
Axel Springer 181

back-to-the-landers 30
Badger, Adam 72
Baidu 23, 93
Baltimore 138, 144
Bangladesh 107
banner ads 40–41, 56
Barcelona 74, 125, 251n58
Barcelona en Comú (Barcelona in Common) 125
Bard/Gemini 171, 205; see also Gemini-powered version of Bard
Barlow, John Perry 29, 31
Barnes & Noble 65
Bateson, Gregory 30
Beer, Stafford 197–98
behavior modification 2, 9–12, 88, 116
behavioral profiling 41, 44
behavioral surplus 7–8, 10–13, 82
Beijing 221
Belgium 168
Belt and Road Initiative (BRI) 209

Benavidez, Nora 191
Beniger, James 48
Benjamin, Ruha 22
Bentham, Jeremy 84
Berlin 74, 103
Berners-Lee, Tim 33
Beware 131
Bezos, Jeff 66, 107
Biden, Pres. Joe 5, 154, 172, 219; analytics 120, 134; big data 1, 10, 67, 92, 95, 128–29, 131–32, 134–35, 140, 151, 225; Executive Order regarding safe use of AI 173; policing 133
Bing 162, 177, 182
biometric categorization 175–76, 219; biometric data 222, 173, 175, 219
Bird 77
Birhane, Abeha 212
black communities 92, 142
Black Lives Matter movement 128, 249n8
blacklisting 93
Blecharczyk, Nathan 73
Bloomberg 13, 129
Bloomberg Philanthropies 120, 123
Blueprint for an AI Bill of Rights 172
Bluesky Social 193
body-cam video 128
boilerplate clauses/legal jargon 54, 86
Botswana 109
Boundless Information 15
Bradford, Anu 186, 207–8, 214, 216–17, 220
Brady, Tye 98
Brand, Stewart 29–32
Brandeis, Louis 153
Bratton, William 132, 240n25
Brayne, Sarah 130, 134–36
Brazil 125
Breaking Things at Work (Mueller) 203
Bria, Francesca 123–24
brick-and-mortar stores/structures 45–46, 66, 198
Brin, Sergey 8, 37–38
British House of Lords communications and digital select committee 181
British North American colonies 17–18
broadband infrastructures 57, 191–92, 208, 249n17, 249n20; American Association for Public Broadband (AAPB) 191; community broadband networks 191; Institute for Local Self-Reliance (ILSR) 191; Internet for All program

191; municipal broadband networks 191; Tribal Nations' fiber networks 191
"broken windows theory" 132, 240n25
Bronx 122, 144–45
Brooklyn 144
Brown, Michael 128
browser(s) 33, 40–42, 111, 155, 157, 164–66, 250n27
Brunei 221
Brussels 166, 186, 217
"brute-force" attacks 159
"Building Global Communities" (Zuckerberg) 62
Bulgaria 168
Burawoy, Michael 103–4
business models 3, 7, 10, 12–13, 37–38, 40–43, 56, 59, 61, 65–66, 70–73, 75, 82, 106, 116, 150, 152, 163, 189–90, 218
Business Process Reengineering 101
ByteDance 166

Cahn, Albert Fox 173
California 5, 102, 144–47, 152, 156, 192, 230n57
California Consumer Privacy Act (CCPA) 152
Canary home security monitor 115
Cantrell, LaToya (mayor of New Orleans) 130
"captured city" 138
Caribbean Islands 17
CCTV (closed circuit television) 120, 127, 138–39, 144; cameras 120, 138–39, 144; footage 175
Cecchinato, Maria 102
Center for Democracy and Technology (CDT) 51, 115, 173
Center for Humane Technology 64
Center for Media Education (CME) 51
Centers for Medicare and Medicaid Services (CMS) 45
Central Intelligence Agency (CIA) 129, 136
Central Political and Legal Affairs Commission (China) 95
Cerf, Vincent 27
CERN 33
Charlotte, North Carolina 138
chatbot Bard 178; see also Bard/Gemini
ChatGPT 171–72, 174, 177–79, 181, 185–86, 205
Chesky, Brian 73–74

Index

child abuse 162
Children's Online Privacy Protection Act (COPPA) 52–54, 160
China 2, 4–5, 10, 20–23, 92–95, 123–24, 127, 139, 148, 152, 177, 186, 207–10, 213–16, 218–21, 234n69, 253n29, 253–54n48, 254n53; absence of citizen protection against state action 207; Anti-Monopoly Law 207; Cyber Security Law (2016) 95, 207; Data Security Law 207; digital authoritarian governance model 215; "Guidelines of Social Credit System Construction" (2014–2020) 93; international involvement in standard setting across technologies 210; Personal Information Protection Law 207
Chinese Communist Party (CCP) 93, 207–9, 212
Chinese higher import tariffs on certain European and U.S made products 221
Chinese-made surveillance technologies or censorship techniques 215
Chinese social credit system 93
Chinese State Council 93
Chinese state-driven regulatory model 207, 214–17, 219–20, 222
Chips and Science Act 221
choice architecture of digital services 8, 192
Chongqing 95
Chrome browser 155
Cisco 81, 118, 121–22, 138
cities as engines of capitalist accumulation 122–23
citizen participation 125–26, 199
Citizen Virtual Patrol 140
city indebtedness rankings 123
city surveillance 4, 127, 137, 143; technocratic concept 137
Civiq Smartscape 121
Clark, Brett 16
Clark, Jennifer 121
Clearview AI 148
Clickworker 106
Clinton administration 39, 51–52
Clinton, Bill 34
"Cloud Empire" 23
cloud services 65–66, 104, 183, 209, 253n29
CMGI 43
CO_2 emissions 187
CoEvolution Quarterly 30–31

cognitive behavioral manipulation 175, 219
Colau, Ada 125
Cold War 14, 29, 34, 197, 212
Cole, G.D.H. 197, 250n45
collusive agreements among AI corporations 185
colonialism (historical) 5, 16–17, 20, 23–24, 44, 80, 208, 211, 232n89; British settler 18; internal 17, 44; *see also* data colonialism
colonization of everyday life 83–84
Colorado 153, 156
Columbus, Christopher 17
Columbus, Ohio 120
commercialization of the internet 3, 34–35, 37–38, 42, 56
Communications Workers of America (Local 7777) 196
Community Control over Police Surveillance (CCOPS) ordinances 139, 147
community media activists 194
community-owned internet providers 191–92
Comparitech 95
COMPSTAT 132
Compton, California 144
computer networks 1, 27–28, 31–32; commercial networks 32–33; *see also* private networks
Connecticut 153, 156
conquest by declaration 18
conquistadores 17
consumer feedback vs. meaningful participation 46–48
Consumer Privacy Protection Act (proposal) 53
consumer surveillance 51, 54–58, 243n2
"consumerforce" 49
contactless payment 161, 167
content moderation 189–90, 202, 214, 218
contextual advertising 55–56, 58, 165, 243n2
control of workers 1, 4, 90, 205
control revolution 48
Cook County Public Defender 142
cooperative ownership of online malls 192
"Cop City" 145–46, 242n82, 242n90, 242n91
copyright 174, 180–81, 187, 194, 248n52; activists 194; breaches 180; copyright law 174; issues and journalistic implications of generative

AI usage 180–81, 187; EU U.S. copyright law 181
CoreLogic 91
coronavirus pandemic 73, 100, 107, 125
Corporate Business Systems (CBS) 101
"Corporation for Digital Public Infrastructure" proposal (2018) 195
Cortana 113
The Costs of Connection (Couldry and Mejias) 16
Couldry, Nick 3, 16–17, 20–24, 79, 95, 150, 204, 210–12
counterterrorism 127, 136, 240n46
CPAP machine 11
Crain, Matthew 38, 151
Creative Commons 201
"credential stuffing" 159
credit scoring systems 4, 91–94; Chinese style 92–95; Social Credit System (SCS) 94–95
Cross Bronx Expressway 122
cryptography 159
The Culting of Brands: Turn Your Customers Into True Believers (Atkin) 74
cybernetics 31, 197; discourse of systems and information 30; feedback 47; management 50, 198
cyberspace community 37
Cybersyn 197–98, 250n47
Czech Republic 168

D. Davidson & Co. 168
Dallas 121
DARPA (Defense Advanced Research Projects Agency) 25–27, 32
data breaches 1, 113, 141, 147, 151–52, 159; privacy-infringing data extraction 217; *see also* privacy breaches; security breaches
data brokers 23, 50, 90–93, 233n51
data centers 67–68, 185, 187, 198, 247n39
data colonialism 3, 7, 16–17, 19–25, 44, 54, 80, 84, 86, 95, 114, 149–52, 211–12, 225n62; resistance against 211–12; *see also* colonialism (historical); digital colonialism
data fetishism 89
data fusion 135
"data-owning democracy" vs. "digital socialism" 213
data portability 162, 218
data rents 72
"data trusts" and "data

commons," proposals for 196
Datalogic 91
"dataveillance" 116
The Death and Life of Great American Cities (Jacobs) 122
Decidim 125
"Declaration for the Future of the Internet" (2020) 220
DECODE (Decentralized Citizen-Owned Data Ecosystems) 126, 201
decolonial struggles against data extraction 210–12; decolonial approach to data and technology advocated by Couldry and Mejias 211–12
decoupling of U.S. and Chinese technological assets 216
"deepfakes" 172–73
DeepMind 178, 185, 248n57
"defunding the police" 128
DeKalb County, Georgia 146
Delaware 153
Deliveroo 59, 102, 111, 199
Deloitte 118, 121
Democratic Party 34, 153
Democrats 31, 34, 202, 214; corporate 31, 202; New Democrats' third way 34
Denmark 168
deprivatization of the internet 5, 189, 212
Dholera 123
Didi 23
Didier, Emmanuel 132
digital capitalism 2–3, 13, 16, 80–82, 85, 87, 88, 112, 149, 189, 197, 204
digital colonialism 208–10, 212; *see also* data colonialism
digital company town 109
digital control systems/technologies 83, 96, 100–2
Digital Disconnect: How Capitalism Is Turning the Internet Against Democracy (McChesney) 194
digital empires 186, 207, 220
Digital Empires: The Global Battle to Regulate Technology (Bradford) 186
digital enclosure of the internet 3–4, 44–46, 61, 63, 69, 85, 150, 192
digital managerial control of workers 4, 50–51, 70, 96–101
digital media literacy 150, 206
digital pat downs 128
digital platform(s) 2–3, 45, 59, 77, 125, 162, 197–201, 203, 207, 210; as publicly-owned utilities 198–99
digital protectionism 186, 214; U.S. accusations of digital protectionism by EU 214
digital rights group Access Now 176
digital rights group EDRi 176
Digital Services Act (DSA) 5, 161, 165, 168, 218
Digital Silk Road 209, 215
digital sovereignty 64, 171, 207, 209–10, 221
digital Taylorism 4, 104–6
digital tools, behavioral requirements of 5, 206
digital tools design 5, 8–9, 54, 63–64, 69, 80–81, 86–87, 89, 97, 103, 112, 116, 140–41; problems inherent to 203–4
"disciplinary gaze" of authorities 84
District of Columbia 155–56, 160
DMV records 137
DNA samples 134
Domain Awareness System (DAS) 137–38
Doordas 77
dot-com investment bubble 40, 42–43, 53–57, 65, 105, 224n33
DoubleClick 3, 40, 43, 51–53, 55–58, 156–58
Drivers Union 101–2
drones 92, 95, 106, 127, 138–39, 147, 170, 238n42
DuckDuckgo 58, 150, 243n2
DVRs 113
Dynamics 365 185
Dyson, Esther 29, 31

EAGL Technology 141
East Asia 107
eBay 63, 65, 92
Echo 80
economic libertarianism 31–32
Economic Policy Institute (EPI) 77
economic recession, early 1990s 38
economies of action 9
8kun 195, 250n34
Eisenhower, Dwight D. 14
Electronic Frontier Foundation (EFF) 28, 114, 139
electronic logging devices (ELDs) 98–99
Electronic Privacy Information Center (EPIC) 51, 173
enclosure movements 17, 44
encryption 22, 35, 236n7
end-user license agreements (EULAs) 19
"engineering of human souls" 94
England 18, 44, 204, 228n29, 232n17
Enterprise Resource Planning (ERP) system 8, 101, 185
Epsilon 91
Equifax 11, 23
equity firm KRR 181
Ericson, Richard 137
Ericsson 121
Ernst & Young 118
EU AI regulations 5, 186, 219; and "Brussels effect" 217
EU Artificial Intelligence Act 5, 170–71, 174; finalized 174–77
EU Data Protection Directive 52
EU Digital Markets Act (DMA) 5, 157, 161–62, 165–67, 170, 218
EU Digital Services Act (DSA) 5, 161–63, 165, 168, 170, 218; platform transparency and accountability 218
EU focus on fundamental rights vs. U.S. focus on national security 214
EU member states 33, 52, 165, 170, 176, 215, 218
EU Parliament 161, 174, 218
EU rights-driven regulatory model 207, 216, 219–20, 222
Eubanks, Virginia 88
European Capital of Innovation (2014) 125
European Center for Digital Rights (NOYB) 164
European Commission 162–63, 165–69, 170, 174, 218
European Council of Ministers 161, 170, 218
European Court of Justice (EJC) 164
European digital privacy and human rights groups 176
European Union (EU) 2, 4–6, 33, 52, 95, 126, 150–52, 154, 157, 161–78, 186–89, 191, 207, 210, 214–22
Ewen, Stuart 49

Facebook 1, 3, 8–10, 13, 23, 40, 57, 59–63, 65–66, 68–69, 80, 105–6, 114, 118, 148, 153, 158, 162–65, 167, 182, 184, 189–90, 193–94, 196, 198, 203, 209, 214, 224n26; as "meaningful communities" 62; as an online directory 61; and sense of belonging 63
Facebook-Cambridge Analytica scandal 1, 151–52
Facebook's "Free Basics" 208–9
facial recognition 70, 95, 100, 105, 114, 127, 130, 144–45; camera 105; drones 238n42
fake news 1, 162, 168

Fasman, Jon 140–41, 143, 147–48
"fauxtomation" 106
Fearless Cities summits 200, 251n58
Federal Bureau of Alcohol, Tobacco, Firearms and Explosives (ATF) 146
Federal Bureau of Investigation (FBI) 15, 22, 127, 129–30, 136, 146
Federal Communications Commission 191
Federal Motor Carrier Safety Administration 99
Federal Trade Commission (FTC) 46, 52–53, 153–54, 158–60, 183, 214, 233n51, 237n15
Federation of Dutch Trade Unions (FNV) 102
FedEx 99; computerized package scanner 99; driver-facing and front-facing cameras 99
"fediverse" model of social networking 193, 195, 202
Ferguson, Andrew 129–30, 140, 144
Ferguson, Missouri 128
FICO credit score 93
FIFA World Cup 120
Financial Times 13, 181
financialization 15, 45, 72–73, 78, 192
fingerprints 134, 138, 232n17
Finland 168, 174
Firefox 157
Fitbit 89
5G networks 209, 216
flash-bang grenades 127
"flawed democracies" 219; U.S. flaws inherent to non-parliamentarian, presidential system 219
Flink 102
Floyd, George 128
FNV Young & United (webpage) 102
"focused deterrence" 130
food delivery platforms 102, 196, 201
Fordist-Keynesian compromise of embedded liberalism 122
Foster, John Bellamy 3, 7, 14–16, 19
Foucault, Michel 84–85
Fourth Amendment to the U.S. Constitution 115
France 162, 171, 174–75, 209
free global cyber commons vs. competing systems of cyber sovereignty 209
Frischmann, Brett 87, 151

From Counterculture to Cyberculture (Turner) 29
Fuchs, Christian 195
Fuller, Buckminster 30
function creep 90, 117, 136
fusion centers 135
Fusus 139

G7 173
Gab 195, 250n34
gamification: as "soft control" 69; of work 69–70, 104
Gandy, Oscar H., Jr. 11
GE (General Electric) 121
Gebbia, Joe 73
Gemini 178
Gemini Pro-powered version of Bard 178
General Data Protection Regulation (GDPR) 5, 151–53, 163–64, 217, 243n4
General Motors (GM) 9, 96
general-purpose AI models 171
general-purpose AI systems 171, 174, 186, 218
generative AI models 181, 186, 190
gentrification 74
Georgia 139, 146
Germany 102–3, 123, 162, 171, 174–75, 185, 209, 231n62, 254n51
Getaround 77
ghost workers 105, 189; see also gig workers
gig economy 1, 3, 69, 71, 103, 196, 230n57
gig workers 5, 102–3, 189; Gig Workers Rising 102
Gilder, George 31
Gingrich, Newt 31
Global Business Network 31–32, 226n32
Global Digital Services Organization (GDSO; proposed new UN allied agency) 201–2
Global North 107–8, 123
Global Smart Cities & Connected Communities Think Tank 123
Global South 107, 109, 208–9, 212
Gmail 58
GMB union (UK) 102
Golden Shield and *Sharp Eyes* projects 95, 234n69
Google 2–3, 5, 8–10, 13, 15, 17, 19, 23, 37–38, 40, 43, 54, 61, 63, 65–69, 71, 80–82, 106, 114, 116, 118–19, 139–48, 152–59, 162–63, 165, 167–68, 170, 178–85, 189, 194, 198, 201, 208–9, 214, 218, 220, 224n27,

229n81; ad exchange ADX 156; Ad manager suite 156; Flights 165; Hotels 165; Play store 156; Plus social network 58, 229n89; shopping 165
Gore, Al 34–35, 227n52
GoTo 5
government-owned computer networks 3, 25
GPT (Generative Pre-Trained Transformer) 172; GPT-4 model 181; GPT4o 179
GPUs (graphics procesing units) 187
Great Recession 128
Green Taxi Cooperative in Denver metro area 196
"grinding" vs. "oppositional play" in work games 104
Guam 155
The Guardian 142, 145
"guild socialism for the digital economy" 199
Guild Socialists in Britain 197
Gurgaon 124

hackers 29, 113–14, 119, 172, 204; culture 29; hacktivism 204, 251n75
Haggerty, Kevin 137
harassment 114, 162, 190
Harris, David Evan 186–87
Harris, Tristan 64
Harvey, David 17, 45
hate crimes 162
hate speech 1, 105, 168, 217
Hayek, Friedrich 198
HBO series *The Wire* 132
health insurance companies 15, 115
Heineken 171
high-altitude balloons creating aerial wireless network 208; "Project Loon" 252n6
"high-impact" GPAIs 174, 219
High-Performance Computing and Communications Act (1991) 35
"high-risk" AI systems 170, 174, 176, 187
Hikvision 95
Hill, Kasmir 148
Hitachi 122
Ho, Daniel 173
Hollings, Sen. Fritz (D-SC) 53
home as "data factory" 112
home automation system(s) 4, 112–13
Home Depot 145
home security surveillance system(s) 4, 112–13, 115
hospitals 85, 91, 118
House Judiciary Committee 154, 243n18

HTTP cookies 41–42
Huawei 208, 220; smart phones 210; struggles in U.S. 208
human intelligence tasks (HITs) 105
Human Rights Watch 95
HunchLab 133, 142
Hungary 168, 174, 215
hyperlink(s) 33, 40

IBM 26, 33, 79, 118, 120, 122, 138
IBM Solutions 120
IBM's Intelligence Operation Center (IOC) 120
Iceland 165
identity theft 1
IG Metall (Germany) 102
illiberal tendencies in U.S. and some EU member states 215
Illinois Biometric Information Privacy Act (BIPA) 148
"image generators" 178, 181
IMSI (international mobile subscriber information) catchers 130, 142
In-Q-Tel 129
"independent" contractors 4, 60, 69–71, 96, 98–99, 230–31n57
India 2, 18, 22, 107, 123–24, 212, 232n17, 252n7
Indianapolis 138
Indo-Pacific 221
industrial capitalism 11–12, 63
industry "self-regulation" 38–39, 51–54, 150, 171, 207
Inflation Reduction Act 221
Inflection AI 185, 248n58
Information Infrastructure Task Force (IITF) 39
"information superhighway" 35
Information Technology: A Luddite Analysis (Webster and Robins) 7
"informed consent" principle 152
Infrastructure Investment and Jobs Act (2023) 191
Instagram 153, 163–65, 167, 193, 208
"instrumentarianism" 10
insurance companies 4, 11, 88, 115–16
insurance firms 122
Intel 121
Intelligence Law Enforcement Centers 120
interactive communication devices 3, 50
interactive digital advertising 3, 39
Intercept 180

International Labour Oganization (ILO) 107
internet advertising companies 42, 55, 57
internet bubble 1, 32
Internet Engineering Task Force (IETF) 42
Internet Explorer 155
Internet of Things (IoT) 23, 79–80, 89, 112–13, 117
internet policy 35, 38–39, 45
internet reform 2, 152–53, 188
internet service providers (ISPs) 15, 64, 214; lobby 191
interoperability 158, 162, 193, 218
Interpublic 58
intertribal broadband collaboration 249–50n21
inverse panopticon effect 85
investment bubble 179
investors 3, 8, 12, 38, 41–42, 56, 59, 71–72, 106, 123, 177, 183, 191, 193, 202, 216
Invite Media 157
iOS system 13, 155, 165, 167
Iowa 153, 249n17
iPad 161
IPOs (Initial Public Offerings) 42, 73
Iraq 129, 238
Ireland 164
Ireland's Data Protection Commission (DPC) 164
iris scan 148
Israel 95, 145, 242n91, 253n29
Israeli-occupied West Bank 95, 145
Italy 171, 174
iTunes platform 9
IWGB (Independent Workers Union of Great Britain) 102
IWW (Industrial Workers of the World) 102

Jacobs, Jane 122
Japan 35, 194, 219, 253n29
Jawbone Up 89
JD.com 92
Jinping, Xi (president of China) 221
Jones, Phil 106, 108, 111
journalism 49, 182, 194; ethos of, as public service 194; independent journalists 194
JP Morgan Chase 129
"juking the stats" 132, 200n30
just-in-time: scheduling 100; techniques 96

Kahn, Robert 27
Kanter, Jonathan 154, 183
Kelly, Kevin 29, 31
Kenya 208, 212

Khan, Lina 154, 158–59, 243n18
Kim, Nancy 54
Kitchin, Rob 124, 135
KPMG 118
Kwet, Michael 208, 212
Kyoto Protocol 118

"labor" of consumption 85
labor platforms 60, 106
labor rights of digital gig workers 5, 196
land enclosures 44
Landrieu, Mitch (mayor of New Orleans) 129
Latin America 20, 107, 209; *see also* South America
Latino communities 92, 142; Latino men 128
Lavasa 124
law enforcement departments 4, 114
lawsuits 157–158, 160, 207;
lean production 96
LeCun, Yann 178
legislation 3, 5, 36, 51–54, 147, 150, 153–54, 158, 170, 173–74, 186–87, 190, 202, 214, 217
Leufer, Daniel 176
Levine, Yasha 26, 171
LexisNexis 91
liberal market societies 21–22, 152
Liberty Mutual 115
libraries 118, 181, 192, 194–95, 227n52
Lichtenstein, Nelson 96
Lieberman, Donna 145
Liechtenstein 165
"lifelogging" 89
Lime 77
Lionsbridge 108
LLMs (large language models) 184
location information/geolocation data 9, 13, 80, 88, 99–100, 103, 133, 137–38, 141, 143, 152, 160, 175, 200, 210
Loeb, Zachary 205
log-off strike 103
logistics 2–3, 65–66, 68, 96, 106
long-term home stays 60
Los Angeles Police Department (LAPD) 129–30, 132
Louisiana State police 130
"low-tier" AI 174
Lucy Parsons Labs 142
Luddism 111, 203–5; machine breaking 111; *see also* New Luddism
Luria, Gil 168
Luxembourg 168
Lyft 7, 101, 103, 201, 230n50; scooters 77

Index

MacArthur Justice Center 142
machine learning 10, 67, 92, 95, 120, 128, 131–32, 134–36, 140, 151, 225n62
Madisonian system of checks and balances 153
Malaysia 221, 253n29
Manufacturing Consent (Burawoy) 103
Maps Google 58
marketers 3, 13, 37, 40–43, 46–47, 50, 55–58, 92, 112, 244n45
marketing complex 38, 51–52, 54
Markoff, John 28
Masdar City 123
Massachusetts Drivers United 102
Mastodon 193, 195, 201–2
McCain, John, Senator (R-AZ) 53
McChesney, Robert 3, 7, 28, 34, 49, 153, 194–95, 215
McGuirk, Justin 112
MCI 33
McLuhan, Marshall 30
Media Group 77
media obsession with user engagement 63–64, 67, 188
media watchdog Free Press 190
Meelaunee forest 145
Mejias, Ulises 3, 16–17, 20–24, 79, 95, 150, 204, 210–12
Memphis 139
Merit 33
Meta 40, 118, 158, 162–67, 169
Metha, Amit 157
Microsoft 8–9, 15, 23, 57–59, 67, 113, 118, 121, 137, 153, 155, 157, 161, 166, 170, 177, 180–85, 209, 218, 247n39, 248n58, 253n29; Azure cloud 59, 177, 184–85, 253n29; CityNext program 121; Software as a Service (SaaS) 183–84
Microwork 4, 70, 104–11; "black box labor" 110; essence of 110–11, microworkers 4, 70, 106–11; nonmonetary rewards 109; obstacles to worker organizing 110
militarized policing 4, 127
military conflicts in Eastern Europe and Middle East 221
"military-digital complex" (Levine) 171
military-industrial-communications complex 3, 127
military-industrial complex 7, 14–15, 30
military-industrial research culture 29–30
military intelligence 15, 171

military Keynesianism 15, 224n31
Minneapolis, Minnesota 128
Minocher, Xerxes 144
Misfit 89
misinformation 1, 180, 187, 190–91
mission creep 127
Missouri 128, 153
Mistral AI 171, 174, 177, 179, 184
MIT Technology Review 121
"modernization of social governance" 22
Modi, Narendra 124
Le Monde 181
monetary rents 72
monopolistic features/practices 2, 5, 39, 153–55, 183, 199
monopolistic practices 2, 5, 154, 183
Montana 153, 249n17
Monthly Review 7, 14, 16
Moody's 123
Morozov, Evgeny 7, 10–11, 123–24, 197–98, 202, 223n15, 223–24n22, 239n67, 250n47
Mosaic 33
Moses, Robert 122
Motorola Solutions 145
Motorola's Command Center Predictive Suite 130
Mozilla 157; 237n18
MTurkGrind 110
Mueller, Gavin 203–4
mug shots 134
Muldoon, James 13, 62, 126, 189, 197–203, 213
municipal networks 191, 249n17
Muscogee Creek people 145
music streaming 168–69; app market 168; services 169
Musk, Elon 178, 191
MySpace 57

National Center for Supercomputing Applications 33
National Information Infrastructure (NII) 39
National Institute of Health 202
National Institute of Standard and Technology (NIST) 172
National Longitudinal Study of Adolescent Health 136
National Longitudinal Survey of Youth 136–37
National Science Foundation (NSF) 27–28, 32–35, 202
National Science Foundation Network (NSFNET) 27–28, 32–35; NSFNET's Acceptable Use Policy (AUP) 32

National Security Agency (NSA) 15, 110, 129, 135, 215
National Telecommunications and Information Administration (NTIA) 39
nationalistic protectionism 221
NATO 25
natural monopolies 153, 198–99
Naughton, John 155, 178–79
Neighbors App 114, 139
neoliberal consensus 34–35
Nest Protect smoke detector 115
Netflix 23, 59
The Netherlands 102, 168, 245n54, 253n29
Netscape 41, 155
network effects 4, 60, 193, 198
network entrepreneur 31
networks of resistance 242n90
Neurath, Otto 197, 250n45
neuroliberalism 8
New Brandeisians 153–54, 188–89, 192
New Communalists 30, 32
New Hampshire 141
New Jersey 140, 153, 156
new Luddism 203–5; *see also* Luddism
new municipalism 191, 199–201, 249n17; projects of autonomy at local level 200; risk of falling into localist trap 200
New Orleans 128–30; New Orleans City Council 130
New School for Social Research 200
New York City 31, 74, 77, 101, 122, 128, 132, 144–45, 196
New York Civil Liberties Union 145
New York Police Department (NYPD) 129, 132, 137, 144, 240n33
New York Stock Exchange 67, 73, 138
New York Supreme Court 144
New York Taxi Workers Alliance (NYTWA) 101
The New York Times 145, 180–81, 195
New York's Freedom of Information Law 144
New York's Local Law 18 77
Newark, New Jersey 140
News Corp 181
NGOs 123
1994 Republican Revolution 34
1998 World Wide Web Conference 37
Nividia 179, 183, 247n39, 253n29
Noble, Safiya 22

NOLA for Life (New Orleans murder reduction program) 129
non–Apple smartwatches 161
Nordense 121
Norway 165, 168
"notice and choice" principles and practices 53–54
Nuance Communications 183
nudges 8, 69–70
Nyabola, Nanjala 212

Oakland, California 144
Oakland's Privacy Advisory Commission 148
Obama, Barack 75; 2008 and 2012 presidential election campaigns 75, 231n85
occupied territories 110
office workers 100
Ongweso, Edward, Jr. 205
online advertising 38, 43, 55–58, 168
online marketplace(s) 3, 5, 65, 77, 157
online platforms 2, 3, 5, 13, 59, 63, 73, 106, 108, 150, 162–63, 178, 182, 189–91, 197, 206; as "inequality machines" 189
Online Privacy Alliance (OPA) 52
online surveillance 2, 54, 58
OpenAI 170, 172, 177–181, 183–84, 246n15; its API (Application Programming Interface) 177; Dall-E 178; Midjourney 178
Open Markets Institute 169
open-source AI models 174, 177, 186
open-source licenses 187
open-source model LLaMA (Large Language Model Meta AI) 175
Operation Shield 145
opt-in default setting 51–54
opt-out default setting 51, 53
Oracle 185
Oregon 142, 153
Oregon Justice Research Center 142
Orlando, Florida 139

Page, Larry 8, 37–38
Pakistan 107, 208, 219
Palantir 23, 128–30, 135
PalTalk 15
panopticon 84, 232n17, 232n19
payment service provider (PSP) 167
PayPal 92, 129
PeekYou 91
pension funds 15, 122; pension fund legislation 202

Pentagon 7, 14, 25
People's Bank of China 94
People's Republic of China 22, 93
Perplexity 247n41
Persistent Surveillance Systems (PSS) 138, 144
personalized advertising 1, 10, 165, 237n10
Philippines 221
Photobucket 181
Pickard, Victor 194–95
piggybacking on existing public infrastructures/media 194–95
Pinochet, Augusto 197
platform capitalism 2, 24, 43, 94–95, 106, 150, 192, 196, 207, 213, 222; global impact of competing regulatory models 213–22
platform cooperatives 5, 196, 199–200, 250n39
Platform Cooperativism Consortium 200, 250n39
platform economy 59, 63, 77, 102, 105, 199, 213, 222, 250n45
platform monopoly 43, 55, 58
platform owners 59–60, 89
platform socialism 5, 188, 197–201, 203, 206, 212–13, 222
Platform Socialism (Muldoon) 197, 201, 203
platform workers 60, 102, 216
Pokémon Go 9
Poland 168, 174, 215
"Police Cloud" systems in China 95
police department(s) 114, 116, 121, 127–31, 133–35, 138–41, 144–45, 147–48
police surveillance 4, 130, 134, 139, 147; five shifts in data-driven techniques 130–36; reproduction of inequality with data-intensive practices 134
political consensus of the 1980s about neoliberalism 62
political economy 11, 14, 37, 124, 188
portability of data 162, 218
Portland, Maine 74
Portland, Oregon 142
Porto Alegre 125
predictive analytics 128, 132
predictive policing 4, 83, 120–21, 127–30, 132, 139, 142–44
PredPol 133
primitive accumulation 45, 228n29
principle of subsidiarity 201
PRISM 15
prisons 84–85, 107–8, 121, 145

privacy advocates 144, 194
privacy and security program 159–60
privacy breaches 2, 4, 151; *see also* data breaches; security breaches
privacy invasions 140–43; *see also* privacy breaches
privacy laws 136, 148, 153, 173, 214, 243n2
privacy of users 2, 4, 114, 158, 163, 209
privacy policy 13, 53, 163
privacy protection 5, 39, 52–53, 115, 150, 170
private computer network(s) 32, 220, 249n18
private equity firms/investors 56–57, 145, 181, 202
private networks 32, 220
private sector 2, 28, 32, 34–35, 38–39, 51, 120, 136, 144, 147, 173
private social scoring systems 175
privatization 2–3, 5, 25, 33–35, 45, 62, 65, 72, 123, 189, 191, 199, 209
proactive strategies to prevent crime 132
programmable manufacturing equipment 3, 50
Progressive American reformers 199
Progressive Era 153
Progressive insurance company 88
Project Astra 179
Project LASER (Los Angeles Strategic Extraction and Restoration) 131
proto-platformization 3, 43–44
psychographic segmentation/profiling 9, 217
public domain 44, 47, 191
public investments in social media 194
"public lane" for cloud (carved out of Amazon Web Services) proposal 195
public libraries 194–95
public media 194, 227n52
public-oriented government 22
public ownership 26, 32, 153, 199
public policy 4, 76, 126, 150, 188
public-private partnership(s) 22, 35, 138
public sector 3, 35, 39, 118
public security bureaus (PSBs) in China 95
The Public Service Media and Public Service Internet Manifesto (2021) 191

public social scoring systems 175
public spaces 45, 115, 176, 192, 219, 234n69
public utilities 118, 198–99
publishers 11, 40–41, 43, 53, 55, 57–58, 157–58, 161–62, 168, 180–82, 248n53
Puerto Rico 155–56
Purpose 74
PwC 118

Qatar 109
Qualcomm 121
Quantified Self movement 4, 88–90
quantitative easing 71, 78, 231n66
Queens 144

Race After Technology (Benjamin) 22
racial modes of social control 22
Racketeer Influenced and Corrupt Organizations (Rico) Act 146, 242n87
Radin, Margaret Jane 86
Randall, Caelyn 144
rare-earth metals 187
Raw Story 180
real-estate developers 118
real-time biometric data 175, 219
Real-Time Crime Centers 120, 139
real-time monitoring of lower Manhattan 137
real-time RBI systems 175–76
Reddit 110, 181, 203, 251n72
redlining 92
Re-engineering Humanity (Frischmann and Selinger) 54, 87
refugee camps 107–8, 110
regulating digital markets and services, as well as AI 207
regulation(s) 2, 4–5, 34, 38–39, 46, 51–54, 62, 68, 71, 76–77, 94, 114, 124, 140–41, 143, 150–54, 161–65, 170–76, 184–86, 189, 194–95, 199, 207, 214, 217–20
remote biometric identification (RBI) systems 175–76, 219
rentier relations 3, 59; rentierism 59
repression of dissidents or minorities in China 95
Republicans of the 1990s 31, 34
residential spaces 114
resistance against city surveillance and smart policing 143–146

resistance against workplace exploitation 206
retailers 43, 82, 91, 96
retrospective (non-real-time) RBI systems 175
return on investment (ROI) 28, 43, 56
Rheingold, Howard 29
Rhode Island 156
Ricaurte, Paola 212
ride-hail drivers 71, 103
ride-hailing companies/platforms 13, 23, 68, 72–73, 77, 103–4
ride hailing services 60, 102, 201
Ride Share Drivers United, California 102
"RideLondon" app (hypothetical) 201
right of free speech 190; Article 10 in European Convention of Human Rights (ECHR) 190; First Amendment to U.S. Constitution 190
"the right to be forgotten" 151
"Right to the City" movement 125
right-wing propaganda, amplification of 190; see also social media and left-wing movements
RightsCon summits 212
Rikap, Cecilia 184
Rio de Janeiro 120
The Rise of Big Data Politics (Ferguson) 129
Robins, Kevin 47–48, 85
robot dogs 145
Rockefeller Foundation 120, 123
Romer, Paul 195
Roomba 203
Royal Dutch Shell 31
Royal Society 18
RTM (Risk Terrain Modeling) 133
Rust Belt regions 108
Rutgers University 133

sabotaging 111; see also Luddites
Safe Street Rebel 204
St. Louis 144
Samasource 107–8
Samsung 23, 157
San Francisco 28–29, 31, 73–74, 156; bohemian 28–29, 31; San Francisco Bay area 28–29
SAP 185
Scale 106
scanner 139; software 70, 95, 100, 127, 130, 148; surveillance 148; technology 114, 144–45
Schrems, Max 164
Schwartz, Peter 31

Scoot 77
Scotland 18
search engine(s) 1, 3, 21, 37–38, 55, 58, 104, 150, 155–57, 162, 165, 177–78, 180, 182, 189, 198, 201, 224n28, 243n2, 243n21, 248n53
Section 230 of 1996 Communications Act 64, 190, 214, 216
security agencies 22, 94, 135, 234n69
security breaches 87, 113–14, 147, 151–52, 159; see also data breaches; privacy breaches
Seeing Machines' computer vision-equipped inward facing cameras 99
self-governing social media communities 193–95, 200
self-help tips to break one's online addiction 206
self-monitoring 20, 84, 89
self-preferencing business practice 218
self-regulation 38–39, 51–54, 150, 171, 207; "mandatory self-regulation" for foundation AI models 171
self-tracking digital tools 88–90
Selinger, Evan 8, 151
Shanghai 95
sharing economy 73, 77
Shenzhen 95
short-term rental accommodations/platforms 1, 3, 73, 75, 77–78, 201; see also Airbnb
Shotspotter 141–42; see also SoundThinking
Shutterstock 181
Sidewalk Labs 125
Siemens 118, 122, 177, 229n1
Silicon Valley 14–15, 2829, 38, 75, 93–94, 128, 202, 205, 209, 212, 226–27n34
Simon & Schuster 181
Simplisafe 113
Simulmatics Corporation 11
Siri, virtual assistant 113, 161
sixties countercultural activities and institutions 29–31
Skype 15
Slack 103
slavery 6, 21, 24
Sloan, Alfred 9
Smart Cities, Civic Hackers, and the Quest for a New Utopia (Townsend) 119
smart city/cities 4, 81, 87, 112, 117–25, 128, 135, 137, 139, 143, 210, 238n37, 238n50

smart home 4, 81, 87, 112–13, 115, 117, 178, 206, 237n8
smart phones as slotmachines 63–64
smart policing 4, 127–28, 130, 132, 134, 137, 143, 149
smart self 4, 81, 87–88, 90, 96
smart society 4, 80, 85
smart spaces 87
smart speakers 80, 160
smart technology 1, 117, 140
SmartCap's EEG-monitoring hats 99
smartphones 1, 64, 68, 71, 79–80, 85, 113, 119, 128, 138, 155, 157, 161, 165, 169, 203, 206, 215, 238n37
Snapshot 88
sniper rifles 127
Snowden, Edward 1, 15, 135, 164, 171
social digital spaces built with taxpayer money 192–93
social knowledge 20–22, 69
social media 1, 3, 5, 21, 61, 63–65, 75, 91–95, 100, 104, 119, 129, 131, 162, 163–65, 168, 188, 190, 192–96, 98, 215, 238n37, 242n90, 249n8, 250n27; and left-wing movements 249n8
social networks 24, 30, 31, 57–58, 61, 129, 140, 158, 189, 195, 201–3, 250n27
social ownership of digital assets 5, 197, 199
social quantification sector 23–25, 206
social Taylorism 7, 48–49, 80
"soft control" by management 69, 96
Songdo 123
Soofa 121
Sora ("text-to-video diffusion" AI model) 178
SoundThinking (formerly ShotSpotter) 142
"sousveillance" 102, 205–6; bottom-up practices of surveillance 206
South Africa 109
South America 123; see also Latin America
South Korea 123, 194, 219, 253n29
South River forest 145
Soviet Union 29, 34
Spain 168, 200
Spin 77
Spokeo 91
SPORH (Stops Per On-Road Hour) 99
Spotify 23, 59, 169, 229n1
Srnicek, Nick 43, 188, 195
Stack Overflow 181

Standard & Poor's 123
Stanford University 31, 171
starter-interrupt device 88, 99
State Farm 115
statements of rights and responsibilities (SRRs) 19
Steps to an Ecology of Mind (Bateson) 30
Stingray devices 128, 130, 134, 142–43
S.T.O.P. (Surveillance Technology Oversight Project) 144–45
stop-and-frisk practics/programs 128, 144
Stop Cop City: activists 146; crackdowns on opponents 146–47; movement 242n87
"strategic partnerships" among AI corporations 185
streaming services 5, 57, 59, 116, 161, 168–69
Streeter, Thomas 28
submachine guns 127
"Summit for Democracy" (2021) 219
Sunstein, Cass 47
surveillance advertising 3, 11, 37–38, 40–42, 44, 51, 54–55, 57–58, 92
surveillance capitalism 3, 7–14, 16–17, 23–24, 81, 87, 215–16, 223–24n22
surveillance practices 44, 134, 210–15, 219
Surveillance Tech Oversight Project 173
surveillance technologies 2, 88, 120, 130, 140, 144, 147, 209–10, 215, 232n17
Surveillance Valley: The Secret Military Mission of the Internet (Levine) 26
Suspicious Activity Reports (SARs) 135
sustainability 118, 124–25, 174, 187
SWAT (special weapons and tactics) teams 127, 146
Sweden 165, 231n62
Switzerland 165, 178

Taiwan 221, 253n29
TalkingData 23
targeted display 56–57
tariff increases on U.S. and EU imports of certain Chinese products 221
Tarnoff, Ben 34, 67, 189, 192, 194–97
Task Rabbit 63
Taylor, Astra 106
Taylor, Frederick Winslow 48, 50

TCP/IP protocols 27, 34
Teamsters Union 102
tech unicorn 177, 246n28
TechCrunch 181
techlash 1, 125
technocratic decoupling between EU, U.S. and China 220
"techno-democracies" vs. "techno-autocracies" 219
techno-libertarianism 216
technological infrastructure 3, 20, 50, 122
technological solutionism 124, 126, 202, 239n67
technologies as possible sites of political struggle 203, 206
techno-nationalism risks 216
technopolitics 80, 83, 87
techno-social engineering 54, 87
Telecommunications Act of 1996 34
telematics 99
television networks 11
Tencent 20, 208
Tennessee 147, 153, 156
terra nullius, legal doctrine of 17–18
Texas 153
textile militants in early 19th-century England 204
Thiel, Peter 129
This Machine Kills (podcast by Ongweso and Sadowski) 205
Thompson, E.P. 18
Thomson Reuters CLEAR 91
Thuisbezorgd 102
TikTok 163, 166, 168, 208, 245n64
time-and-motion studies 102–3
Toffler, Alvin 31
Too Smart 80
Toronto 125
Townsend, Anthony 119, 238n37
Toyota 96
tracking technologies 41, 88–90, 97, 99, 103, 122, 130
transfer of ownership of computer network 3, 33, 35
Tribal Broadband Connectivity Program (TBCP) 192; Acorn project at Hoopa Valley Reservation, California 192; see also intertribal broadband collaboration
trucking industry 98–99
Trump, Donald 190, 251n68
Tumbler 181
TurkerNation 110
Turker-themed Reddit threads 110–11
Turner, Fred 29–30

Turner, Maurice 115
Twitter 57, 61, 74, 105; *see also* X
two-factor authentication 113, 237*n*10
"tyranny of convenience" 45

Uber 2–3, 13, 23, 59–60, 68–73, 77, 93, 96, 101–4, 111, 118, 196, 201, 204, 230*n*57, 231*n*62; Uber Eats 102
"undressing" apps 186
United Arab Emirates 123
United Kingdom 17–18, 44, 102, 123, 178, 197, 209, 219
United States 2, 4–5, 15, 17, 20, 22–23, 29–30, 49, 51–52, 70, 76, 92–95, 99, 107, 114, 120–23, 145, 151–52, 172, 186, 188, 192, 195–96, 207, 209, 214, 219, 224*n*33, 231*n*62, 253*n*29
U.S.-China tensions centered on South China Sea and Taiwan 221
U.S. Congress 5, 31, 38, 51, 114, 153, 173, 186, 213–14
U.S. Defense Department 14, 25, 31
U.S. Department of Commerce 52
U.S. Department of Defense 14, 25, 31, 110
U.S. Department of Energy 172
U.S. Department of Homeland Security (DHS) 94, 127, 129, 135, 172, 240–41*n*46
U.S. Department of Transportation 120
U.S. Immigration and Customs Enforcement (ICE) 110, 129
U.S. market-driven regulatory model 207, 213–14, 216, 222
U.S. restrictions on exports of strategic technologies to China 221, 253*n*29
U.S. Supreme Court 153, 160, 219
U.S. tech companies fostering path dependency of next generation of users 209
Up&Go, New York City and Philadelphia 196

UPS 99
Urban Footprint 121
urban issues, holistic approach to 124–25
urban planners 118, 121–22, 238*n*50
USSA (United Services Automobile Association) 116
Utah 153
Uyghurs 95

value generation 81–83
van Doorn, Niels 72
venture capital investments 12, 38, 42, 71–72, 106, 110, 129, 184–85, 202, 253*n*68
The Verge 129
Vestager, Margrethe, 165
Vietnam 221
Vietnam War 30, 224*n*33
Virgin Islands 156
Virginia 153, 156
virtual community 63, 251*n*72
virtual private networks (VPNs) 220
Vivint 113
voice recognition technology 117, 243*n*3
von Thun, Max 169
Vrbo 77

Wales 18
Wall Street 14, 35, 56, 73
Walmart 96–97, 109
Washington State 102
We See It All: Liberty and Justice in an Age of Perpetual Surveillance (Fasman) 143
wearable devices 89, 117
weblining 92
Webster, Frank 7, 47–48, 85
WeChat 23
Weelaunee forest 145
Weizenbaum, Joseph 87
Wells Fargo 145
West Baltimore, Maryland 144
West Virginia 147
Western Union 34
WhatsApp 153, 158, 162, 208
White House 5, 35, 38, 46, 52, 154, 172–73

whitelisting 93
Whole Earth Catalog 29–31
Whole Earth Electronic Link (the Well) 29
Whole Earth network 29, 31
Whole Earth Software Review 30
Wi-Fi 119, 125
Wide Area Networks (WANs) 70
Wiener, Norbert 30
Wikipedia 201, 251*n*62
Winston, Ali 129
Wired 29, 31
work games 103–4
Work Without the Worker (Jones) 108
worker resistance to digital management practices 4, 101–4, 110–11
"workfare" 108
workplace democracy 199, 250*n*45
WorkSmart 100
World Bank 107
World Social Forums 125
World Trade Organization 210
World Wide Web 33, 37–38, 40
Wozniak, Steve 28
WPP 57
Wyze 113

X (formerly Twitter) 105, 163, 190, 193, 198, 201, 203
Xinjiang 95

Yahoo 15, 55, 57–58
Young, Arthur 18
Your Face Belongs to Us (Hill) 148
YouTube 15, 57–58, 61, 190, 193

Zabasearch 91
Zhima Credit score 93
Zuboff, Shoshana 2, 7–14, 16–17, 21–23, 81–82, 223–24*n*22, 224*n*28
Zuckerberg, Mark 20, 61–62, 71, 191, 194
Zuckerman, Ethan 192–93, 195

www.ingramcontent.com/pod-product-compliance
Ingram Content Group UK Ltd.
Pitfield, Milton Keynes, MK11 3LW, UK
UKHW050540150426
5217IPUK00026B/2013